AF646602

SOLVING ENGINEERING MECHANICS PROBLEMS WITH MATLAB

SOLVING

ENGINEERING MECHANICS PROBLEMS

WITH

MATLAB

Rao V. Dukkipati
Ph.D., P.E.
Fellow of ASME & CSME
Professor, Chair and Graduate Programme Director
Department of Mechanical Engineering
Fairfield University, Fairfield, Connecticut, USA

J. Srinivas
M.E., Ph.D.
Professor
Department of Mechanical Engineering
Chaitanya Engineering College
Visakhapatnam, India
&
Visiting Faculty
Kyungpook National University
Taegu, South Korea.

PUBLISHING FOR ONE WORLD

NEW AGE INTERNATIONAL (P) LIMITED, PUBLISHERS

New Delhi • Bangalore • Chennai • Cochin • Guwahati • Hyderabad
Jalandhar • Kolkata • Lucknow • Mumbai • Ranchi
Visit us at www.newagepublishers.com

Copyright © 2009, New Age International (P) Ltd., Publishers
Published by New Age International (P) Ltd., Publishers
First Edition: 2009

All rights reserved.

No part of this book may be reproduced in any form, by photostat, microfilm, xerography, or any other means, or incorporated into any information retrieval system, electronic or mechanical, without the written permission of the copyright owner.

Branches:

- 36, Malikarjuna Temple Street, Opp. ICWA, Basavanagudi, **Bangalore**. ✆ (080) 26677815
- 26, Damodaran Street, T. Nagar, **Chennai**. ✆ (044) 24353401
- Hemsen Complex, Mohd. Shah Road, Paltan Bazar, Near Starline Hotel, **Guwahati**. ✆ (0361) 2543669
- No. 105, 1st Floor, Madhiray Kaveri Tower, 3-2-19, Azam Jahi Road, Nimboliadda, **Hyderabad**. ✆ (040) 24652456
- RDB Chambers (Formerly Lotus Cinema) 106A,1st Floor, S.N. Banerjee Road, **Kolkata**. ✆ (033) 22275247
- 18, Madan Mohan Malviya Marg, **Lucknow**. ✆ (0522) 2209578
- 142C, Victor House, Ground Floor, N.M. Joshi Marg, Lower Parel, **Mumbai**. ✆ (022) 24927869
- 22, Golden House, Daryaganj, **New Delhi**. ✆ (011) 23262370, 23262368

ISBN : 978-81-224-2601-4

Rs. 695.00

C-08-10-3168

Printed in India at Glorious Printers, Delhi.
Typeset at Inhouse.

PUBLISHING FOR ONE WORLD
NEW AGE INTERNATIONAL (P) LIMITED, PUBLISHERS
4835/24, Ansari Road, Daryaganj, New Delhi-110002
Visit us at **www.newagepublishers.com**

We dedicate this book

to

Lord Sri Venkateswara

PREFACE

Engineering mechanics is an exciting and challenging field and considers the motion of bodies under the action of forces and the effects of forces on that motion. Many times, the computations in Mechanics are difficult to solve manually. Sometimes, the output results are very interesting with variation in the input parameters like coefficient of friction or angle of inclined plane. So it requires to repeat the computations several times to understand their influence. In comparison to many computational software, MATLAB provides a unique platform for conducting numerical and graphical parametric studies in a very concise way. This book is designed and organized around the concepts of engineering mechanics: statics and dynamics, and mechanics of materials, as they have been developed in an introductory undergraduate course for engineering students of all disciplines.

This book includes the coverage of basic methods of engineering mechanics: statics and dynamics, using MATLAB. With this foundation of basic principles, the book provides opportunities to explore advanced topics in engineering mechanics.

Chapter 1 presents a brief introduction to engineering mechanics: statics and dynamics and mechanics of materials. An introduction to MATLAB basics is presented in Chapter 2. MATLAB is considered as the software of choice. MATLAB can be used interactively and has an inventory of routines, called as functions, which minimize the task of programming even more. Further information on MATLAB can be obtained from: The MathWorks, Inc., 3 Apple Hill Drive, Natick, MA 01760. In the computational aspects, MATLAB has emerged as a very powerful tool for numerical computations involved in engineering. The idea of computer-aided design and analysis using MATLAB with the Symbolic Math Tool box has been incorporated. Chapter 3 consists of many solved problems that demonstrate the application of MATLAB to the analysis of engineering mechanics: statics and dynamics, and mechanics of materials. Presentations are limited to very basic topics to serve as an introduction to advanced topics in those areas of discipline. Chapters 2 and 3 include a great number of worked examples to guide the student in understanding the basic principles, concepts in mechanics: statics and dynamics. Chapter 4 presents some more solved applications of commonly used MATLAB functions such as differentiation and integration, symbolic calculations etc. This chapter is simply an extension to Chapter 2. In Chapter 5, additional classified examples of MATLAB for statics and dynamics problems are elaborately given. Background for understanding all these examples is a first course in mathematics along with concepts of physics.

Some of the topics not covered under the book have been presented concisely as appendices. Centroids and moment of inertia of common geometrical shapes are given in Appendix-I and fundamental principles of vibration analysis derived from Mechanics are presented in Appendix-II of the book. A short bibliography to guide the student to further sources of information on Engineering Mechanics and Mechanics of Materials are provided at the end of the book. We sincerely hope that the final outcome of this book will help the students in developing an appreciation for the topic of Engineering Mechanics.

We are grateful to all those who have had a direct impact on this work. Many people working in the general areas of engineering mechanics have influenced the format of this book. We greatly owe our indebtedness to all the authors of the articles listed in the bibliography of this book. Finally, we would very much like to acknowledge the encouragement, patience, and support provided by all our family members who have also shared in all the pain, frustration, and fun of producing a manuscript. We also acknowledge New Age Publishers, New Delhi in bringing this book in a nice format over a short span of time.

We would appreciate being informed of errors, or receiving other comments about the book. Please write to the author's Fairfield University address or send e-mail to **Rdukkipati@mail.fairfield.edu** or **Professordukkipati@yahoo.com**.

Rao V. Dukkipati
J. Srinivas

CONTENTS

Preface vii

CHAPTER 1: ENGINEERING MECHANICS 1–24

1.1 Introduction 1
1.2 Newtonian Mechanics 1
1.3 Newton's Laws of Motion 1
1.4 Resultants of Coplanar Force Systems 2
1.5 Resultants of Non-coplanar Force Systems 3
1.6 Equilibrium of Coplanar Force Systems 4
1.7 Equilibrium of Non-coplanar Force System 6
1.8 Trusses 6
1.9 Analysis of Beams 7
1.10 Friction 7
1.11 First Moments and Centroids 8
1.12 Virtual Work 9
1.13 Kinematics of a Particle 10
1.14 D'Alembert's Principle 14
1.15 Kinematics of a Rigid Body in Plane Motion 14
1.16 Moments of Inertia 16
1.17 Dynamics of a Rigid Body in Plane Motion 18
1.18 Work and Energy 20
1.19 Impulse and Momentum 21
1.20 Three-dimensional Mechanics 23
1.21 Summary 24

CHAPTER 2: MATLAB BASICS 25–102

2.1 Introduction 25
- 2.1.1 Starting and Quitting MATLAB 26
- 2.1.2 Display Windows 26
- 2.1.3 Entering Commands 26
- 2.1.4 MATLAB Expo 26
- 2.1.5 Abort 26
- 2.1.6 The Semicolon (;) 26
- 2.1.7 Typing % 26
- 2.1.8 The clc Command 26
- 2.1.9 Help 26
- 2.1.10 Statements and Variables 27

2.2 Arithmetic Operations 27

2.3 Display Formats 27

2.4 Elementary Math Built-in Functions 28

2.5 Variable Names 30

2.6 Predefined Variables 30

2.7 Commands for Managing Variables 31

2.8 General Commands 31

2.9 Arrays 33
- 2.9.1 Row Vector 33
- 2.9.2 Column Vector 33
- 2.9.3 Matrix 33
- 2.9.4 Addressing Arrays 34
  - 2.9.4.1 *Colon for a vector* 34
  - 2.9.4.2 *Colon for a matrix* 34
- 2.9.5 Adding Elements to a Vector or a Matrix 34
- 2.9.6 Deleting Elements 34
- 2.9.7 Built-in Functions 34

2.10 Operations with Arrays 35
- 2.10.1 Addition and Subtraction of Matrices 35
- 2.10.2 Dot Product 36
- 2.10.3 Array Multiplication 36
- 2.10.4 Array Division 36
- 2.10.5 Identity Matrix 36
- 2.10.6 Inverse of a Matrix 36

2.10.7 Transpose 36
2.10.8 Determinant 37
2.10.9 Array Division 37
2.10.10 Left Division 37
2.10.11 Right Division 37
2.10.12 Eigenvalues and Eigenvectors 37
2.11 Element-by-Element Operations **38**
2.11.1 Built-in Functions for Arrays 39
2.12 Random Numbers Generation **40**
2.12.1 The Random Command 40
2.13 Polynomials **41**
2.14 System of Linear Equations **42**
2.14.1 Matrix Division 42
2.14.2 Matrix Inverse 42
2.15 Script Files **47**
2.15.1 Creating and Saving a Script File 47
2.15.2 Running a Script File 47
2.15.3 Input to a Script File 47
2.15.4 Output Commands 48
2.16 Programming in MATLAB **48**
2.16.1 Relational and Logical Operators 48
2.16.2 Order of Precedence 49
2.16.3 Built-in Logical Functions 49
2.16.4 Conditional Statements 50
2.16.5 Nested if Statements 51
2.16.6 else AND elseif Clauses 51
2.16.7 MATLAB while Structures 51
2.17 Graphics **53**
2.17.1 Basic 2-D Plots 53
2.17.2 Specialized 2-D Plots 54
2.17.2.1 *Overlay plots* 55
2.17.3 3-D Plots 55
2.17.4 Saving and Printing Graphs 61
2.18 Input/Output in MATLAB **62**
2.18.1 The fopen Statement 62

2.19 Symbolic Mathematics **63**
2.19.1 Symbolic Expressions 63
2.19.2 Solution to Differential Equations 65
2.19.3 Calculus 65
2.20 The Laplace Transforms **67**
2.21 Additional Examples **68**
2.22 Summary **102**

CHAPTER 3: MATLAB TUTORIAL FOR MECHANICS **103–217**
3.1 Introduction **103**
3.2 Example Problems and Solutions **104**
3.2.1 Statics 104
3.2.2 Dynamics 162
3.3 Summary **217**

CHAPTER 4: MATLAB FOR ENGINEERING APPLICATIONS **219–253**
4.1 Numerical Calculations **219**
4.2 Writing Script Files and Functions **232**
4.3 Graphics **234**
4.4 Symbolic Calculations **239**
4.5 Differentiation and Integration **248**
4.6 Summary **253**

CHAPTER 5: MATLAB–A COMPUTATIONAL TOOL FOR MECHANICS **255–377**
5.1 Statics **255**
5.2 Dynamics 325
5.2.1 Kinematics of Particles 325
5.2.2 Particle Kinetics 338
5.2.3 Planar Kinematics of Rigid Bodies 348
5.2.4 Rigid Body Kinetics in Plane Motion 356
5.2.5 Three-dimensional Motion of Rigid Bodies 371
5.3 Summary **377**

APPENDIX **379-392**
I Centroids and Moments of Inertia **379**
II Vibration Engineering–Application of Dynamics **381**

BIBLIOGRAPHY **393-394**

CHAPTER 1

ENGINEERING MECHANICS

1.1 INTRODUCTION

Engineering mechanics is the science that considers the motion of bodies under the action of forces and the effects of forces on that motion. Mechanics includes *statics* and *dynamics*. Statics deals with the special case of a body at rest or a body that moves with constant velocity. A body at rest or moving with constant velocity is said to be in *equilibrium*. This is sometimes also called as static equilibrium. When the body moves with finite velocity or acceleration, the principles of statics are no longer applicable. The mechanics of such a system is called *dynamics*. When the body has no rotational motion, it is called as particle. Usually, each point on a rigid-body is always at a constant distance from any other point in the body. Dynamics is further divided into *kinematics* and *kinetics*. *Kinematics* defines the relationships among displacement, velocity, and acceleration of a moving body. *Kinetics* defines the relationship between the forces that act on a body and the motion of the body. Analysis of the bodies can either be conducted in plane or in three-dimensions. A rigid-body in space has six degree of freedom. This chapter provides a brief review of various topics in Mechanics.

1.2 NEWTONIAN MECHANICS

Collectively, the study of statics and dynamics is called *classical mechanics*. Classical mechanics treats the motion of bodies of ordinary size that move at speeds that are small compared to the speed of light. Newton (1642-1727) formulated the law of universal gravitation and the mathematics of calculus. Newton introduced the concepts of force and mass. Newton formulated the three laws of motion that are the basis of engineering applications of mechanics. The classical mechanics is often called *Newtonian mechanics*.

1.3 NEWTON'S LAWS OF MOTION

Newton's laws are fundamentally applicable directly to a particle that is a body which may be treated as having point mass. Newton's laws may be stated in terms of a particle parameters as follows:

First Law: In the absence of applied forces, a particle originally at rest or moving with constant speed in a straight line will remain at rest or continue to move with constant speed in a straight line.

Second Law: If a particle is subjected to a force, the particle will accelerate. The acceleration of the particle will be in the direction of the force, and the magnitude of the acceleration will be proportional to the force and inversely proportional to the mass of the particle.

Newton's second law can be expressed in the form:

$$a = \frac{F}{m} \qquad \text{...(1.1)}$$

where F = denotes force

m = mass

a = acceleration

Newton's second law is the basis for the study of kinetics of a particle.

Third Law: For every action, there is an equal and opposite reaction, or the mutual forces exerted by two particles on each other are always equal and oppositely directed.

1.4 RESULTANTS OF COPLANAR FORCE SYSTEMS

In statics, it is often necessary to find the state of equilibrium or the resultant of several forces acting in plane. In coplanar system, the resultant R for a concurrent forces is given by

$$R = \sqrt{(\Sigma F_x)^2 + (\Sigma F_y)^2}$$

and

$$\theta_x = \tan^{-1}\left[\frac{\Sigma F_y}{\Sigma F_x}\right] \qquad \text{...(1.2)}$$

where ΣF_x, ΣF_y = algebraic sums of the x and y components of the forces of the system respectively.

θ_x = the angle that the resultant R makes with the x-axis.

For a parallel system, the resultant R is given by

$$R = \Sigma F$$

and

$$R\bar{a} = \Sigma M_O \qquad \text{...(1.3)}$$

where ΣF = algebraic sum of the forces of the system

O = any moment center in the plane

$\bar{a}$ = perpendicular distance from the moment center O to the resultant R

$R\bar{a}$ = moment of R with respect to O

ΣM_O = algebraic sum of the moments of the forces of the system with respect to O

The resultant R for a non-concurrent and non-parallel system is given by

$$R = \sqrt{(\Sigma F_x)^2 + (\Sigma F_y)^2}$$

and $$\theta_x = \tan^{-1}\left[\frac{\Sigma F_y}{\Sigma F_x}\right] \qquad ...(1.4)$$

where ΣF_x, ΣF_y = algebraic sums of the x and y components of the forces of the system respectively

θ_x = the angle that the resultant R makes with the x-axis.

The action of the line of the resultant force is given by

$$R\bar{a} = \Sigma M_o \qquad ...(1.5)$$

where O = any moment center in the plane

$\bar{a}$ = perpendicular distance from the moment center O to the resultant R

$R\bar{a}$ = moment of R with respect to O

ΣM_o = algebraic sum of the moments of the forces of the system with respect to O

1.5 RESULTANTS OF NON-COPLANAR FORCE SYSTEMS

In non-coplanar force systems, the forces are neither all concurrent nor in parallel. The resultant R for a concurrent system is given by

$$R = \sqrt{(\Sigma F_x)^2 + (\Sigma F_y)^2 + (\Sigma F_z)^2} \qquad ...(1.6)$$

with the direction cosines

$$\cos\theta_x = \frac{\Sigma F_x}{R}, \cos\theta_y = \frac{\Sigma F_y}{R} \text{ and } \cos\theta_z = \frac{\Sigma F_z}{R}$$

where ΣF_x, ΣF_y, ΣF_z = algebraic sums of the x, y and z components of the forces of the system respectively.

For a parallel system, the resultant R is given by

$$R = \Sigma F$$
$$R\bar{x} = \Sigma M_z \qquad ...(1.7)$$
$$R\bar{z} = \Sigma M_x$$

where ΣF = algebraic sum of the forces of the system

$\bar{x}$ = perpendicular distance from the yz plane to the resultant

$\bar{z}$ = perpendicular distance from the xy plane to the resultant

ΣM_x, ΣM_z = algebraic sum of the moments of the forces of the system about the x and z axes respectively

If $\Sigma F = 0$, the resultant couple C when exists is given by

$$C = \sqrt{(\Sigma M_x)^2 + (\Sigma M_z)^2}$$

and $$\phi = \tan^{-1}\left[\frac{\Sigma M_z}{\Sigma M_x}\right] \quad ...(1.8)$$

where ϕ is the angle that the vector representing the resultant couple makes with the x-axis.

The magnitude of the resultant R of the non-concurrent system at the origin (x, y and z axes is placed with their origin at the base point) is given by

$$R = \sqrt{(\Sigma F_x)^2 + (\Sigma F_y)^2 + (\Sigma F_z)^2} \quad ...(1.9)$$

with the direction cosines

$$\cos\theta_x = \frac{\Sigma F_x}{R}, \cos\theta_y = \frac{\Sigma F_y}{R} \text{ and } \cos\theta_z = \frac{\Sigma F_z}{R} \quad ...(1.10)$$

The magnitude of the resulting couple C is given by

$$C = \sqrt{(\Sigma M_x)^2 + (\Sigma M_y)^2 + (\Sigma M_z)^2} \quad ...(1.11)$$

with the direction cosines

$$\cos\phi_x = \frac{\Sigma M_x}{C}, \cos\phi_y = \frac{\Sigma M_y}{C} \text{ and } \cos\phi_z = \frac{\Sigma M_z}{C} \quad ...(1.12)$$

where ΣM_x, ΣM_y, ΣM_z = algebraic sums of the moments of the forces of the system about x, y and z-axes respectively.

ϕ_x, ϕ_y, ϕ_z = angles which the vector representing the couple C makes with the x, y and z-axes respectively.

1.6 EQUILIBRIUM OF COPLANAR FORCE SYSTEMS

The necessary and sufficient conditions for the equilibrium of a coplanar force system are:

$$R = \Sigma F = 0$$

and $$C = \Sigma M = 0 \quad ...(1.13)$$

where ΣF = vector sum of all forces of the system

ΣM = vector sum of the moments of all the forces of the system.

For concurrent system, any of the following sets of equations ensures equilibrium (the resultant is zero). The concurrency is assumed at the origin.

Set 1: $\Sigma F_x = 0$

$\Sigma F_y = 0$

Set 2: $\Sigma F_x = 0$

$\Sigma M_A = 0$ ...(1.14)

(A may be chosen any place in the plane except on the y-axis)

Set 3: $\Sigma M_A = 0$

$\Sigma M_B = 0$

(A and B may be chosen any place in the plane except A, B and the origin do not lie on the same straight line)

where ΣF_x, ΣF_y = algebraic sum of the x and y components of the forces of the system respectively.

ΣM_A, ΣM_B = algebraic sum of the moments of the forces of the system about A and B respectively.

For parallel system, any of the following sets of equations ensures equilibrium (the resultant is neither a force nor a couple).

Set 1: $\Sigma F = 0$

$\Sigma M_A = 0$ ...(1.15)

Set 2: $\Sigma M_A = 0$

$\Sigma M_B = 0$

(A and B may chosen any place in the plane provided in the line joining A and B is not parallel to the forces of the system)

where ΣF = algebraic sum of the forces of the system parallel to the action lines of the forces

ΣM_A, ΣM_B = algebraic sum of the moments of the forces of the system about A and B respectively.

For non-concurrent and non-parallel system, any of the following sets of equations ensures equilibrium (the resultant is neither a force nor a couple).

Set 1: $\Sigma F_x = 0$

$\Sigma F_y = 0$

$\Sigma M_A = 0$ ...(1.16)

Set 2: $\Sigma F_x = 0$

$\Sigma M_A = 0$

$\Sigma M_B = 0$ ...(1.17)

(Provided that the line joining A and B is not perpendicular to the x-axis)

Set 3: $\Sigma M_A = 0$

$\Sigma M_B = 0$

$\Sigma M_C = 0$ ...(1.18)

(Provided that A, B and C do not lie on the same straight line)

where ΣF_x, ΣF_y, ΣF_z = algebraic sum of the x, y and z components of the forces respectively

ΣM_A, ΣM_B, ΣM_C = algebraic sum of the moments of the forces of the system about any three points A, B and C in the plane respectively.

1.7 EQUILIBRIUM OF NON-COPLANAR FORCE SYSTEM

The necessary and sufficient conditions that R and C be zero vectors are given by

$$R = \Sigma F = 0$$

and $$C = \Sigma M = 0 \qquad ...(1.19)$$

where ΣF = vector sum of all the forces of the system

ΣM = vector sum of the moments of all the forces of the system relative to any point.

For a concurrent and non-coplanar system, the following set of equations must be satisfied:

$$\Sigma F_x = 0$$
$$\Sigma F_y = 0 \qquad ...(1.20)$$
$$\Sigma F_z = 0$$

where ΣF_x, ΣF_y, ΣF_z = algebraic sums of the x, y and z components of the forces of the system respectively.

For a parallel non-coplanar system, the set of equations to be satisfied for equilibrium are:

$$\Sigma F_y = 0$$
$$\Sigma M_x = 0 \qquad ...(1.21)$$
$$\Sigma M_z = 0$$

where ΣF_y = algebraic sum of the forces of the system along the y-axis which is selected parallel to the system.

ΣM_x, ΣM_z = algebraic sums of the moments of the forces of the system about the x and z axes respectively.

The necessary and sufficient conditions required for equilibrium for a non-concurrent non-coplanar system are given by

$$\Sigma F_x = 0$$
$$\Sigma F_y = 0$$
$$\Sigma F_z = 0$$
$$\Sigma M_x = 0$$
$$\Sigma M_y = 0$$
$$\Sigma M_z = 0 \qquad ...(1.22)$$

where ΣF_x, ΣF_y, ΣF_z = algebraic sums of the x, y and z components of the forces of the system respectively.

ΣM_x, ΣM_y, ΣM_z = algebraic sums of the components of the forces of the system about the x, y and z axes respectively.

1.8 TRUSSES

A *truss* is a system of slender members that are pinned together. The members are free to rotate at the pinned joints and carry forces only. All the external forces act at the joints.

Trusses are examples of coplanar force systems in equilibrium. Trusses are assumed to be rigid members all located in one plane. The weights of the truss members are neglected. Forces are transmitted from one member to another through pin joints. These members are called two-force members. A two-force member is in equilibrium under the effect of two resultant forces one at each end. The two-force members will be either in tension or compression. There are two methods available for the analysis of trusses.

For a stable planar truss, the following condition applies:

$$2n = m + p \qquad ...(1.23)$$

where n = number of joints

m = number of members

p = number of unknown external forces

In a stable three-dimensional truss, the condition that holds is

$$3n = m + p \qquad ...(1.24)$$

Method of Joints:

In this method, a free-body diagram of any pin in the truss is drawn. Maximum of two unknown forces act on that pin. Proceed from one pin to another until all unknowns have been obtained.

Method of Sections:

A free-body diagram of a section of the truss is drawn. The forces in the members cut act as external forces. The system is a non-concurrent and non-parallel one. In any one section no more than three unknown forces are to be found.

1.9 ANALYSIS OF BEAMS

A *beam* is a long member that is subjected to external lateral forces, F and external lateral moments, T. These cause internal lateral forces V called *shear forces* and internal lateral moments M known as *bending moments*. The shear forces and the bending moments induce lateral deformations, also called *bending*.

Shear and Moment Diagrams:

Shear and moment diagrams show the variation of V and M across a beam. Summation of the forces and moments of forces are used to plot V and M directly for the beams.

The slope of the shear diagram at any section along the beam is the negative of the load per unit length at that point. The change in shear between two sections of a beam carrying a distributed load equals the negative area of the load diagram between the two sections. The slope of the moment diagram at any section along the beam is the value of the shear at that section. The change in the moment between two sections of a beam equals the area of the shear diagram between the two sections.

1.10 FRICTION

The tangential force that opposes the sliding of one body relative to the other is the *static friction* between the two bodies. *Kinetic friction* is the tangential force between two bodies after motion begins. The body begins to slide only when the applied force exceeds the static frictional force at the floor. Referring to Fig.1.1, the following terms are defined:

Limiting friction F' is the maximum value of static friction that occurs when motion is impending.

Coefficient of static friction is the ratio of the limiting friction F' to the normal force N, or $\mu = \left(\dfrac{F'}{N}\right)$

Coefficient of kinetic friction is the ratio of the kinetic friction to the normal force. *Angle of reponse*, α is the angle to which an inclined plane may be raised before an object resting on it will move under the action of the force of gravity and the reaction of the plane. In Fig.1.1, R is the resultant of F' and N and $\alpha = \phi$.

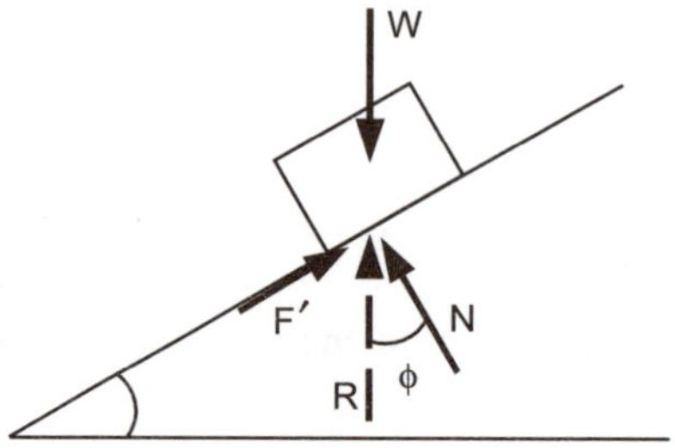

Fig. 1.1 Sliding friction

Belt Friction: When a belt passes over a pulley, the tensions in the belt on the two sides of the pulley will differ. When slip is about to occur, the tensions T_1 and T_2 are given by

$$T_1 = T_2 e^{\mu\beta} \quad ...(1.25)$$

where T_1 = larger tension

T_2 = smaller tension

μ = coefficient of friction

β = angle of wrap (radian)

e = 2.718 (base of natural logarithms)

Screw frictional force occurs when a nut is screwed over a bolt.

1.11 FIRST MOMENTS AND CENTROIDS

The centroidal position vector $\bar{\mathbf{r}}$ of a geometry composed of n areas $d_1, d_2, d_3, ..., d_n$ located at points $P_1, P_2, P_3, ..., P_n$ represented by position vectors $\mathbf{r}_1, \mathbf{r}_2, \mathbf{r}_3, ..., \mathbf{r}_n$ respectively is defined as

$$\bar{\mathbf{r}} = \frac{\sum_{i=1}^{n} \mathbf{r}_i d_i}{\sum_{i=1}^{n} d_i}$$

where d_i = area of ith element

$\mathbf{r}_i$ = position vector of the ith element

$\sum_{i=1}^{n} d_i$ = total area of all n elements

$\sum_{i=1}^{n} \mathbf{r}_i d_i$ = first moment of area of all the elements relative to selected point O.

The centroid can be written in terms of x, y and z coordinates as

$$\bar{x} = \frac{\sum_{i=1}^{n} x_i d_i}{\sum_{i=1}^{n} d_i}$$

$$\bar{y} = \frac{\sum_{i=1}^{n} y_i d_i}{\sum_{i=1}^{n} d_i}$$

$$\bar{z} = \frac{\sum_{i=1}^{n} z_i d_i}{\sum_{i=1}^{n} d_i}$$

where d_i = magnitude of area of the ith element

$\bar{x}, \bar{y}, \bar{z}$ = coordinates of centroid of the assemblage

x_i, y_i, z_i = coordinates of P_i at which d_i is concentrated.

The centroid of a continuous quantity of mass m is given by

$$\bar{\mathbf{r}} = \frac{\int \mathbf{r} dm}{\int dm}$$

or

$$\bar{x} = \frac{\int x dm}{\int dm} = \frac{Q_{yz}}{m}$$

$$\bar{y} = \frac{\int y dm}{\int dm} = \frac{Q_{xz}}{m}$$

$$\bar{z} = \frac{\int z dm}{\int dm} = \frac{Q_{xy}}{m}$$

where Q_{xy}, Q_{yz}, Q_{xz} = first moments with respect to xy, yz and xz planes respectively.

1.12 VIRTUAL WORK

A *virtual displacement* δs of a particle is defined as any infinitesimal change in the position of the particle consistent with the constraints imposed on the particle. *Virtual work* δU done by a force, F_t is defined as $F_t\, \delta s$, where F_t is the magnitude of the component of the force along the virtual displacement δs. Similarly, virtual work δU done by a couple of moment M is defined as $M\delta\theta$, where $\delta\theta$ is the virtual angular displacement.

Equilibrium: The necessary and sufficient condition for the *equilibrium of a particle* is zero virtual work done by all the forces acting on the particle during any virtual displacement δs. The necessary and sufficient condition for the *equilibrium of a rigid-body* is zero virtual work done by all the external forces acting on the body during any virtual displacement consistent with the constraints imposed on the body. The *equilibrium* of a system of rigid bodies exists if the potential energy V has a stationary value. *Stable equilibrium* occurs if the potential energy V is a minimum. *Unstable equilibrium* occurs if the potential energy V is a maximum. *Neutral equilibrium* exists if a system remains in any position in which it is placed.

If x is the variable and V the potential energy of the system, then we have

$$\frac{d^2V}{dx^2} > 0 \quad \text{(stable equilibrium)}$$

$$\frac{d^2V}{dx^2} < 0 \quad \text{(unstable equilibrium)}$$

$$\frac{d^2V}{dx^2} = 0 \quad \text{(neutral equilibrium)}$$

1.13 KINEMATICS OF A PARTICLE

Kinematics is the study of motion without considering the forces or other factors that influence the motion.

Rectilinear motion: The general expression for displacement(s), velocity(v) and acceleration(a) are derived from the following three differential relations:

$$a = dv/dt$$

$$v = ds/dt$$

$$ads = vdv$$

For motion of a point along a straight line, the following formulas are valid for constant acceleration $a = k$:

$$v = v_0 + kt$$

$$v^2 = v_0^2 + 2ks$$

$$s = v_0 t + \frac{1}{2}kt^2 \qquad \text{...(1.25a)}$$

$$s = \frac{1}{2}(v + v_0)t$$

where v_0 = initial velocity

v = final velocity

k = constant acceleration

t = time

s = displacement.

Curvilinear motion: In a plane, curvilinear motion is motion along a plane curve (path). The velocity and acceleration of a point on such a curve can be expressed either as:

(*a*) rectangular components

(*b*) tangential and normal components

(*c*) radial and transverse components

(a) Rectangular components

The position vector **r**(*t*), the velocity vector **v**(*t*), and the acceleration vector **a**(*t*) are given by

$$\mathbf{r}(t) = x(t)i + y(t)j + z(t)k$$

$$\mathbf{v}(t) = v_x(t)i + v_y(t)j + v_z(t)k = \dot{x}(t)i + \dot{y}(t)j + \dot{z}(t)k$$

$$\mathbf{a}(t) = a_x(t)i + a_y(t)j + a_z(t)k = \ddot{x}(t)i + \ddot{y}(t)j + \ddot{z}(t)k \qquad ...(1.26)$$

where the over-dot represents time differentiation.

The rectangular components of velocity and acceleration are given by

$$\mathbf{v}\,(t) = \dot{x}(t), v_y(t) = \dot{y}(t), v_z(t) = \dot{z}(t)$$

$$\mathbf{a}\,(t) = \ddot{x}(t), a_y(t) = \ddot{y}(t),\ a_z(t) = \ddot{z}(t) \qquad ...(1.27)$$

The components of position are given by

$$\mathbf{x}\,(t) = x(t_1) + \int_{t_1}^{t} v_x(s)\,ds$$

$$\mathbf{y}\,(t) = y(t_1) + \int_{t_1}^{t} v_y(s)\,ds$$

$$\mathbf{z}\,(t) = z(t_1) + \int_{t_1}^{t} v_z(s)\,ds \qquad ...(1.28)$$

The components of velocity are given by

$$\mathbf{v}_x\,(t) = v_x(t_1) + \int_{t_1}^{t} a_x(s)\,ds$$

$$\mathbf{v}_y\,(t) = v_y(t_1) + \int_{t_1}^{t} a_y(s)\,ds$$

$$\mathbf{v}_z\,(t) = v_z(t_1) + \int_{t_1}^{t} a_z(s)\,ds \qquad ...(1.29)$$

Also

$$v_x^2\,(x) = v_x^2(x_1) + 2\int_{x_1}^{x} a_x(x)\,dx \qquad ...(1.30)$$

(b) Tangential and normal components

Referring to Fig.1.2, the velocity vector ***v*** can be written as

$$\mathbf{v} = u\mathbf{n}_t \qquad ...(1.31)$$

where v = magnitude of the velocity vector

$\mathbf{n}_t$ = tangential unit vector directed along the velocity vector.

Also $$\dot{\mathbf{n}}_t = \dot{\theta}\mathrm{n}_n \qquad ...(1.32)$$

and $$\dot{\mathbf{n}}_n = -\dot{\theta}\mathbf{n}_t$$

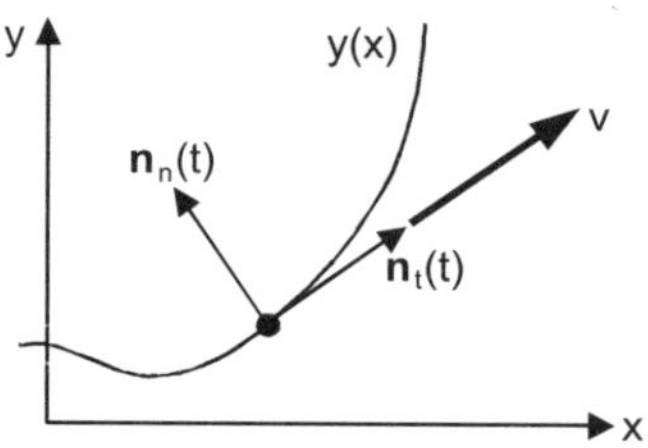

Fig. 1.2 Velocity vectors

where $\mathbf{n}_n$ = the normal unit vector defined to be perpendicular to the tangential unit vector.

In Fig. 1.3, the radius of curvature ρ of the path at time t is shown which is obtained by the intersection of the lines extending from $\mathbf{n}_t(t)$ and $\mathbf{n}_t(t + \Delta t)$ where Δt is time increment. The angle θ changes an incremental amount Δθ and the point moves an incremental amount Δs.

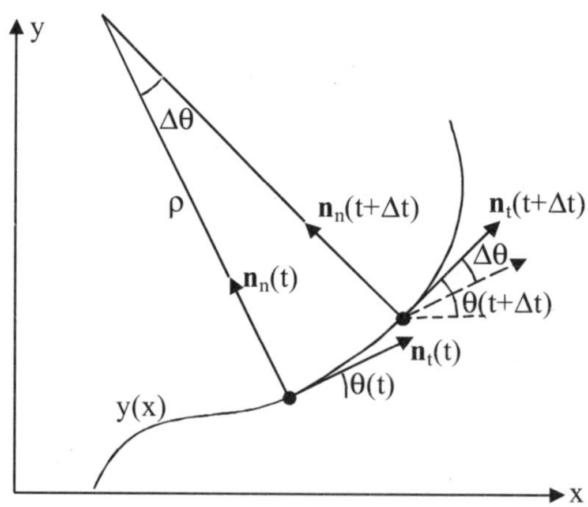

Fig. 1.3 Radius of curvature of path

Now $$\rho\Delta\theta = \Delta s$$

or $$\rho\dot{\theta} = v \qquad ...(1.33)$$

where $\dot{\theta} = \dfrac{d\theta}{dt}$

$v = \dfrac{ds}{dt}$

Differentiating Eq. (1.31) with respect to time gives

$$\mathbf{a} = a_t\mathbf{n}_t + a_n\mathbf{n}_n \qquad ...(1.34)$$

where $a_t = \dot{v}$

$$a_n = \frac{v^2}{\rho} \qquad ...(1.35)$$

(c) Radial and transverse components

The polar form of a position vector is

$$\mathbf{r} = r\mathbf{n}_r \qquad ...(1.36)$$

where $r = |\mathbf{r}|$ is the magnitude of r

$\mathbf{n}_r$ = unit vector in the direction of $\mathbf{r}$

$\mathbf{n}_r$ = the radial unit vector

$\mathbf{n}_\theta$ = the circumferential unit vector

The circumferential unit vector is perpendicular to $\mathbf{n}_r$. Hence, $\mathbf{n}_r$ and $\mathbf{n}_\theta$ are related to i and j in Fig.1.4 as

$$\mathbf{n}_r(t) = \cos\theta(t)i + \sin\theta\,(t)j$$

$$\mathbf{n}_\theta(t) = -\sin\theta(t)i + \cos\theta\,(t)j \qquad ...(1.37)$$

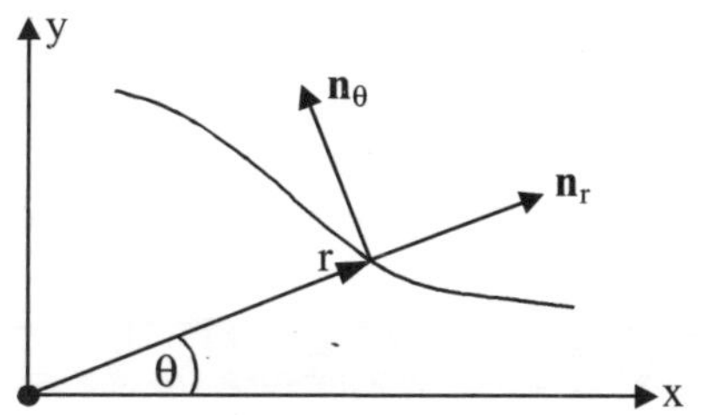

Fig. 1.4 Radial and transverse components

Differentiating Eq. (1.37) with respect to time gives

$$\dot{\mathbf{n}}_r = -\dot{\theta}\sin\theta i + \dot{\theta}\cos\theta j = \dot{\theta}(-\sin\theta i + \cos\theta j) = \dot{\theta}\mathbf{n}_\theta \qquad ...(1.38)$$

and $$\dot{\mathbf{n}}_\theta(t) = -\dot{\theta}\mathbf{n}_r \qquad ...(1.39)$$

Hence, the derivatives of the unit vectors are given by

$$\dot{\mathbf{n}}_r = -\dot{\theta}\mathbf{n}_\theta$$

$$\dot{\mathbf{n}}_\theta = -\dot{\theta}\mathbf{n}_r \qquad ...(1.40)$$

Similarly, from Eqs. (1.36) and (1.40), we get

$$\mathbf{v} = v_r\mathbf{n}_r + v_\theta\mathbf{n}_\theta \qquad ...(1.41)$$

where $v_r = \dot{r}$

$v_\theta = r\dot{\theta}$

Differentiating Eq. (1.41) with respect to time gives

$$\mathbf{a} = a_r \mathbf{n}_r + a_\theta \mathbf{n}_\theta \qquad ...(1.42)$$

where $a_r = \ddot{r} - r\dot{\theta}^2$

$$a_\theta = r\ddot{\theta} + 2\dot{r}\dot{\theta} \qquad ...(1.43)$$

1.14 D'ALEMBERT'S PRINCIPLE

Kinetics of particles begins with Newton's second law, which relates forces, accelerations and time. The second form of writing Newton's law is called D'Alembert's principle which gives the condition for dynamic equilibrium. It states that

$$\Sigma F - ma = 0 \qquad ...(1.44)$$

where ΣF = vector sum of all the forces acting on the particle

m = mass of the particle

a = acceleration of the particle.

Hence, an imaginary force (also known as inertia force) which is collinear with ΣF but opposite in sense and of magnitude ma would cause it to be in equilibrium if applied to the particle. All the equations of equilibrium are then applicable. Kinetics of particles can also be attempted with work-energy principle and impulse-momentum principle, just as a rigid-body. When velocities and displacements are given instead of acceleration as in spring problems, the problem should be attempted with work-energy principle. On the other hand when velocity and time are given as in recoil or impact problems, the equations of motion are formulated from impulse-momentum relations.

1.15 KINEMATICS OF A RIGID BODY IN PLANE MOTION

A rigid-body is configured by specified dimensions and rotational motion should be considered in addition to translation. In plane motion of a rigid body, every point in the body remains at a constant distance from a fixed plane. As shown in Fig.1.15, B is an arbitrary point in the body and x-y-z is a non-rotating reference frame.

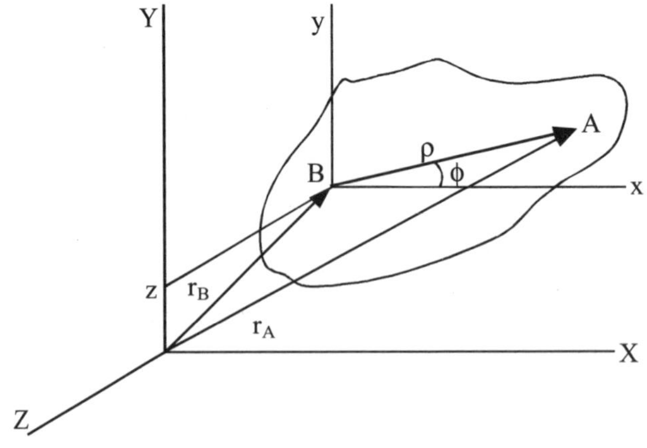

Fig. 1.5 Representation of plane motion of rigid-body

The position vector $\mathbf{r}_A$ of any point A (fixed or moving) in the lamina is given by

$$\mathbf{r}_A = \mathbf{r}_B + \boldsymbol{\rho} \quad \text{...(1.45)}$$

where $\mathbf{r}_B$ = position vector of B

$\boldsymbol{\rho}$ = vector BA

Differentiating Eq. (1.45) gives

$$\mathbf{v}_A = \dot{\mathbf{r}}_A = \dot{\mathbf{r}}_B + \dot{\boldsymbol{\rho}} = \dot{\mathbf{r}}_B + \boldsymbol{\rho}\omega\mathbf{e}_\phi \quad \text{...(1.46)}$$

where $\dot{\mathbf{r}}_B$ = linear velocity of B relative to the fixed axes X, Y and Z

ω = angular velocity (magnitude) of $\boldsymbol{\rho}$ about any line parallel to Z-axis

$\mathbf{e}_\phi$ = unit vector perpendicular to $\boldsymbol{\rho}$ (in the direction of increasing ϕ)

The acceleration $\mathbf{a}_A$ is given by

$$a_A = \dot{v}_A = \ddot{r}_A = \ddot{r}_B - \rho\omega^2 e_\rho + \rho\alpha e_\phi \quad \text{...(1.47)}$$

where $\ddot{\mathbf{r}}_B$ = acceleration of B relative to fixed axes X, Y and Z

e_ρ = unit vector along ρ directed from B toward A

e_ϕ = unit vector perpendicular to $\boldsymbol{\rho}$ (in the direction of increasing ϕ)

α = angular acceleration (magnitude) of $\boldsymbol{\rho}$ about any line parallel to the Z-axis.

Equations (1.46) and (1.47) can also be written in vector form as follows:

$$\mathbf{v_A} = \mathbf{v_B} + \boldsymbol{\omega} \times \boldsymbol{\rho}$$

or

$$\mathbf{v_A} = \mathbf{v_B} + \mathbf{v_{A/B}} \quad \text{...(1.48)}$$

and

$$\mathbf{a_A} = \mathbf{a_B} + \boldsymbol{\omega} \times (\boldsymbol{\omega} \times \boldsymbol{\rho}) + \boldsymbol{\alpha} \times \boldsymbol{\rho}$$

$$\mathbf{a_A} = \mathbf{a_B} + \mathbf{a_{A/B}} \quad \text{...(1.49)}$$

Here $\times$ is a vector (cross) product.

also

$$\boldsymbol{\omega} = \dot{\phi}\mathbf{k} = \omega\mathbf{k}$$

$$\boldsymbol{\alpha} = \ddot{\phi}\mathbf{k} = \dot{\omega}\mathbf{k} = \alpha\mathbf{k} \quad \text{...(1.50)}$$

$$\rho\alpha e_\rho = \omega \times \rho$$

$$\rho\alpha e_\phi = \alpha \times \rho$$

$$-\rho\omega^2 e_\rho = \omega \times (\omega \times \rho)$$

in which $\mathbf{v}_{A/B}$ = relative velocity of A as it rotates around B.

$\mathbf{a}_{A/B}$ = relative acceleration of A as it rotates around B.

Thus in plane motion the term '$\boldsymbol{\omega}\times(\boldsymbol{\omega}\times\boldsymbol{\rho})$' becomes $-\boldsymbol{\rho}\boldsymbol{\omega}^2$

Another method of computing velocities and accelerations in rigid bodies is to draw the vector diagrams. These are drawn based on the concept that relative velocity is always perpendicular to the line joining the two points and the two components of relative acceleration are perpendicular to each other. Sometimes a rotating frame of reference is used to represent the motion of a body translating with velocity **v**, and acceleration **a**, on another rotating body as in case of a crank and slotted lever linkage. Here the acceleration expression becomes

$$\mathbf{v_A} = \mathbf{v_B} + (\boldsymbol{\omega} \times \boldsymbol{\rho}) + \mathbf{v}$$

and

$$\mathbf{a_A} = \mathbf{a_B} + \boldsymbol{\omega} \times (\boldsymbol{\omega} \times \boldsymbol{\rho}) + \boldsymbol{\alpha} \times \boldsymbol{\rho} + 2\boldsymbol{\omega} \times \mathbf{v_{AB}} + \mathbf{a}$$

Here the term '$2\boldsymbol{\omega}\times\mathbf{v_{AB}}$' is called coriolis component of acceleration.

1.16 MOMENTS OF INERTIA

The axial moment of inertia or the second moment of inertia I, of an element of area about an axis in its plane (Fig.1.6) is given by

$$dI_x = y^2 dA$$

$$dI_y = x^2 dA \quad ...(1.51)$$

The polar moment of inertia J, of an element about an axis perpendicular to its plane is the product of the area of the element and square of its distance from the axis. Referring to Fig.1.6, the polar moment of inertia is

$$dJ = \rho^2 dA = (x^2 + y^2)dA = dI_y + dI_x \quad ...(1.52)$$

The product of inertia of an element of area in Fig.1.6 is given by

$$dI_{xy} = xy\,dA \quad ...(1.53)$$

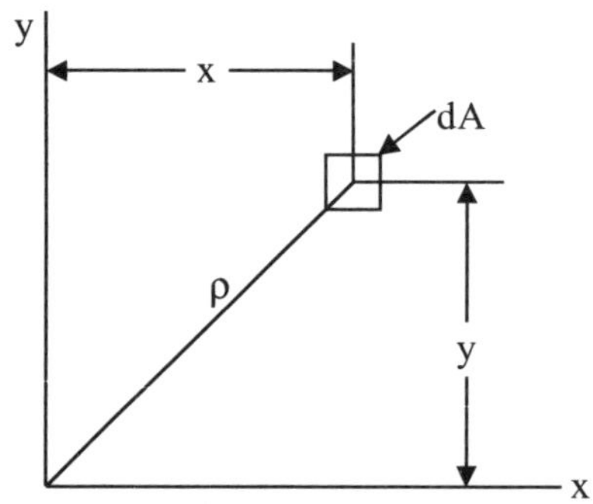

Fig. 1.6 Definition of product of inertia

The axial moment of inertia of an area is the sum of the axial moments of inertia of its elements

$$I_x = \int y^2 dA$$

$$I_y = \int x^2 dA \quad ...(1.54)$$

The radius of gyration of an area with respect to an axis is given by

$$k = \sqrt{I/A} \quad ...(1.55)$$

The polar moment of inertia of an area is the sum of the polar moments of inertia of its elements

$$J = \int \rho^2 \, dA \quad ...(1.56)$$

The product of inertia of an area is the sum of the products of inertia of its elements

$$I_{xy} = \int xy dA \quad ...(1.57)$$

The parallel-axis theorem states that the axial or polar moment of inertia of an area about any axis equals the axial or polar moment of inertia of the area about a parallel axis through the centroid of the area plus the product of the area and the square of the distance between the two parallel axes. Referring to Fig.1.7, we have

$$I_x = I_{x'} + Am^2$$
$$I_y = I_{y'} + An^2$$
$$I_z = J + Ar^2$$
$$I_{xy} = I_{xy} + Amn \quad \text{...(1.58)}$$

where A = area

I_{xy} = product of inertia of the area with respect to x and y axes

$I_{x'y'}$ = product of inertia about two parallel centroid axes x' and y'

m, n = coordinates of G relative to the (x, y) axes through O or the coordinates of O relative to the (x', y') axes through G.

x, y = any axes through O

$x'y'$ = coplanar parallel axes through the centroid G.

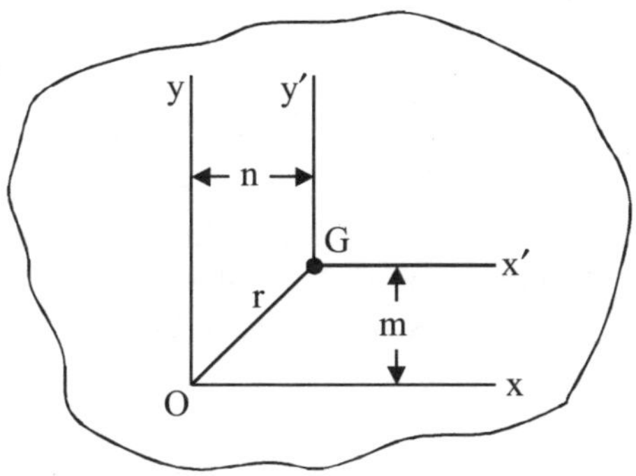

Fig. 1.7 Parallel-axis theorem

The axial or polar moment of inertia or product of inertia of a composite area is the sum of the axial or polar moments of inertia, or products of inertia, of the component areas of the whole area.

Mass moment of inertia of a rigid-body is defined as: $I = \int r^2\, dm$, where r is perpendicular distance from the z-axis to the arbitrary element dm. Thus the value of I differs for each axis about which it is computed. Usually in planar kinetics, the axis which is generally chosen for analysis passes through body's mass center G and is always perpendicular to the plane of motion. It's units are kgm^2. Sometimes I is given as radius of gyration about center G. The radius of gyration k of a body wiıh respect to an axis is given by

$$k = \sqrt{I/m} \quad \text{...(1.59)}$$

Consider Fig.1.8, the axial moment of inertia of a mass dm is given by

$$I_{xx} = I_x = \int (y^2 + z^2)\, dm \text{ and } I_{xy} = I_{yx} = \int xy\, dm$$
$$I_{yy} = I_y = \int (x^2 + z^2)\, dm \text{ and } I_{yz} = I_{zy} = \int yz\, dm$$
$$I_{zz} = I_z = \int (x^2 + y^2)\, dm \text{ and } I_{zx} = I_{xz} = \int xz\, dm \quad \text{...(1.60)}$$

where dm = mass of element

I_x, I_y, I_z = axial moments of inertia with respect to the x, y and z axes respectively.

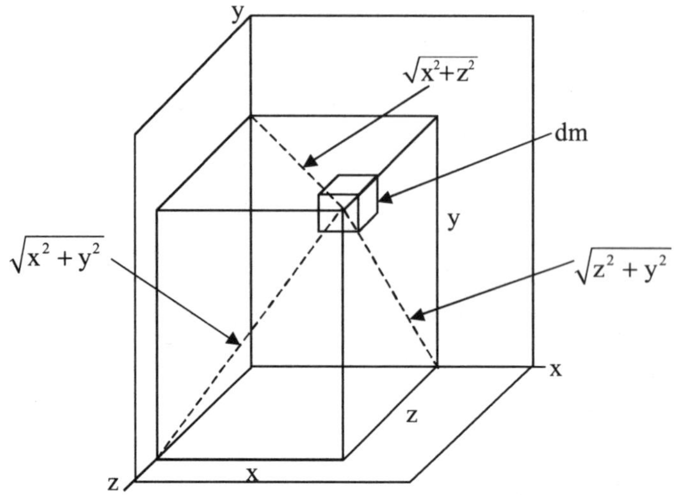

Fig. 1.8 Polar moment of inertia of composite area

In three dimensions, a body has six components of inertia for any specified x, y, z axes. Three of these are moments of inertia about each of the axes I_x, I_y, I_z and three are products of inertia each defined from two orthogonal planes I_{xy}, I_{yz}, I_{xz}. If either one or both of the planes are planes of symmetry, then the product of inertia with respect to these planes will be zero. Such axes are principal axes of inertia.

The product of inertia of a mass is given by

$$I_{xy} = \int xy\, dm$$

The parallel axis theorem states that the moment of inertia of a body about an axis is equal to the moment of inertia I about a parallel axis through the center of gravity of the body plus the product of the mass of the body and the square of the distance between the two parallel axes. For example, moment of inertia of slender rod about one end can be computed in terms of moment of inertia about mass center G according to the following relation: $I_C = I_G + m\left(\frac{\ell}{2}\right)^2$. A summary of moments of inertia of some geometrical shapes are given in Appendix-1.

1.17 DYNAMICS OF A RIGID BODY IN PLANE MOTION

When the motion is specified in terms of acceleration and forces, Newton's second law or D'Alembert's principle can be directly used. For a rigid-body in plane motion , there are three equations describing dynamic equilibrium: Two force relations for translation along x and y direction and one moment equation about a point usually the center of mass G. In vector form the equations of plane motion are given by

$$\Sigma\mathbf{F} = m\,\bar{\mathbf{a}} \qquad \text{...(1.61)}$$

$$\Sigma\mathbf{M}_O = I_O\alpha\mathbf{k} + m\mathbf{r}_{G/O} \times \mathbf{a}_O = \left(I_O\alpha + m\bar{x}a_{Oy} - m\bar{y}a_{Ox}\right)\mathbf{k} \qquad \text{...(1.62)}$$

where $\Sigma\mathbf{F}$ = resultant of the external forces acting on the body

$\Sigma\mathbf{M}_O$ = resultant of the external moments acting on the body

m = mass of the body

$\bar{\mathbf{a}}$ = acceleration of the mass center of the body

$\mathbf{a}_O$ = acceleration of reference point O

α = angular acceleration of the body

I_O = moment of inertia of the body relative to the reference point O

$\bar{x}, \bar{y}$ = coordinates of the mass center relative to the reference point O

$\mathbf{r}_{G/O}$ = position vector of the mass center relative to the reference point O

a_{Ox}, a_{Oy} = magnitude of the components of the acceleration of the reference point O along the x and y axes.

The scalar equations of the plane motion are given by

$$\Sigma F_x = m\bar{a}_x$$

$$\Sigma F_y = m\bar{a}_y$$

$$\Sigma \bar{M} = \bar{I}\alpha \qquad \text{...(1.62}a\text{)}$$

where $\Sigma F_x, \Sigma F_y$ = algebraic sums of the magnitudes of the components of the external forces along the x and y axes respectively.

m = mass of the body

$\bar{a}_x, \bar{a}_y$ = components of the linear acceleration of the mass center in x and y directions respectively.

$\Sigma \bar{M}$ = algebraic sum of the moments of the external forces about the mass center.

$\bar{I}$ = moment of inertia of the body about the mass center.

α = magnitude of the angular acceleration of the body.

The scalar equations for translation of a rigid body are given by

$$\Sigma F_x = ma_x$$

$$\Sigma F_y = ma_y \qquad \text{...(1.63)}$$

and $$\Sigma M = 0$$

where $\Sigma F_x, \Sigma F_y$ = algebraic sums of the components of the external forces in the x and y directions respectively.

m = mass of the body

a_x, a_y = acceleration components in the x and y directions respectively.

ΣM = sum of the moments of the external forces about the mass center of the body.

The scalar equations of motion of a rigid body under the action of an unbalanced force system for a body with a plane of symmetry and rotates about a fixed axis perpendicular to the plane are given by

$$\Sigma F_n = m\bar{r}\omega^2$$

$$\Sigma F_t = m\bar{r}\alpha \qquad \text{...(1.64)}$$

$$\Sigma M_O = I_O\alpha$$

where ΣF_n = algebraic sum of the components of all external forces (which are the applied forces F_1, F_2, F_3, etc., the gravitational force on the body, and the reaction R of the axis on the body) along the n axis, which is the line drawn between the center of rotation O and the mass center G; note that the positive sense is from G toward O because $\bar{a}_n = \bar{r}\omega^2$ has that sense

ΣF_t = algebraic sum of the components of the external forces along the t axis, which is perpendicular to the n axis at O; note that the positive sense along this axis agrees with that of $\bar{a}_t = \bar{r}\alpha$

ΣM_O = algebraic sum of the moments of the external forces about the axis of rotation through O; note that positive sense agrees with the assumed sense of the angular acceleration α

m = mass of the body

G = center of mass of the body

$\bar{r}$ = distance from the center of rotation O to the mass center G

I_O = moment of inertia of the body about the axis of rotation

ω = angular speed of the body

α = magnitude of the angular acceleration of the body

This type of rotation is called *non-centroidal rotation*. Example: a lever oscillating about a point of suspension. When G and O coincide (rotation about a fixed axis through G) and $F = 0$, the equations of motion are given by

$$\Sigma F_x = 0,\ \Sigma F_y = 0, \Sigma \bar{M} = \bar{I}\alpha \quad \text{...(1.65)}$$

where ΣF_x = algebraic sum of the components of the external forces along any axis chosen as the x-axis.

ΣF_y = algebraic sum of the components of the external forces along the y-axis

$\Sigma \bar{M}$ = algebraic sum of the moments of the external forces about the axis of rotation through the mass center G (axis of symmetry)

$\bar{I}$ = moment of inertia of the body about the axis of rotation through the mass center G

α = magnitude of the angular acceleration of the body

This type of rotation is called *centroidal rotation*.

1.18 WORK AND ENERGY

The work done to move particle or a body from point 1 to point 2 by the resultant force F acting on the particle is given by

$$U_{1-2} = \int_1^2 F \bullet dr \quad \text{...(1.66)}$$

The kinetic energy T of a particle with mass m and moving with speed v is defined as $\frac{1}{2}mv^2$. In connected rigid bodies the total kinetic energy of all the bodies at a configuration is computed.

The work done on a particle by the resultant force as it moves from point 1 to point 2 is equal to the change in kinetic energy.

$$U_{1-2} = T_2 - T_1 \qquad ...(1.67)$$

where T_1, T_2 = initial and final kinetic energy respectively at points 1 and 2.

The kinetic energy T of a rigid-body in translation is

$$T = \frac{1}{2}mv^2.$$

The kinetic energy T of a rigid-body in rotation is

$$T = \frac{1}{2}I_O\omega^2 \qquad ...(1.68)$$

where I_O = mass moment of inertia of the body about the axis of rotation

ω = angular speed

The kinetic energy T of a body in plane motion is given by

$$T = \frac{1}{2}\bar{I}\omega^2 \qquad ...(1.69)$$

where ω = angular speed

$\bar{I}$ = moment of inertia about an axis through the mass center parallel to the z-axis.

The change in potential energy may be defined as the negative of the work done by the conservative force acting on the body in bringing it from the datum to a final position. The selection of the datum is arbitrary. The principle of work and energy states that 'the work done by the external forces acting on a rigid body during a displacement is equal to the change in kinetic energy of the body during the same displacement'. The sum of the work done by the non-conservative external forces such as friction and the work done by the internal forces acting on a system of particles is equal to the change in the total (kinetic and potential) energy of the system of the particles over the time interval of the action

or $$E = T + V \qquad ...(1.70)$$

where E = total energy (kinetic and potential) of conservative system

T = kinetic energy

V = potential energy

The law of conservation of energy states that if a particle (or body) is acted upon by a conservation force system, the sum of the kinetic energy and potential energy is a constant. Thus for non-conservative system, principle of work and energy is used and for conservative system, principle of conservation of energy can be employed.

1.19 IMPULSE AND MOMENTUM

In particles, there is only linear momentum since it has no angular motion.

The *linear momentum* of the ith particle is defined as

$$\mathbf{L}_i = m_i\mathbf{v}_i \qquad ...(1.71)$$

where m_i = mass of the ith particle

$\mathbf{v}_i$ = ith particle's velocity.

The linear momentum of the system of particles is the sum of the linear momentum of the particles.

or $$\mathbf{L} = \sum_{i=1}^{n} \mathbf{L}_i \qquad ...(1.72)$$

The linear impulse acting on the system imparted over a time interval is given by

$$\mathbf{G}_{1-2} = \int_{t_1}^{t_2} \mathbf{F}dt \qquad ...(1.73)$$

Then the principle of impulse and momentum can be written as:

$$\mathbf{G}_{1-2} = \mathbf{L}_2 - \mathbf{L}_1 \qquad ...(1.74)$$

where $\mathbf{L}_p$ = linear momentum of the system at a state p.

Equation (1.74) states that the linear impulse acting on the system is equal to the change in the system's linear momentum over the time interval.

For a rigid body, there is angular momentum and angular impulse in addition to linear counterparts. Angular impulse is created by a moment of a force while the angular momentum is due to inertia of the body and angular velocity of rotation. Angular momentum sometimes called as the moment of the linear momentum is defined as:

$$\mathbf{H}_o = \mathbf{r} \times \mathbf{L} \qquad ...(1.75)$$

where $\mathbf{H}_O$ is the angular momentum about the point O.

The angular momentum of a system of n rigid bodies about a point A is given by

$$\mathbf{H}_A = \sum_{i=1}^{n} \mathbf{H}_{Ai} = \sum_{i=1}^{n} \mathbf{r}_{i/A} \times \mathbf{L}_i = \sum_{i=1}^{n} \mathbf{r}_{i/A} \times m_i v_i \qquad ...(1.76)$$

where A is any point.

The resultant external moment about the point A is given by

$$\mathbf{M}_A = \sum_{i=1}^{n} M_{Ai} = \sum_{i=1}^{n} r_{i/A} \times \mathbf{F}_i \qquad ...(1.77)$$

Differentiating Eq. (1.76) with respect to time yields

$$\dot{\mathbf{H}}_A = -\mathbf{v}_A \times L + \mathbf{M}_A \qquad ...(1.78)$$

If the point A is fixed then $\mathbf{v}_A = 0$ and if the point A is the mass center C then $\mathbf{v}_A \times \mathbf{L} = \mathbf{v}_C \times m\mathbf{v}_C = 0$. Thus, whether the point A represents a fixed point O or the mass center C:

$$\dot{\mathbf{H}}_A = \mathbf{M}_A \qquad ...(1.79)$$

Equation (1.79) states that the time derivative of the angular momentum about a fixed point O (or the mass center O) is equal to the resultant external moment about the fixed point O (or the mass center C). Integrating Eq. (1.79) yields

$$\mathbf{N}_{A1-2} = \int_{t_1}^{t_2} \mathbf{M}_A dt = \int_{t_1}^{t_2} \dot{\mathbf{H}}_A dt = \mathbf{H}_A(t_2) - \mathbf{H}_A(t_1) = \mathbf{H}_{A2} - \mathbf{H}_{A1} \qquad ...(1.80)$$

where $\mathbf{N}_{A1-2} = \int_{t_1}^{t_2} \mathbf{M}_A dt$ angular impulse imparted over the time interval

Hence $$\mathbf{N}_{A1-2} = \mathbf{H}_{A2} - \mathbf{H}_{A1} \qquad ...(1.81)$$

Equation (1.81) states that the angular impulse acting on the system about the fixed point O (or the mass center G) is equal to the change in the system's angular momentum about the point O (or the mass center G) over the time interval. This is illustrated in Fig.1.9, where initial and final velocities are interest are taken at mass center G.

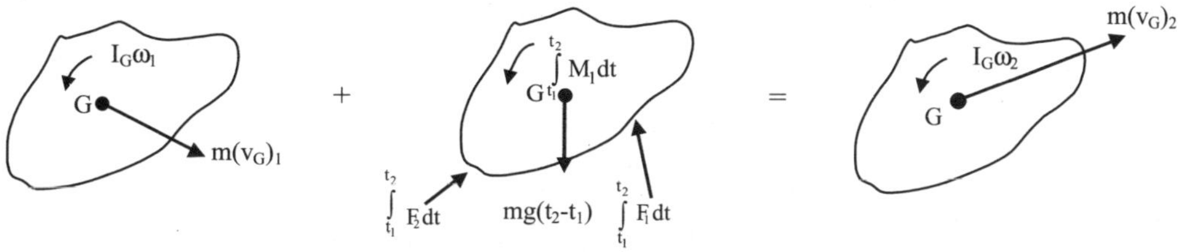

Fig.1.9 Principle of Impulse-Momentum of rigid body

Conservation of linear momentum in a given direction occurs if the sum of the external forces in that direction is zero. For examples in case of two balls (particles) colliding head-on with each other either centrally or obliquely, the linear momentum is conserved along the line of collision. Mathematically, it is written as $\Sigma L_p = 0$. Conservation of angular momentum about an axis occurs if the sum of the moments of the external forces about that axis is zero. Mathematically it is written as: $\Sigma H_p = 0$. This occurs in cases where the rotating or oscillating body suffers different speeds when the point of suspension changes. As an example, a diver jumping from a height into water maintains the constant angular momentum during his motion.

1.20 THREE-DIMENSIONAL MECHANICS

In three dimensional motion, the angular velocity and acceleration vector has components in more than one axis, unlike in plane motion where for example $\boldsymbol{\omega} = \omega\hat{k}$ a single component parallel to z-axis. Similar to two dimensional motion the motion of two points A and B on a body, or a series of connected bodies can be related using relative motion analysis with rotating and translating axes at B. If a body undergoes general motion, then the motion of a point A in the body can be related to the motion of another point B using a relative motion analysis along with translating axes at B:

$$\mathbf{v_A} = \mathbf{v_B} + \boldsymbol{\omega} \times \mathbf{r}_{AB}$$

$$\mathbf{a_A} = \mathbf{a_B} + \boldsymbol{\alpha} \times \mathbf{r}_{AB} + \boldsymbol{\omega} \times (\boldsymbol{\omega} \times \mathbf{r}_{AB})$$

There are three scalar equations of translational motion for a rigid-body that moves in three dimensions.

$$\Sigma F_x = m(a_{Gx})$$

$$\Sigma F_y = m(a_{Gy})$$

$$\Sigma F_z = m(a_{Gz})$$

The three scalar equations of rotational motion depend upon the location of the x, y, z reference.

$$\Sigma M_x = I_x \dot{\omega}_x - (I_y - I_z)\,\omega_y\omega_z$$

$$\Sigma M_y = I_y \dot{\omega}_y - (I_z - I_x)\,\omega_z\omega_x$$

$$\Sigma M_z = I_z \dot{\omega}_z - (I_x - I_y)\,\omega_x\omega_y$$

Most often, these axes are oriented so that they are the principal axes of inertia. If the axes are fixed in and move with the rotation ω of the body, then the equations are referred to as Euler equations of motion.

The angular motion of a gyroscope is best described using the changes in motion of the three Euler angles. These angular velocity components are the precision $\dot{\phi}$, nutation $\dot{\theta}$ and spin $\dot{\psi}$. If $\psi = 0$ and ϕ and θ are constant then the motion is referred to as steady precision. The angular velocity of the body is specified only in terms of Euler angle θ as:

$$\omega = \omega_x \mathbf{i} + \omega_y \mathbf{j} + \omega_z \mathbf{k} = \dot{\theta}\mathbf{i} + \dot{\phi}\sin\theta\,\mathbf{j} + \left(\dot{\phi}\cos\theta + \dot{\psi}\right)\mathbf{k}$$

The spin velocity is given by $\Omega = \Omega_x \mathbf{i} + \Omega_y \mathbf{j} + \Omega_z \mathbf{k} = \dot{\theta}\,\mathbf{i} + \dot{\phi}\sin\theta\,\mathbf{j} + \dot{\phi}\cos\theta\,\mathbf{k}$

It is spin of a gyro rotor that is responsible for holding the rotor from falling downward and instead causing it to precess about a vertical axis. This phenomenon is called the 'gyroscopic effect'.

1.21 SUMMARY

This chapter has briefly presented the fundamental concepts of engineering mechanics. Starting with equilibrium relations, topics such as trusses and beams, centroids, moment of inertia, friction, kinematics of particle, kinetics of particle, kinematics of rigid bodies and planar kinetics of bodies along with dynamics of bodies in three dimensional motions were briefly presented with simple relations. Student undergone Mechanics course in earlier semesters can easily understand the concepts presented in this chapter. Next chapter highlights the important features of MATLAB.

CHAPTER 2

MATLAB BASICS

2.1 INTRODUCTION

This chapter is a brief introduction to **MATLAB** (an abbreviation of **MAT**rix **LAB**oratory) basics, registered trademark of computer software, version 4.0 or later developed by the Math Works Inc. The software is widely used in many of science and engineering fields. MATLAB is an interactive program for numerical computation and data visualization. MATLAB is supported on Unix, Macintosh, and Windows environments. For more information on MATLAB, contact **the mathworks.com**. A Windows version of MATLAB is assumed here. The syntax is very similar for the DOS version.

MATLAB integrates mathematical computing, visualization, and a powerful language to provide a flexible environment for technical computing. The open architecture makes it easy to use MATLAB and its companion products to explore data, create algorithms, and create custom tools that provide early insights and competitive advantages.

Known for its highly optimized matrix and vector calculations, MATLAB offers an intuitive language for expressing problems and their solutions both mathematically and visually. Typical uses include:

- Numeric computation and algorithm development.
- Symbolic computation (with the built-in Symbolic Math functions).
- Modeling, simulation, and prototyping.
- Data analysis and signal processing.
- Engineering graphics and scientific visualization.

In this chapter, we will introduce the MATLAB environment. We will learn how to create, edit, save, run, and debug M-files (ASCII files with series of MATLAB statements). We will see how to create arrays (matrices and vectors), and explore the built-in MATLAB linear algebra functions for matrix and vector multiplication, dot and cross products, transpose, determinants, and inverses, and for the solution of linear equations. MATLAB is based on the language C, but is generally much easier to use. We will also see how to program logic constructs and loops in MATLAB, how to use subprograms and functions, how to use comments (%) for explaining the programs and tabs for easy readability, and how to print and plot graphics both two and three dimensional. MATLAB's functions for symbolic mathematics are presented. Use of these functions to perform symbolic operations, to develop closed form expressions for solutions to algebraic equations, ordinary differential equations, and system of equations was presented. Symbolic mathematics can also be used to determine analytical expressions for the derivative and integral of an expression.

2.1.1 Starting and Quitting MATLAB

To start MATLAB click on the MATLAB icon or type in MATLAB, followed by pressing the *enter* or *return* key at the system prompt. The screen will produce the MATLAB **prompt** >> (or EDU >>), which indicates that MATLAB is waiting for a command to be entered.

In order to quit MATLAB, type **quit** or **exit** after the prompt, followed by pressing the *enter* or *return* key.

2.1.2 Display Windows

MATLAB has three display windows. They are

1. *A Command Window* which is used to enter commands and data to display plots and graphs.
2. *A Graphics Window* which is used to display plots and graphs.
3. *An Edit Window* which is used to create and modify *M*-files. *M*-files are files that contain a program or script of MATLAB commands.

2.1.3 Entering Commands

Every command has to be followed by a carriage return <cr> (enter key) in order that the command can be executed. MATLAB commands are case sensitive and *lower case* letters are used throughout.

To execute an *M*-file (such as Project_1.m), simply enter the name of the file without its extension (as in Project_1).

2.1.4 MATLAB Expo

In order to see some of the MATLAB capabilities, enter the *demo* command. This will initiate the *MATLAB EXPO*. *MATLAB Expo* is a graphical demonstration environment that shows some of the different types of operations which can be conducted with MATLAB.

2.1.5 Abort

In order to *abort* a command in MATLAB, hold down the control key and press *c* to generate a local abort with MATLAB.

2.1.6 The Semicolon (;)

If a semicolon (;) is typed at the end of a command the output of the command is not displayed.

2.1.7 Typing %

When per cent symbol (%) is typed in the beginning of a line, the line is designated as a comment. When the *enter* key is pressed the line is not executed.

2.1.8 The clc Command

Typing *clc* command and pressing *enter* cleans the command window. Once the *clc* command is executed a clear window is displayed.

2.1.9 Help

MATLAB has a host of built-in functions. For a complete list, refer to MATLAB user's guide or refer to the *on-line Help*. To obtain help on a particular topic in the list, *e.g.*, inverse, type *help inv*.

2.1.10 Statements and Variables

Statements have the form

```
>> variable = expression
```

The equals ("=") sign implies the assignment of the expression to the variable. For instance, to enter a 2 × 2 matrix with a variable name *A*, we write

```
>> A == [1 2 ; 3 4] ⟨ret⟩
```

The statement is executed after the carriage return (or enter) key is pressed to display

```
A =
     1    2
     3    4
```

2.2 ARITHMETIC OPERATIONS

The symbols for arithmetic operations with scalars are summarized below in Table 2.1.

Table 2.1

Arithmetic operation	Symbol	Example
Addition	+	$6+3=9$
Subtraction	−	$6-3=3$
Multiplication	*	$6*3=18$
Right division	/	$6/3=2$
Left division	\	$6\backslash 3=3/6=1/2$
Exponentiation	^	$6\wedge 3=6^3=216$

2.3 DISPLAY FORMATS

MATLAB has several different screen output formats for displaying numbers. These formats can be found by typing the help command: help format in the Command Window. A few of these formats are shown in Table 2.2 for 2π.

Table 2.2 Display formats

Command	Description	Example
format short	Fixed-point with 4 decimal digits	>> 351/7 ans = 50.1429
format long	Fixed-point with 14 decimal digits	>> 351/7 ans = 50.14285714285715
format short e	Scientific notation with 4 decimal digits	>> 351/7 ans = 5.0143e + 001
format long e	Scientific notation with 15 decimal digits	>> 351/7 ans = 5.014285714285715e001
format short g	Best of 5 digit fixed or floating point	>> 351/7 ans = 50.143

format long g	Best of 15 digit fixed or floating point	>> 351/7 ans = 50.1428571428571
format bank	Two decimal digits	>> 351/7 ans = 50.14
format compact	Eliminates empty lines to allow more lines with information displayed on the screen	
format loose	Adds empty lines (opposite of compact)	

2.4 ELEMENTARY MATH BUILT-IN FUNCTIONS

MATLAB contains a number of functions for performing computations which require the use of logarithms, elementary math functions, and trigonometric math functions. List of these commonly used elementary MATLAB mathematical built-in functions are given in Tables 2.3 to 2.8.

Table 2.3 Common math functions

Function	Description
abs(x)	Computes the absolute value of **x**.
sqrt(x)	Computes the square root of **x**.
round(x)	Rounds **x** to the nearest integer.
fix(x)	Rounds (or truncates) **x** to the nearest integer toward 0.
floor(x)	Rounds **x** to the nearest integer toward $-\infty$.
ceil(x)	Rounds **x** to the nearest integer toward ∞.
sign(x)	Returns a value of –1 if **x** is less than 0, a value of 0 if **x** equals 0, and a value of 1 otherwise.
rem(x,y)	Returns the remainder of x/y. for example, **rem(25, 4)** is 1, and **rem(100, 21)** is 16. This function is also called a **modulus** function.
exp(x)	Computes e^x, where e is the base for natural logarithms, or approximately 2.718282.
log(x)	Computes ln **x**, the natural logarithm of **x** to the base e.
log10(x)	Computes $\log_{10}$ **x**, the common logarithm of **x** to the base 10.

Table 2.4 Exponential functions

Function	Description
exp(x)	Exponential (e^x)
log(x)	Natural logarithm
log10(x)	Base 10 logarithm
sqrt(x)	Square root

Table 2.5 Trigonometric and hyperbolic functions

Function	Description
sin(x)	Computes the sine of **x**, where **x** is in radians.
cos(x)	Computes the cosine of **x**, where **x** is in radians.
tan(x)	Computes the tangent of **x**, where **x** is in radians.
asin(x)	Computes the arcsine or inverse sine of **x**, where **x** must be between –1 and 1. The function returns an angle in radians between $-\pi/2$ and $\pi/2$.
acos(x)	Computes the arccosine or inverse cosine of **x**, where **x** must be between –1 and 1. The function returns an angle in radians between 0 and π.
atan(x)	Computes the arctangent or inverse tangent of **x**. The function returns an angle in radians between $-\pi/2$ and $\pi/2$.
atan2(y,x)	Computes the arctangent or inverse tangent of the value y/x. The function returns an angle in radians that will be between $-\pi$ and π, depending on the signs of **x** and **y**.
sinh(x)	Computes the hyperbolic sine of **x**, which is equal to $\frac{e^{\mathbf{x}} - e^{-\mathbf{x}}}{2}$.
cosh(x)	Computes the hyperbolic cosine of **x**, which is equal to $\frac{e^{\mathbf{x}} + e^{-\mathbf{x}}}{2}$.
tanh(x)	Computes the hyperbolic tangent of **x**, which is equal to $\frac{\sinh \mathbf{x}}{\cosh \mathbf{x}}$.
asinh(x)	Computes the inverse hyperbolic sine of **x**, which is equal to $\ln\left(\mathbf{x} + \sqrt{\mathbf{x}^2 + 1}\right)$.
acosh(x)	Computes the inverse hyperbolic cosine of **x**, which is equal to $\ln\left(\mathbf{x} + \sqrt{\mathbf{x}^2 - 1}\right)$.
atanh(x)	Computes the inverse hyperbolic tangent of **x**, which is equal to $\ln\sqrt{\frac{1+\mathbf{x}}{1-\mathbf{x}}}$ for $\lvert\mathbf{x}\rvert \le 1$.

Table 2.6 Round-off functions

Function	Description	Example
round(x)	Round to the nearest integer	>> round(20/6) ans = 3
fix(x)	Round towards zero	>> fix(13/6) ans = 2
ceil(x)	Round towards infinity	>> ceil(13/5) ans = 3
floor(x)	Round towards minus infinity	>> floor(–10/4) ans = –3
rem(x,y)	Returns the remainder after **x** is divided by **y**	>> rem(14,3) ans = 2
sign(x,y)	Signum function. Returns 1 if $\mathbf{x} > 0$, –1 if $\mathbf{x} < 0$, and 0 if $\mathbf{x} = 0$.	>> sign(7) ans = 1

Table 2.7 Complex number functions

Function	Description
conj(x)	Computes the complex **conjugate** of the complex number **x**. Thus, if **x** is equal to $a + ib$, then **conj(x)** will be equal to $a - ib$.
real(x)	Computes the real portion of the complex number **x**.
imag(x)	Computes the imaginary portion of the complex number **x**.
abs(x)	Computes the absolute value of **magnitude** of the complex number **x**.
angle(x)	Computes the angle using the value of **atan2(imag(x), real(x))**; thus, the angle value is between $-\pi$ and π.

Table 2.8 Arithmetic operations with complex numbers

Operation	Result
$c_1 + c_2$	$(a_1 + a_2) + i(b_1 + b_2)$
$c_1 + c_2$	$(a_1 - a_2) + i(b_1 - b_2)$
$c_1 \bullet c_2$	$(a_1a_2 - b_1b_2) + i(a_1b_2 - a_2b_1)$
$\frac{c_1}{c_2}$	$\left(\frac{a_1a_2 + b_1b_2}{a_2^2 + b_2^2}\right) + i\left(\frac{a_2b_1 - b_2a_1}{a_2^2 + b_2^2}\right)$
$\lvert c_1 \rvert$	$\sqrt{a_1^2 + b_1^2}$ (magnitude or absolute value of c_1)
c_1*	$a_1 - ib_1$ (conjugate of c_1)
(Assume that $c_1 = a_1 + ib_1$ and $c_2 = a_2 + ib_2$.)	

2.5 VARIABLE NAMES

A variable is a name made of a letter or a combination of several letters and digits. Variable names can be up to 63 (in MATLAB 7) characters long (31 characters on MATLAB 6.0). MATLAB is case sensitive. For instance, *XX*, *Xx*, *xX*, and *xx* are the names of four different variables. It should be noted here that not to use the names of a built-in functions for a variable. For instance, avoid using: sin, cos, exp, sqrt, ..., etc. Once a function name is used to define a variable, the function cannot be used.

2.6 PREDEFINED VARIABLES

MATLAB includes a number of predefined variables. Some of the predefined variables that are available to use in MATLAB programs are summarized in Table 2.9.

Table 2.9 Predefined variables

Predefined variable in MATLAB	Description
ans	Represents a value computed by an expression but not stored in variable name.
pi	Represents the number π.
eps	Represents the floating-point precision for the computer being used. This is the smallest difference between two numbers.
inf	Represents infinity which for instance occurs as a result of a division by zero. A warning message will be displayed or the value will be printed as ∞.
i	Defined as $\sqrt{-1}$, which is: 0 + 1.0000i.
j	Same as i.
NaN	Stands for Not a Number. Typically occurs as a result of an expression being undefined, as in the case of division of zero by zero.
clock	Represents the current time in a six-element row vector containing year, month, day, hour, minute, and seconds.
date	Represents the current date in a character string format.

2.7 COMMANDS FOR MANAGING VARIABLES

Table 2.10 lists commands that can be used to eliminate variables or to obtain information about variables that have been created. The procedure is to enter the command in the Command Window and the *Enter* key is to be pressed.

Table 2.10 Commands for managing variables

Command	Description
clear	Removes all variables from the memory.
clear x, y, z	Clears/removes only variables **x, y,** and **z** from the memory.
who	Lists the variables currently in the workspace.
whos	Displays a list of the variables currently in the memory and their size together with information about their bytes and class.

2.8 GENERAL COMMANDS

In Tables 2.11 to 2.15 the useful general commands on on-line help, workspace information, directory information, and general information are given.

Table 2.11 On-line help

Function	Description
help	Lists topics on which help is available.
helpwin	Opens the interactive help window.
helpdesk	Opens the web browser based help facility.
help *topic*	Provides help on *topic*.
lookfor *string*	Lists help topics containing *string*.
demo	Runs the demo program.

Table 2.12 Workspace information

Function	Description
who	Lists variables currently in the workspace.
whos	lsts variables currently in the workspace with their size.
what	Lists m-, mat-, and mex-files on the disk.
clear	Clears the workspace, all variables are removed.
clear *x y z*	Clears only variables *x*, *y*, and *z*.
clear all	Clears all variables and functions from workspace.
mlock *fun*	Locks function *fun* so that **clear** cannot remove it.
munlock *fun*	Unlocks function *fun* so that **clear** can remove it.
clc	Clears command window, command history is lost.
home	Same as **clc**.
clf	Clears figure window.

Table 2.13 Directory information

Function	Description
pwd	Shows the current working directory.
cd	Changes the current working directory.
dir	Lists contents of the current directory.
ls	Lists contents of the current directory, same as **dir**.
path	Gets or sets MATLAB search path.
editpath	Modifies MATLAB search path.
copyfile	Copies a file.
mkdir	Creates a directory.

Table 2.14 General information

Function	Description
computer	Tells you the computer type your are using.
clock	Gives you wall clock time and date as a vector.
date	Tells you the date as a string.
more	Controls the paged output according to the screen size.
ver	Gives the license and the version information about MATLAB installed on your computer.
bench	Benchmarks your computer on running MATLAB compared to other computers.

Table 2.15 Termination

Function	Description
c (Control-c)	Local abort, kills the current command execution.
quit	Quits MATLAB.
exit	Same as **quit**.

2.9 ARRAYS

An array is a list of numbers arranged in rows and/or columns. A one-dimensional array is a row or a column of numbers and a two-dimensional array has a set of numbers arranged in rows and columns. An array operation is performed *element-by-element*.

2.9.1 Row Vector

A vector is a row or column of elements.

In a row vector the elements are entered with a space or a comma between the elements inside the square brackets. For example, $x = [7\ -1\ 2\ -5\ 8]$

2.9.2 Column Vector

In a column vector the elements are entered with a semicolon between the elements inside the square brackets. For example, $x = [7;\ -1;\ 2;\ -5;\ 8]$

2.9.3 Matrix

A matrix is a two-dimensional array which has numbers in rows and columns. A matrix is entered row-wise with consecutive elements of a row separated by a space or a comma, and the rows separated by emicolons or carriage returns. The entire matrix is enclosed within square brackets. The elements of the matrix may be real numbers or complex numbers. For example, to enter the matrix,

$$A = \begin{bmatrix} 1 & 3 & -4 \\ 0 & -2 & 8 \end{bmatrix}$$

The MATLAB input command is

```
A = [1 3 -4 ; 0 -2 8]
```

Similarly for complex number elements of a matrix B

$$B = \begin{bmatrix} -5x & \ln 2x + 7\sin 3y \\ 3i & 5 - 13i \end{bmatrix}$$

The MATLAB input command is

```
B = [-5*x  log(2*x) + 7*sin(3*y) ; 3i 5 - 13i]
```

2.9.4 Addressing Arrays

A colon can be used in MATLAB to address a range of elements in a vector or a matrix.

2.9.4.1 Colon for a vector

Va(:) – refers to all the elements of the vector **Va** (either a row or a column vector).
Va(*m*:*n*) – refers to elements *m* through *n* of the vector **Va**.
For instance

```
>> V=[2 5 –1 11 8 4 7 –3 11]
>> u=V(2:8)
   u=
      5 –1 11 8 4 7 –3 11
```

2.9.4.2 Colon for a matrix

Table 2.16 gives the use of a colon in addressing arrays in a matrix.

Table 2.16 Colon use for a matrix

Command	Description
A(:, *n*)	Refers to the elements in all the rows of a column *n* of the matrix *A*.
A(*n*, :)	Refers to the elements in all the columns of row *n* of the matrix *A*.
A(:, *m*:*n*)	Refers to the elements in all the rows between columns *m* and *n* of the matrix *A*.
A(*m*:*n*, :)	Refers to the elements in all the columns between rows *m* and *n* of the matrix *A*.
A(*m*:*n*, *p*:*q*)	Refers to the elements in rows *m* through *n* and columns *p* through *q* of the matrix *A*.

2.9.5 Adding Elements to a Vector or a Matrix

A variable that exists as a vector, or a matrix, can be changed by adding elements to it. Addition of elements is done by assigning values of the additional elements, or by appending existing variables. Rows and/or columns can be added to an existing matrix by assigning values to the new rows or columns.

2.9.6 Deleting Elements

An element, or a range of elements, of an existing variable can be deleted by reassigning blanks to these elements. This is done simply by the use of square brackets with nothing typed in between them.

2.9.7 Built-in Functions

Some of the built-in functions available in MATLAB for managing and handling arrays as listed in Table 2.17.

Table 2.17 Built-in functions for handling arrays

Function	Description	Example
length(A)	Returns the number of elements in the vector **A**.	>> *A* = [5 9 2 4]; >> length(*A*) ans = 4
size(A)	Returns a row vector [*m*, *n*], where *m* and *n* are the size $m \times n$ of the array **A**.	>>*A* = [2 3 0 8 11 ; 6 17 5 7 1] *A* = 2 3 0 8 11 6 17 5 7 1 >> size(*A*) ans = 2 5
reshape(A, m, n)	Rearrange a matrix *A* that has *r* rows and *s* columns to have *m* rows and *n* columns. *r* times *s* must be equal to *m* times *n*.	>> *A* = [3 1 4 ; 9 0 7] *A* = 3 1 4 9 0 7 >> *B* = reshape(A, 3, 2) *B* = 3 0 9 4 1 7
diag(v)	When *v* is a vector, creates a square matrix with the elements of *v* in the diagonal	>> *v* = [3 2 1]; >> *A* = diag(*v*) *A* = 3 0 0 0 2 0 0 0 1
diag(A)	When *A* is a matrix, creates a vector from the diagonal elements of *A*.	>> *A* = [1 8 3 ; 4 2 6 ; 7 8 3] *A* = 1 8 3 4 2 6 7 8 3 >> vec = diag(*A*) vec = 1 2 3

2.10 OPERATIONS WITH ARRAYS

We consider here matrices that have more than one row and more than one column.

2.10.1 Addition and Subtraction of Matrices

The addition (the sum) or the subtraction (the difference) of the two arrays is obtained by adding or subtracting their corresponding elements. These operations are performed with arrays of identical size (same number of rows and columns).

For example, if A and B are two arrays (2×3 matrices).

$$A = \begin{bmatrix} a_{11} & a_{12} & a_{13} \\ a_{21} & a_{22} & a_{23} \end{bmatrix} \text{ and } B = \begin{bmatrix} b_{11} & b_{12} & b_{13} \\ b_{21} & b_{22} & b_{23} \end{bmatrix}$$

Then, the matrix addition ($A + B$) is obtained by adding A and B is

$$\begin{bmatrix} a_{11} + b_{11} & a_{12} + b_{12} & a_{13} + b_{13} \\ a_{21} + b_{21} & a_{22} + b_{22} & a_{23} + b_{23} \end{bmatrix}$$

2.10.2 Dot Product

The dot product is a scalar computed from two vectors of the same size. The scalar is the sum of the products of the values in corresponding positions in the vectors.
For n elements in the vectors A and B:

$$\text{dot product} = A \bullet B = \sum_{i=1}^{n} a_i \, b_i$$

dot(A, B): Computes the dot product of A and B. If A and B are matrices, the dot product is a row vector containing the dot products for the corresponding columns of A and B.

2.10.3 Array Multiplication

The value in position $c_{i,j}$ of the product C of two matrices, A and B, is the dot product of row i of the first matrix and column of the second matrix:

$$c_{i,j} = \sum_{k=1}^{n} a_{i,k} \, b_{k,j}$$

2.10.4 Array Division

The division operation can be explained by means of the identity matrix and the inverse matrix operation.

2.10.5 Identity Matrix

An identity matrix is a square matrix in which all the diagonal elements are 1's, and the remaining elements are 0's. If a matrix A is square, then it can be multiplied by the identity matrix, I, from the left or from the right:

$$AI = IA = A$$

2.10.6 Inverse of a Matrix

The matrix B is the inverse of the matrix A if when the two matrices are multiplied the product is the identity matrix. Both matrices A and B must be square and the order of multiplication can be AB or BA.

$$AB = BA = I$$

2.10.7 Transpose

The transpose of a matrix is a new matrix in which the rows of the original matrix are the columns of the new matrix. The transpose of a given matrix A is denoted by A^T. In MATLAB, the transpose of the matrix A is denoted by A'.

2.10.8 Determinant

A determinant is a scalar computed from the entries in a square matrix. For a 2 × 2 matrix A, the determinant is

$$|A| = a_{11}\, a_{22} - a_{21}\, a_{12}$$

MATLAB will compute the determinant of a matrix using the **det** function:

det(A): Computes the determinant of a square matrix A.

2.10.9 Array Division

MATLAB has two types of array division, which are the left division and the right division.

2.10.10 Left Division

The left division is used to solve the matrix equation $Ax = B$ where x and B are column vectors. Multiplying both sides of this equation by the inverse of A, A^{-1}, we have

$$A^{-1}Ax = A^{-1}B$$

or $$Ix = x = A^{-1}B$$

Hence $$x = A^{-1}B$$

In MATLAB, the above equation is written by using the left division character:

$$x = A \backslash B$$

2.10.11 Right Division

The right division is used to solve the matrix equation $xA = B$ where x and B are row vectors. Multiplying both sides of this equation by the inverse of A, A^{-1}, we have

$$x \bullet AA^{-1} = B \bullet A^{-1}$$

or $$x = B \bullet A^{-1}$$

In MATLAB, this equation is written by using the right division character:

$$x = B / A$$

2.10.12 Eigenvalues and Eigenvectors

Consider the following equation:

$$AX = \lambda X \qquad ...(2.1)$$

where A is an $n \times n$ square matrix, X is a column vector with n rows and λ is a scalar.

The values of λ for which X are nonzero are called the *eigenvalues* of the matrix A, and the corresponding values of X are called the *eigenvectors* of the matrix A.

Equation (2.1) can also be used to find the following equation:

$$(A - \lambda I)X = 0 \qquad ...(2.2)$$

where I is an $n \times n$ identity matrix. Equation (2.2) corresponding to a set of homogeneous equations and has nontrivial solutions only if the determinant is equal to zero, or

$$|A - \lambda I| = 0 \qquad ...(2.3)$$

Equation (2.3) is known as the *characteristic equation* of the matrix A. The solution to Eq. (2.3) gives the eigenvalues of the matrix A.

MATLAB determines both the eigenvalues and eigenvectors for a matrix A.

eig(A): Computes a column vector containing the eigenvalues of A.

[Q, d] = eig(A) Computes a square matrix Q containing the eigenvectors of A as columns and a square matrix d containing the eigenvlaues (λ) of A on the diagonal. The values of Q and d are such that Q^*Q is the identity matrix and A^*X equals λ times X.

***Triangular factorization or lower-upper factorization*:** Triangular or lower-upper factorization expresses a square matrix as the product of two triangular matrices—a lower triangular matrix and an upper triangular matrix. The **lu** function in MATLAB computes the LU factorization.

[L, U] = lu(A): Computes a permuted lower triangular factor in L and an upper triangular factor in U such that the product of L and U is equal to A.

***QR factorization*:** The **QR** factorization method factors a matrix A into the product of an orthonormal matrix and an upper-triangular matrix. The **qr** function is used to perform the **QR** factorization in MATLAB.

[Q, R] = qr(A): Computes the values of Q and R such that $A = QR$. Q will be an orthonormal matrix, and R will be an upper triangular matrix.

For a matrix A of size $m \times n$, the size of Q is $m \times m$, and the size of R is $m \times n$.

***Singular Value Decomposition (SVD)*:** Singular value decomposition decomposes a matrix A (size $m \times n$) into a product of three matrix factors.

$$A = USV$$

where U and V are orthogonal matrices and S is a diagonal matrix. The size of U is $m \times m$, the size of V is $n \times n$, and the size of S is $m \times n$. The values on the diagonal matrix S are called singular values. The number of nonzero singular values is equal to the rank of the matrix.

The *SVD* factorization can be obtained using the **svd** function.

[U, S, V] = svd(A): Computes the factorization of A into the product of three matrices, USV, where U and V are orthogonal matrices and S is a diagonal matrix.

svd(A): Returns the diagonal elements of S, which are the singular values of A.

2.11 ELEMENT-BY-ELEMENT OPERATIONS

Element-by-element operations can only be done with arrays of the same size. Element-by-element multiplication, division, and exponentiation of two vectors or matrices is entered in MATLAB by typing a period in front of the arithmetic operator. Table 2.18 lists these operations.

Table 2.18 Element-by-element operations

Arithmetic operators			
Matrix operators		Array operators	
+	Addition	+	Addition
–	Subtraction	–	Subtraction
*	Multiplication	•*	Array multiplication
^	Exponentiation	•^	Array exponentiation
/	Left division	•/	Array left division
\	Right division	•\	Array right division

2.11.1 Built-in Functions for Arrays

Table 2.19 lists some of the many built-in functions available in MATLAB for analyzing arrays.

Table 2.19 MATLAB built-in array functions

Function	Description	Example
mean(A)	If *A* is a vector, returns the mean value of the elements	>> *A* = [3 7 2 16]; >> mean(*A*) ans = 14
C = max(A)	If *A* is a vector, *C* is the largest element in *A*. If *A* is a matrix, *C* is a row vector containing the largest element of each column of *A*.	>> *A* = [3 7 2 16 9 5 18 13 0 4]; >> *C* = max(*A*) *C* = 18
[d, n] = max(A)	If *A* is a vector, *d* is the largest element in *A*, *n* is the position of the element (the first if several have the max value).	>> [*d*, *n*] = max(*A*) *d* = 18 *n* = 7
min(A)	The same as **max(A)**, but for the smallest element.	>> *A* = [3 7 2 16]; >> min(*A*) ans = 2
[d, n] = min(A)	The same as **[d, n] = max(A)**, but for the smallest element.	
sum(A)	If *A* is a vector, returns the sum of the elements of the vector.	>> *A* = [3 7 2 16]; >> sum(*A*) ans = 28
sort(A)	If *A* is a vector, arranges the elements of the vector in ascending order.	>> *A* = [3 7 2 16]; >> sort(*A*) ans = 2 3 7 16
median(A)	If *A* is a vector, returns the median value of the elements of the vector.	>> *A* = [3 7 2 16]; >> median(*A*) ans =5
std(A)	If *A* is a vector, returns the standard deviation of the elements of the vector.	>> *A* = [3 7 2 16]; >> std(*A*) ans = 6.3770
det(A)	Returns the determinant of a square matrix *A*.	>> *A* = [1 2 ; 3 4]; >> det(*A*) ans = 2
dot(a, b)	Calculates the scalar (dot) product of two vectors *a* and *b*. The vector can each be row or column vectors.	>> *a* = [5 6 7]; >> *b* = [4 3 2]; >> dot(*a*,*b*) ans = 52
cross(a, b)	Calculates the cross product of two vectors *a* and *b*, (*a* × *b*). The two vectors must have 3 elements	>> *a* = [5 6 7]; >> *b* = [4 3 2]; >> cross(*a*, *b*) ans = –9 18 –9

inv(A)	Returns the inverse of a square matrix *A*.	>> *a* = [1 2 3; 4 6 8; –1 2 3]; >> inv(*A*) ans = –0.5000 0.0000 –0.5000 –5.0000 1.5000 1.0000 3.5000 –1.0000 –0.5000

2.12 RANDOM NUMBERS GENERATION

There are many physical processes and engineering applications that require the use of *random numbers* in the development of a solution.

MATLAB has two commands *rand* and *rand n* that can be used to assign random numbers to variables.

The *rand* command: The *rand* command generates uniformly distributed over the interval [0, 1]. A *seed value* is used to initiate a random sequence of values. The seed value is initially set to zero. However, it can be changed with the *seed* function.

The command can be used to assign these numbers to a scalar, a vector, or a matrix, as shown in Table 2.20.

Table 2.20 The *rand* command

Command	Description	Example
rand	Generates a single random number between 0 and 1.	>> rand ans = 0.9501
rand(1, n)	Generates an *n* elements row vector of random numbers between 0 and 1.	>> *a* = rand(1, 3) *a* = 0.4565 0.0185 0.8214
rand(n)	Generates an *n* × *n* matrix with random numbers between 0 and 1.	>> *b* = rand(3) *b* = 0.7382 0.9355 0.8936 0.1763 0.9165 0.0579 0.4057 0.4103 0.3529
rand(m, n)	Generates an *m* × *n* matrix with random numbers between 0 and 1.	>> *c* = rand(2, 3) *c* = 0.2028 0.6038 0.1988 0.1987 0.2722 0.0153
randperm(n)	Generates a row vector with *n* elements that are random permutation of integers 1 through *n*.	>> randperm(7) ans = 5 2 4 7 1 6 3

2.12.1 The Random Command

MATLAB will generate Gaussian values with a mean of zero and a variance of 1.0 if a normal distribution is specified. The MATLAB functions for generating Gaussian values are as follows:

randn(n): Generates an $n \times n$ matrix containing Gaussian (or normal) random numbers with a mean of 0 and a variance of 1.
randn(m, n): Generates an $m \times n$ matrix containing Gaussian (or normal) random numbers with a mean of 0 and a variance of 1.

2.13 POLYNOMIALS

A *polynomial* is a function of a single variable that can be expressed in the following form:

$$f(x) = a_0x^n + a_1x^{n-1} + a_2x^{n-2} + \cdots + a_{n-1}x^1 + a_n$$

where the variable is x and the coefficients of the polynomial are represented by the values a_0, a_1, ... and so on. The *degree* of a polynomial is equal to the largest value used as an exponent.

A vector represents a polynomial in MATLAB. When entering the data in MATLAB, simply enter each coefficient of the polynomial into the vector in descending order. For example, consider the polynomial

$$5s^5 + 7s^4 + 2s^2 - 6s + 10$$

To enter this into MATLAB , we enter this as a vector as

```
>>x = [5 7 0 2 -6 10]
  x =
        5 7 0 2 -6 10
```

It is necessary to enter the coefficients of all the terms.
MATLAB contains functions that perform polynomial multiplication and division, which are listed below:
conv(a, b): Computes a coefficient vector that contains the coefficients of the product of polynomials represented by the coefficients in **a** and **b**. The vectors **a** and **b** do not have to be the same size.
[q, r] = deconv(n, d): Returns two vectors. The first vector contains the coefficients of the quotient and the second vector contains the coefficients of the remainder polynomial.
The MATLAB function for determining the roots of a polynomial is the **roots** function:
root(a): Determines the roots of the polynomial represented by the coefficient vector **a**.
The roots function returns a column vector containing the roots of the polynomial; the number of roots is equal to the degree of the polynomial. When the roots of a polynomial are known, the coefficients of the polynomial are determined when all the linear terms are multiplied, we can use the **poly** function:
poly(r) : Determines the coefficients of the polynomial whose roots are contained in the vector **r**.
The output of the function is a row vector containing the polynomial coefficients.
The value of a polynomial can be computed using the *polyval* function, **polyval (a, x).** It evaluates a polynomial with coefficients **a** for the values in **x**. The result is a matrix the same size ad **x**. For instance, to find the value of the above polynomial at $s = 2$,

```
>>x = polyval([5 7 0 2 -6 10], 2)
  x =
        278
```

To find the roots of the above polynomial, we enter the command **roots (a)** which determines the roots of the polynomial represented by the coefficient vector **a**.

```
>>roots([5 7 0 2 -6 10])
ans =
       -1.8652
       -0.4641 + 1.0832i
       -0.4641 - 1.0832i
        0.6967 + 0.5355i
        0.6967 - 0.5355i
```

```
% or
    >> x = [5 7 0 2 -6 10]
       x =
           5 7 0 2 -6 10
    >> r = roots(x)
    r =
         -1.8652
         -0.4641 + 1.0832i
         -0.4641 - 1.0832i
          0.6967 + 0.5355i
          0.6967 - 0.5355i
```

To multiply two polynomials together, we enter the command *conv*.
The polynomials are: $x = 2x + 5$ and $y = x^2 + 3x + 7$

```
    >>x = [2 5];
    >>y = [1 3 7];
    >>z = conv(x, y)
      z =
          2 11 29 35
```

To divide two polynomials, we use the command *deconv*.

```
      z = [2 11 29 35]; x = [2 5]
      >> [g, t] = deconv (z, x)
      g = 1 3 7
      t = 0 0 0 0
```

2.14 SYSTEM OF LINEAR EQUATIONS

A system of equations is nonsingular if the matrix **A** containing the coefficients of the equations is nonsingular. A system of nonsingular simultaneous linear equations (**AX** = **B**) can be solved using two methods:

(*a*) Matrix Division Method.
(*b*) Matrix Inversion Method.

2.14.1 Matrix Division

The solution to the matrix equation **AX** = **B** is obtained using matrix division, or **X** = **A/B.** The vector *X* then contains the values of **x.**

2.14.2 Matrix Inverse

For the solution of the matrix equation **AX** = **B**, we premultiply both sides of the equation by $\mathbf{A}^{-1}$.

$$\mathbf{A}^{-1}\mathbf{AX} = \mathbf{A}^{-1}\mathbf{B}$$

or

$$\mathbf{IX} = \mathbf{A}^{-1}\mathbf{B}$$

where **I** is the identity matrix.

Hence

$$\mathbf{X} = \mathbf{A}^{-1}\mathbf{B}$$

In MATLAB, we use the command x = inv (**A**)***B**. Similarly, for **XA** = **B**, we use the command **x** = **B***inv (**A**). The basic computational unit in MATLAB is the matrix. A matrix expression is enclosed in square brackets, [ ]. Blanks or commas separate the column elements, and semicolons or carriage returns separate the rows.

```
>>A = [1 2 3 4; 5 6 7 8; 9 10 11 12]
A =
   1    2    3    4
   5    6    7    8
   9   10   11   12
```

The transpose of a simple matrix or a complex matrix is obtained by using the *apostrophe* key

```
>>B = A'
B =
   1    5    9
   2    6   10
   3    7   11
   4    8   12
```

Matrix multiplication is accomplished as follows:

```
>>C = A * B
C =
    30    70   110
    70   174   278
   110   278   446
>>C = B * A
C =
   107   122   137   152
   122   140   158   176
   137   158   179   200
   152   176   200   224
```

The inverse of a matrix D is obtained as

```
>>D = [1 2 ; 3 4]
D =
   1    2
   3    4
>>E = inv (D)
E =
  -2.0000 1.0000
   1.5000-0.5000
```

Similarly, its eigenvalue is

```
>>eig (D)
ans =
   -0.3723
    5.3723
```

Matrix operations require that the matrix dimensions be compatible. If A is an $n \times m$ and B is a $p \times r$ then $A \pm B$ is allowed only if $n = p$ and $m = r$. Similarly, matrix product $A * B$ is allowed only if $m = p$.

Example E2.1: Consider the two matrices;

$$A = \begin{bmatrix} 1 & 0 & 1 \\ 2 & 3 & 4 \\ -1 & 6 & 7 \end{bmatrix} \quad \text{and} \quad B = \begin{bmatrix} 7 & 4 & 2 \\ 3 & 5 & 6 \\ -1 & 2 & 1 \end{bmatrix}$$

Using MATLAB, determine the following:

(*a*) $\mathbf{A}+\mathbf{B}$

(*b*) $\mathbf{AB}$

(*c*) $\mathbf{A}^2$

(*d*) $\mathbf{A}^T$

(*e*) $\mathbf{B}^{-1}$

(*f*) $\mathbf{B}^T\mathbf{A}^T$

(*g*) $\mathbf{A}^2+\mathbf{B}^2-\mathbf{AB}$

(*h*) determinant of **A**, determinant of **B** and determinant of **AB**.

Solution:

```
>> A = [1 0 1; 2 3 4; -1 6 7]
A =
     1     0     1
     2     3     4
    -1     6     7
>> B = [7 4 2; 3 5 6; -1 2 1]
B =
     7     4     2
     3     5     6
    -1     2     1
```

(*a*)
```
>> C = A + B
C =
     8     4     3
     5     8    10
    -2     8     8
```

(*b*)
```
>> D = A*B
D =
     6     6     3
    19    31    26
     4    40    41
```

(*c*)
```
>> E = A^2
E =
     0     6     8
     4    33    42
     4    60    72
```

(*d*)
```
>> % Let F= transpose of A
>> F=A
F =
    1    2   -1
    0    3    6
    1    4    7
```

(*e*)
```
>> H = inv (B)
H =
    0.1111  0.0000  -0.2222
    0.1429 -0.1429    .5714
   -0.1746  0.2857  -0.3651
```

(*f*)
```
>> J = B*A
J =
   6   19    4
   6   31   40
   3   26   41
```

(*g*)
```
>> K = A^2 + B^2 - A * B
K =
   53   52   45
   15   51   58
   -2   28   42
```

(*h*)
```
det (A) = 12
det (B) = -63
det (A*B) = -756
```

Example E2.2: Determine the eigenvalues and eigenvectors of *A* and *B* using MATLAB

$$\mathbf{A} = \begin{bmatrix} 4 & 2 & -3 \\ -1 & 1 & 3 \\ 2 & 5 & 7 \end{bmatrix} \quad \text{and} \quad \mathbf{B} = \begin{bmatrix} 1 & 2 & 3 \\ 8 & 7 & 6 \\ 5 & 3 & 1 \end{bmatrix}$$

Solution:

```
% Determine the eigenvalues and eigenvectors
A = [4 2 -3 ; -1 1 3 ; 2 5 7]
   A =
      4    2   -3
     -1    1    3
      2    5    7
eig (A)
   ans = 0.5949
         3.0000
         8.4051
```

```
    lamda = eig(A)
    lamda = 0.5949
            3.0000
            8.4051
  [V, D] =eig (A)
      V  = -0.6713       0.9163     -0.3905
            0.6713      -0.3984      0.3905
           -0.3144       0.0398      0.8337

      D  =  0.5949       0           0
            0            3.0000      0
            0            8.4051      8.4051
```

Example E2.3: Determine the values of x, y, and z for the following set of linear algebraic equations:

$$x_2 - 3x_3 = -5$$
$$2x_1 + 3x_2 - x_3 = 7$$
$$4x_1 + 5x_2 - 2x_3 = 10$$

Solution:

Here

$$\mathbf{A} = \begin{bmatrix} 0 & 1 & -3 \\ 2 & 3 & -1 \\ 4 & 5 & -2 \end{bmatrix}, \mathbf{B} = \begin{bmatrix} 5 \\ 7 \\ 10 \end{bmatrix} \text{ and } \mathbf{X} = \begin{bmatrix} x_1 \\ x_2 \\ x_3 \end{bmatrix}$$

$$\mathbf{AX} = \mathbf{B}$$
$$\mathbf{A^{-1}AX} = \mathbf{A^{-1}B}$$
$$\mathbf{IX} = \mathbf{A^{-1}B}$$

or $$\mathbf{X} = \mathbf{A^{-1}B}$$

```
>> A = [0 1 -3; 2 3 -1; 4 5 -2];
>> B = [-5; 7; 10]
>> x = inv (A) * B
x     =
        -1.0000
         4.0000
         3.0000
>> check = A * x
check =
           -5
            7
           10
check =
% Alternative method
>> x = A\B
x     =
           -1
            4
            3
```

2.15 SCRIPT FILES

A *script* is a sequence of ordinary statements and functions used at the command prompt level. A script is invoked the command prompt level by typing the file-name or by using the pull down menu. Scripts can also invoke other scripts.

The commands in the Command Window cannot be saved and executed again. Also, the Command Window is not interactive. To overcome these difficulties, the procedure is first to create a file with a list of commands, save it, and then run the file. In this way the commands contained are executed in the order they are listed when the file is run. In addition, as the need arises, one can change or modify the commands in the file, the file can be saved and run again. The files that are used in this fashion are known as *script files*. Thus, a script file is a text file that contains a sequence of MATLAB commands. Script file can be edited (corrected and/or changed) and executed many times.

2.15.1 Creating and Saving a Script File

Any text editor can be used to create script files. In MATLAB script files are created and edited in the Editor/Debugger Window. This window can be opened from the Command Window. From the Command Window, select *File*, *New*, and then *M*-file. Once the window is open, the commands of the script file are typed line by line. The commands can also be typed in any text editor or word processor program and then copied and pasted in the Editor/Debugger Window. The second type of *M*-files is the *function file*. Function file enables the user to extend the basic library functions by adding ones own computational procedures. Function *M*-files are expected to return one or more results. Script files and function files may include reference to other MATLAB toolbox routines.

MATLAB function file begins with a header statement of the form:

function (name of result or results) = name (argument list)

Before a script file can be executed it must be saved. All script files must be saved with the extension ".m". MATLAB refers to them as *M*-files. When using MATLAB *M*-files editor, the files will automatically be saved with a ".m" extension. If any other text editor is used, the file must be saved with the ".m" extension, or MATLAB will not be able to find and run the script file. This is done by choosing *Save As*... from the *File* menu, selecting a location, and entering a name for the file. The names of user defined variables, predefined variables, MATLAB commands or functions should not be used to name script files.

2.15.2 Running a Script File

A script file can be executed either by typing its name in the Command Window and then pressing the *Enter* key, directly from the Editor Window by clicking on the *Run* icon. The file is assumed to be in the current directory, or in the search path.

2.15.3 Input to a Script File

There are three ways of assigning a value to a variable in a script file.

1. The variable is defined and assigned value in the script file.
2. The variable is defined and assigned value in the Command Window.
3. The variable is defined in the script file, but a specified value is entered in the Command Window when the script file is executed.

2.15.4 Output Commands

There are two commands that are commonly used to generate output. They are the *disp* and *fprintf* commands.

1. The *disp* command:

The *disp* command displays the elements of a variable without displaying the name of the variable, and displays text.

disp(name of a variable) or disp('text as string')

```
>>A = [1 2 3 ; 4 5 6 ];
>> disp(A)
          1 2 3
          4 5 6
>> disp('Solution to the problem.')
  Solution to the problem.
```

2. The *fprintf* command:

The *fprintf* command displays output (text and data) on the screen or saves it to a file. The output can be formatted using this command.

Example E2.4: Write a function file Veccrossprod to compute the cross product of two vectors **a**, and **b**, where $\mathbf{a} = (a_1, a_2, a_3)$, $b = (b_1, b_2, b_3)$, and $a \times b = (a_2b_3 - a_3b_2, a_3b_1 - a_1b_3, a_1b_2 - a_2b_1)$. Verify the function by taking the cross products of pairs of unit vectors: (**i**, **j**), (**j**, **k**), etc.

Solution:

```
    function c = Veccrossprod(a, b);
% Veccrossprod : function to compute c = a × b where a and b are 3D vectors
% call syntax:
% c = Veccrossprod(a, b);
c= [a(2) * b(3) - a(3) * b(2); a(3) * b(1) - a(1) * b(3); a(1) * b(2) - a(2) * b(1)];
```

2.16 PROGRAMMING IN MATLAB

One most significant feature of MATLAB is its extendibility through user-written programs such as the *M*-files. *M*-files are ordinary ASCII text files written in MATLAB language. A function file is a subprogram.

2.16.1 Relational and Logical Operators

A relational operator compares two numbers by finding whether a comparison statement is true or false. A logical operator examines true/false statements and produces a result which is true or false according to the specific operator. Relational and logical operators are used in mathematical expressions and also in combination with other commands, to make decision that control the flow a computer program.

MATLAB has six relational operators as shown in Table 2.21.

Table 2.21 Relational operators

Relation operator	Interpretation
<	Less than
<=	Less than or equal
>	Greater than
>=	Greater than or equal
==	Equal
~=	Not equal

The logical operators in MATLAB are shown in Table 2.22.

Table 2.22 Logical operators

Logical operator	Name	Description
& Example: *A*&*B*	AND	Operates on two operands (*A* and *B*). If both are true, the result is true (1), otherwise the result is false (0).
| Example: *A*|*B*	OR	Operates on two operands (*A* and *B*). If either one, or both are true, the result is true (1), otherwise (both are false) the result is false (0).
~ Example: ~*A*	NOT	Operates on one operand (*A*). Gives the opposite of the operand. True (1) if the operand is false, and false (0) if the operand is true.

2.16.2 Order of Precedence

The following Table 2.23 shows the order of precedence used by MATLAB.

Table 2.23

Precedence	Operation
1 (highest)	Parentheses (If nested parentheses exist, inner have precedence).
2	Exponentiation.
3	Logical NOT (~).
4	Multiplication, Division.
5	Addition, Subtraction.
6	Relational operators (>, <, >=, <=, = =, ~=).
7	Logical AND (&).
8 (lowest)	Logical OR (|).

2.16.3 Built-in Logical Functions

The MATLAB built-in functions which are equivalent to the logical operators are:

and (A, B) Equivalent to *A* & *B*
or (A, B) Equivalent to *A* | *B*
not (A) Equivalent to ~*A*

List the MATLAB logical built-in functions are described in Table 2.24.

Table 2.24 Additional logical built-in functions

Function	Description	Example
xor(a, b)	Exclusive or. Returns true (1) if one operand is true and the other is false.	>>xor(8, –1) ans = 0 >>xor(8, 0) ans = 1
all(A)	Returns 1 (true) if all elements in a vector *A* are true (nonzero). Returns 0 (false) if one or more elements are false (zero). If *A* is a matrix, treats columns of A as vectors, returns a vector with 1's and 0's.	>>*A* = [5 3 11 7 8 15] >>all(*A*) ans = 1 >>*B* = [3 6 11 4 0 13] >>all(*B*) ans = 0
any(A)	Returns 1 (true) if any element in a vector *A* is true (nonzero). Returns 0 (false) if all elements are false (zero). If *A* is a matrix, treats columns of *A* as vectors, returns a vector with 1's and 0's.	>>*A* = [5 0 14 0 0 13] >>any(*A*) ans = 1 >>*B* = [0 0 0 0 0 0] >>any(*B*) ans = 0
find(A) **find(A > d)**	If *A* is a vector, returns the indices of the nonzero elements. If *A* is a vector, returns the address of the elements that are larger than *d* (any relational operator can be used).	>>*A* = [0 7 4 2 8 0 0 3 9] >>find(*A*) ans = 2 3 4 5 8 9 >>find(*A* > 4) ans = 4 5 6

The truth table for the operation of the four logical operators, and, or, Xor, and not are summarized in Table 2.25.

Table 2.25 Truth table

INPUT		OUTPUT				
A	*B*	AND *A*&*B*	OR *A*|*B*	XOR (*A*,*B*)	NOT ~*A*	NOT ~*B*
false	false	false	false	false	true	true
false	true	false	true	true	true	false
true	false	false	true	true	false	true
true	true	true	true	false	false	false

2.16.4 Conditional Statements

A conditional statement is a command that allows MATLAB to make a decision of whether to execute a group of commands that follow the conditional statement or to skip these commands.

if conditional expression consists of relational and/or logical operators

```
if  a < 30
    count = count + 1
    disp a
    end
```

The general form of a simple **if** statement is as follows:

```
if  logical expression
          statements
end
```

If the logical expression is true, the statements between the **if** statement and the **end** statement are executed. If the logical expression is false, then it goes to the statements following the **end** statement.

2.16.5 Nested if Statements

Following is an example of *nested if* statements:

```
if  a < 30
    count = count + 1;
    disp(a);
    if  b > a
        b = 0;
    end
end
```

2.16.6 else AND elseif Clauses

The else clause allows to execute one set of statements if a logical expression is true and a different set if the logical expression is false.

```
%   variable name inc
    if  inc < 1
        x_inc = inc/10;
        else
        x_inc = 0.05;
        end
```

When several levels of **if-else** statements are nested, it may be difficult to find which logical expressions must be true (or false) to execute each set of statements. In such cases, the **elaseif** clause is used to clarify the program logic.

2.16.7 MATLAB while Structures

There is a structure in MATLAB that combines the for loop with the features of the if block. This is called the *while loop* and has the form:

while *logical expression*

This set of statements is executed repeatedly as long as the logical expressions remain true (equals +1) or if the expression is a matrix rather than a simple scalar variable, as long as *all* the elements of the matrix remain nonzero.

end

In addition to the normal termination of a loop by means of the **end** statement, there are additional MATLAB commands available to interrupt the calculations. These commands are listed in Table 2.26:

Table 2.26

Command	Description
break	Terminates the execution of MATLAB **for** and **while** loops. In nested loops, **break** will terminate only the innermost loop in which it is placed.
return	Primarily used in MATLAB functions, **return** will cause a normal return from a function from the point at which the **return** statement is executed.
error ('*text*')	Terminates execution and displays the message contained in *text* on the screen. Note, the text must be enclosed in single quotes.

The MATLAB functions used are summarized in Table 2.27 below:

Table 2.27

Function	Description
Relational operators	A MATLAB *logical relation* is a comparison between two variables **x** and **y** of the same size effected by one of the six operators, <, <=, >, >=, = =, ~=. The comparison involves corresponding elements of **x** and **y**, and yields a matrix or scalar of the same size with values of "true" or "false" for each of its elements. In MATLAB, the value of "false" is zero, and "true" has a value of one. Any nonzero quantity is interpreted as "true".
Combinatorial operators	The operators **&** (AND) and | (OR) may be used to combine two logical expressions.
all, any	If **x** is a vector, **all(x)** returns a value of one if *all* of the elements of **x** are nonzero, and a value of zero otherwise. When **X** is a matrix, all(**X**) returns a row vector of ones or zeros obtained by applying all to each of the columns of **X**. The function **any** operates similarly if *any* of the elements of **x** are nonzero.
find	If **x** is a vector, **i** = **find(x)** returns the indices of those elements of **x** that are nonzero (*i.e.*, true). Thus, replacing all the negative elements of **x** by zero could be accomplished by **i = find(x < 0);** **x(i) = zeros(size(i));** If **X** is a matrix, **[i, j]** = **find(X)**operates similarly and returns the row-column indices of nonzero elements.
if, else, elseif	The several forms of MATLAB **if** blocks are as follows: **if** *variable* **if** *variable* 1 **if** *variable* 1 block of statements block of statements block of statements

	executed if *variable* executed if *variable* 1 executed if *variable* 1 is "true", *i.e.*, nonzero is "true", *i.e.*, Nonzero is "true", **end** **else** **else if** *variable* 2 block of statements block of statements executed if *variable* 1executed if *variable* 2 is "false", *i.e.*, zero is "true", **end** **else** **end** block of statements executed if neither *variable* is "true"
break	Terminates the execution of a **for** or **while** loop. Only the innermost loop in which **break** is encountered will be terminated.
return	Causes the function to return at that point to the calling routine. MATLAB *M-file* functions will return normally without this statement.
error (*'text'*)	Within a loop or function, if the statement **error** (*'text'*) is encountered, the loop or function is terminated, and the *text* is displayed.
while	The form of the MATLAB **while** loop is **while** *variable* block of statements executed as long as the value of *Variable* is "true"; i.e., nonzero **end**
	Useful when a function *F* itself calls a second "dummy" function "*f*". For example, the function *F* might find the root of an arbitrary function identified as a generic *f*(*x*). Then, the name of the actual *M-file* function, say **fname**, is passed as a *character string* to the function *F* either through its argument list or as a global variable, and the function is evaluated within *F* by means of **feval.** The use of **feval(name, x1, x2, ..., xn)**, where **fname** is a variable containing the name of the function as a character string; i.e., enclosed in single quotes, and **x1, x2, ..., xn** are the variables needed in the argument list of function **fname**.

2.17 GRAPHICS

MATLAB has many commands that can be used to create basic 2-*D* plots, overlay plots, specialized 2-*D* plots, 3-*D* plots, mesh, and surface plots.

2.17.1 Basic 2-D Plots

The basic command for producing a simple 2-*D* plot is

plot(*x* values, *y* values, 'style option')

where

x values and *y* values are vectors containing the *x*- and *y*-coordinates of points on the graph.

Style option is an optional argument that specifies the color, line-style, and the point-marker style.

The style option in the plot command is a character string that consists of 1, 2, or 3 characters that specify the color and/or the line style. The different color, line-style and marker-style options are summarized in Table 2.28.

Table 2.28 Color, line-style, and marker-style options

Color style-option		Line style-option		Marker style-option	
y	yellow	–	solid	+	plus sign
m	magenta	– –	dashed	o	circle
c	cyan	:	dotted	*	asterisk
r	red	–.	dash-dot	x	x-mark
g	green			.	point
b	blue			^	up triangle
w	white			s	square
k	black			d	diamond, etc.

2.17.2 Specialized 2-D Plots

There are several specialized graphics functions available in MATLAB for 2-*D* plots. The list of functions commonly used in MATLAB for plotting *x*-*y* data are given in Table 2.29.

Table 2.29 List of functions for plotting x–y data

Function	Description
area	Creates a filled area plot.
bar	Creates a bar graph.
barh	Creates a horizontal bar graph.
comet	Makes an animated 2-*D* plot.
compass	Creates arrow graph for complex numbers.
contour	Makes contour plots.
contourf	Makes filled contour plots.
errorbar	Plots a graph and puts error bars.
feather	Makes a feather plot.
fill	Draws filled polygons of specified color.
fplot	Plots a function of a single variable.
hist	Makes histograms.
loglog	Creates plot with log scale on both *x* and *y* axes.
pareto	Makes pareto plots.
pcolor	Makes pseudo color plot of matrix.
pie	Creates a pie chart.
plotyy	Makes a double *y*-axis plot.
plotmatrix	Makes a scatter plot of a matrix.
polar	Plots curves in polar coordinates.
quiver	Plots vector fields.
rose	Makes angled histograms.
scatter	Creates a scatter plot.
semilogx	Makes semilog plot with log scale on the *x*-axis.
semilogy	Makes semilog plot with log scale on the *y*-axis.
stairs	Plots a stair graph.
stem	Plots a stem graph.

2.17.2.1 Overlay plots

There are three ways of generating overlay plots in MATLAB, they are:

(*a*) Plot command.

(*b*) Hold command.

(*c*) Line command.

(*a*) *Plot command*

Example E2.7(*a*) shows the use of plot command used with matrix argument, each column of the second argument matrix plotted against the corresponding column of the first argument matrix.

(*b*) *Hold command*

Invoking hold on at any point during a session freezes the current plot in the graphics window. All the next plots generated by the plot command are added to the exiting plot. See Example E2.7(*a*).

(*c*) *Line command*

The line command takes a pair of vectors (or a triplet in 3-*D*) followed by a parameter name / parameter value pairs as argument. For instance, the command: *line* (*x data, y data, parameter name, parameter value*) adds lines to the existing axes. See Example E2.7(*a*).

2.17.3 3-D Plots

MATLAB provides various options for displaying three-dimensional data. They include line and wire, surface, mesh plots, among many others. More information can be found in the Help Window under Plotting and Data visualization. Table 2.30 lists commonly used functions.

Table 2.30 Functions used for 3-D graphics

Command	Description
plot3	Plots three-dimensional graph of the trajectory of a set of three parametric equations $x(t)$, $y(t)$, and $z(t)$ can be obtained using **plot3 (x,y,z)**.
meshgrid	If **x** and **y** are two vectors containing a range of points for the evaluation of a function, **[X,Y] = meshgrid (x, y)** returns two rectangular matrices containing the x and y values at each point of a two-dimensional grid.
mesh(X,Y,z)	If **X** and **Y** are rectangular arrays containing the values of the x and y coordinates at each point of a rectangular grid , and if **z** is the value of a function evaluated at each of these points, **mesh (X,Y,z)** will produce a three-dimensional perspective graph of the points. The same results can be obtained with **mesh (x,y,z)** can also be used.
meshc, meshz	If the xy grid is rectangular, these two functions are merely variations of the basic plotting program **mesh**, and they operate in an identical fashion. **meshc** will produce a corresponding contour plot drawn on the xy plane below the three-dimensional figure, and **meshz** will add a vertical wall to the outside features of the figures drawn by **mesh**.
surf	Produces a three-dimensional perspective drawing. Its use is usually to draw surfaces, as opposed to plotting functions, although the actual tasks are quite similar. The output of **surf** will be a *shaded* figure. If row vectors of length n are defined by $x = r \cos \theta$ and $y = r \sin \theta$, with $0 \leq \theta \leq 2\pi$, they correspond to a circle of radius r. If $\vec{r}$ is a *column* vector equal to **r = [0 1 2]'**; then **z = r*ones (size(x))** will be a rectangular, $3 \times n$, arrays of 0's and 2's, and **surf (x, y, z)** will produce a shaded surface bounded by three circles; i.e., a cone.

surfc	This function is related to **surf** in the same way that **meshc** is related to mesh.
colormap	Used to change the default coloring of a figure. See the MATLAB reference manual or the help file.
shading	Controls the type of color shading used in drawing figures. See the MATLAB reference manual or the help file.
view	**view(az,el)** controls the perspective view of a three-dimensional plot. The view of the figure is from angle "**el**" above the *xy* plane with the coordinate axes (and the figure) rotated by an angle "**az**" in a clockwise direction about the *z* axis. Both angles are in degrees. The default values are **az** = 37½° and **el** = 30°.
axis	Determines or changes the scaling of a plot. If the coordinate axis limits of a two-dimensional or three-dimensional graph are contained in the row vector $r = [x_{min}, x_{max}, y_{min}, y_{max}, z_{min}, z_{max}]$, **axis** will return the values in this vector, and **axis(r)** can be used to alter them. The coordinate axes can be turned *on* and *off* with **axis ('on')** and **axis ('off')**. A few other string constant inputs to **axis** and their effects are given below: **axis('equal')** *x* and *y* scaling are forced to be the same. **axis('square')** The box formed by the axes is square. **axis('auto')** Restores the scaling to default settings. **axis('normal')** Restoring the scaling to full size, removing any effects of *square* or *equal* settings. **Axis('image')** Alters the aspect ratio and the scaling so the screen pixels are square shaped rather than rectangular.
contour	The use is **contour(x, y, z)**. A default value of N = 10 contour lines will be drawn. An optional fourth argument can be used to control the number of contour lines that are drawn. **contour(x, y, z, N)**, if **N** is a positive integer, will draw **N** contour lines, and **contour(x, y, z, V)**, if **V** is a vector containing values in the range of *z* values, will draw contour lines at each value of $z = V$.
plot3	Plots lines or curves in three dimensions. If **x, y,** and **z** are vectors of equal length, **plot3 (x, y, z)** will draw, on a three-dimensional coordinate axis system, the lines connecting the points. A fourth argument, representing the color and symbols to be used at each point, can be added in exactly the same manner as with **plot**.
grid	**grid on** adds grid lines to a two-dimensional or three-dimensional graph; **grid off** removes them.
slice	Draws "slices" of a volume at a particular location within the volume.

Example E2.5:

(*a*) Generate an overlay plot for plotting three lines

$$y_1 = \sin t$$

$$y_2 = t$$

$$y_3 = t - \frac{t^3}{3!} + \frac{t^5}{5!} + \frac{t^7}{7!}; \qquad 0 \le t \le 2\pi$$

Use (*i*) the plot command
(*ii*) the hold command
(*iii*) the line command

(*b*) Use the functions for plotting *x-y* data for plotting the following functions:

(*i*) $f(t) = t\cos t$
$0 \le t \le 10\pi$

(*ii*) $x = e^t$
$y = 100 + e^{3t}$
$0 \le t \le 2\pi$

Solution:

(*a*) Overlay plot

(*i*)
```
% using the plot command
t = linspace(0, 2*pi, 100);
y1 = sin(t); y2 = t;
y3 = t - (t.^3)/6 + (t.^5)/120 - (t.^7)/5040;
plot(t, y1, t, y2, '-', t, y3, 'o')
axis([0 5 -1 5])
xlabel('t')
ylabel('sin(t) approximation')
title('sin(t) function')
text(3.5,0, 'sin(t)')
gtext('Linear approximation')
gtext('4-term approximation')
```

Output is shown in Fig. E2.5(a).

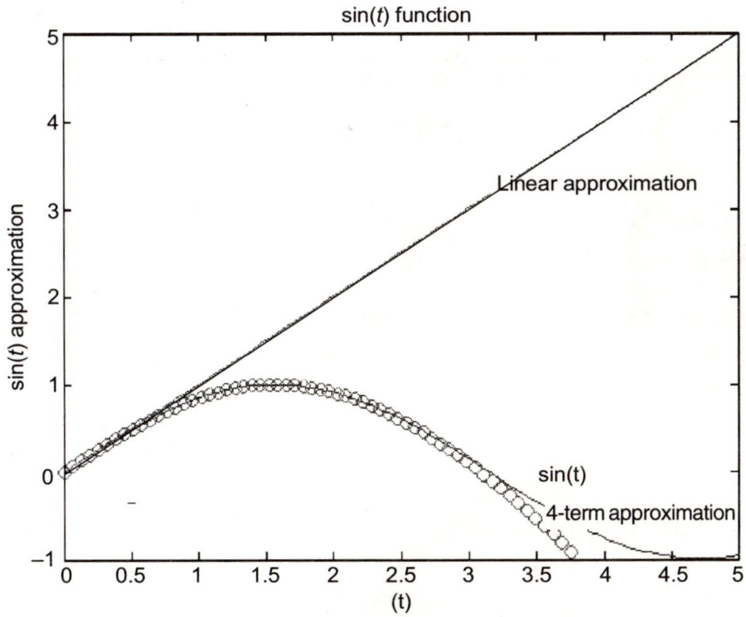

Fig. E2.5 (a)

(*ii*)
```
% using the hold command
x = linspace(0, 2*pi, 100); y1 = sin(x);
plot(x, y1)
hold on
y2 = x; plot(x, y2, '-' )
y3 = x - (x.^3)/6 + (x.^5)/120 - (t.^7)/5040;
plot(x, y3, 'o')
axis([0 5 -1 5])
hold off
```

Output is shown in Fig.E2.5(*b*).

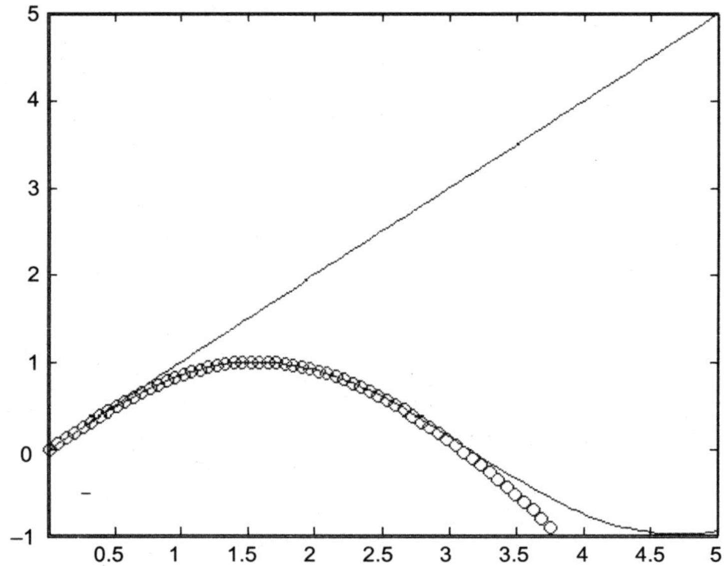

Fig.E2.5 (b)

(*iii*)
```
% using the line command
t = linspace(0, 2*pi, 100);
y1 = sin(t);
y2 = t;
y3 = t - (t.^3)/6 + (t.^5)/120 - (t.^7)/5040;
plot(t, y1)
line(t, y2, 'linestyle', '-')
line(t, y3, 'marker', 'o')
axis([0 5 -1 5])
xlabel('t')
ylabel('sin(t) approximation')
title('sin(t) function')
legend('sin(t)', 'linear approx', '7th order approx')
```

Output is shown in Fig. E2.5(*c*).

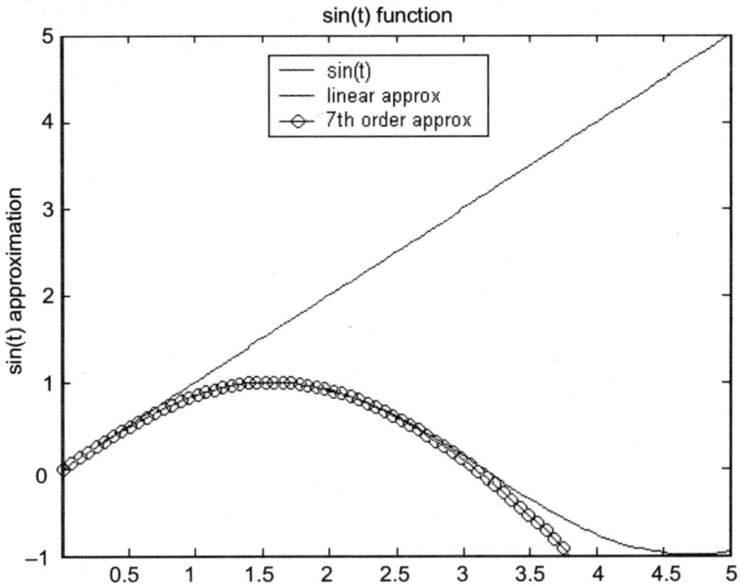

Fig. E2.5 (c)

(*b*) Using Table 2.29 functions

(*i*) `fplot('x.*cos(x)', [0 10*pi])`

This will give the following figure (Fig.E2.5 (*d*)).

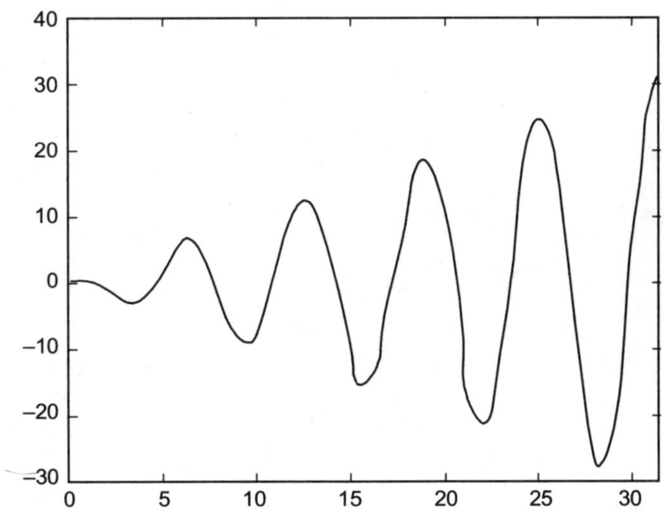

Fig.E2.5 (d)

(*ii*)
```
t = linspace(0, 2*pi, 200);
x = exp(t);
y = 100 + exp(3*t);
loglog(x, y), grid
```

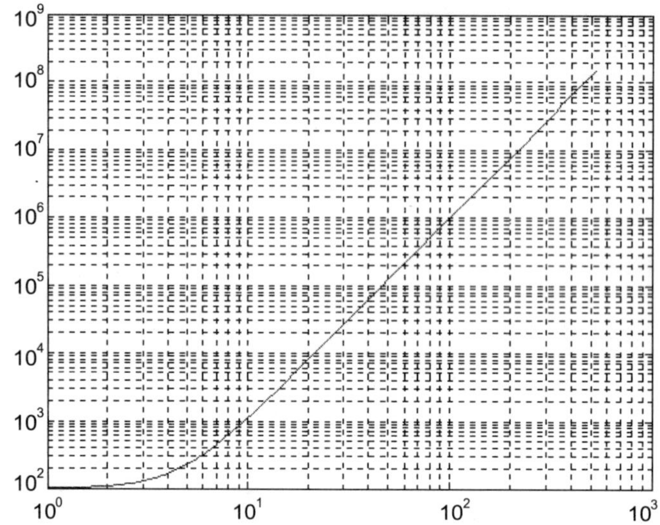

Fig. E2.5 (e)

Example E2.6

(*a*) Plot the parametric space curve of

$x(t) = t$

$y(t) = t^2$

$z(t) = t^3; 0 \leq t \leq 2.0$

(*b*) $z = -7/(1+x^2+y^2); |x| \leq 5, |y| \leq 5$

Solution:

(*a*)
```
>> t = linspace(0, 2,100);
>> x =t; y = t. ^2; z = t. ^3;
>> plot 3 (x, y, z), grid
```

The plot is shown in Figure E2.6 (*a*).

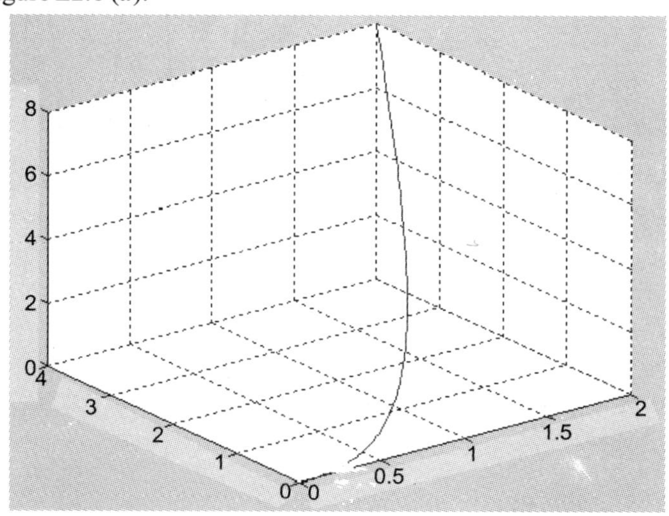

Fig. E2.6 (a)

(*b*)
```
>> t = linspace(0,  2, 100);
>> x = t; y = t. ^2;  z = t. ^3;
>> plot3 (x, y, z),  grid
>> t = linspace(-5, 5, 50); y = x;
>> z = -7./(1+x.^2+y.^2);
>> mesh(z)
```

The plot is shown in Figure E2.6(*b*).

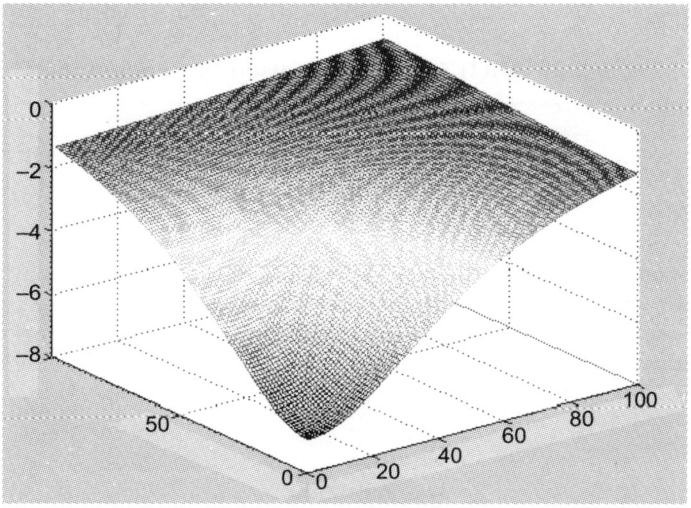

Fig. E2.6(b)

2.17.4 Saving and Printing Graphs

To obtain a hardcopy of a graph, type *print* in the Command Window after the graph appears in the Figure Window. The figure can also be saved into a specified file in the PostScripter or Encapsulated PostScript (EPS) format. The command to save graphics to a file is

print – d devicetype – options filename

where devicetype for PostScript printers are listed in the following Table 2.31.

Table 2.31 Devicetype for PostScript printers

Devicetype	Description	Devicetype	Description
ps	Black and white PostScript	**eps**	Black and white EPSF
psc	Color PostScript	**epsc**	Color EPSF
ps2	Level 2 BW PostScript	**eps2**	Level 2 black and white EPSF
psc2	Level 2 color PostScript	**epsc2**	Level 2 color EPSF

MATLAB can also generate a graphics file in the following popular formats among others:

–dill : Saves file in Adobe Illustrator format.

–djpeg : Saves file as a JPEG image.

–dtiff : Saves file as a compressed TIFF image.

–dmfile : Saves file as an M-file with graphics handles.

2.18 INPUT/OUTPUT IN MATLAB

In this section, we present some of the many available commands in MATLAB for reading data from an external file into a MATLAB matrix, or writing the numbers computed in MATLAB into such an external file.

2.18.1 The fopen Statement

To have the MATLAB read or write a separate data file of numerical values, we need to *connect* the file to the executing MATLAB program. The MATLAB functions used are summarized in Table 2.32.

Table 2.32 MATLAB functions used for input/output

Function	Description
fopen	Connects an existing file to MATLAB or to create a new file from MATLAB. **fid** = **fopen** (*'Filename'*, *permission code*); where, if **fopen** is successful, **fid** will be returned as a positive integer greater than 2. When unsuccessful, a value of –1 is returned. Both the file name and the permission code are string constants enclosed in single quotes. The permission code can be a variety of flags that specify whether or not the file can be written to, read from, appended to, or a combination of these. Some common codes are: Code — Meaning **'r'** — read only **'w'** — write only **'r+'** — read and write **'a+'** — read and append The **fopen** statement positions the file at the beginning.
fclose	Disconnects a file from the operating MATLAB program. The use is **fclose(fid)**, where **fid** is the *file identification number* of the file returned by **fopen.fclose('all')** will close all files.
fscanf	Reads opened files. The use is **A** = **fscanf(fid**, *FORMAT, SIZE*) where *FORMAT* specifies the types of numbers (integers, reals with or without exponent, character strings) and their arrangement in the data file, and optional *SIZE* determines how many quantities are to be read and how they are to be arranged into the matrix **A**. If *SIZE* is omitted, the entire file is read. The *FORMAT* field is a string (enclosed in single quotes) specifying the form of the numbers in the file. The *type* of each number is characterized by a percent sign (%), followed by a letter (**i** or **d** for integers, **e** or **f** for floating-point numbers with or without exponents). Between the percent sign and the type code, one can insert an integer specifying the maximum width of the field.
fprintf	Writes files previously opened. **fprintf(fid**, *FORMAT*, **A)** where **fid** and *FORMAT* have the same meaning as for **fscanf**, with the exception that for output formats the string must end with **\n**, designating the end of a line of output.

2.19 SYMBOLIC MATHEMATICS

In Secs.2.1 to 2.18, the capability of MATLAB for numerical computations have been described. In this section some of MATLAB's capabilities for symbolic manipulations will be presented. Specifically, the symbolic expressions, symbolic algebra, simplification of mathematical expressions, operations on symbolic expressions, solution of a single equation or a set of linear algebraic equations, solutions to differential equations, differentiation and integration of functions using MATLAB are presented.

2.19.1 Symbolic Expressions

A symbolic expression is stored in MATLAB as a *character string*. A single quote marks are used to define the symbolic expression. For instance:

'sin(*y*/*x*)'; '*x*^4 + 5**x*^3 + 7**x*^2 – 7'

The independent variable in many functions is specified as an additional function argument. If an independent variable is not specified, then MATLAB will pick one. When several variables exist, MATLAB will pick the one that is a single lower case letter (except *i* and *j*), which is closest to *x* alphabetically.

The independent variable is returned by the function *symvar*,
symvar(s): Returns the independent variable for the symbolic expression *s*.
For example:

Expression *s*	**symvar(*s*)**
'5 * c * *d* + 34'	*d*
'sin(*y*/*x*)'	*x*

In MATLAB, a number of functions are available to simplify mathematical expressions by expanding the terms, factoring expressions, collecting coefficients, or simplifying the expression. For instance:
expand(s): Performs an expansion of *s*.

A summary of these expressions is given in Table 2.33. A summary of basic operations is given in Table 2.34. The standard arithmetic operation (Table 2.35) is applied to symbolic expressions using symbolic functions. These symbolic expressions are summarized in Table 2.36.

Table 2.33

Simplification	
collect	Collect common terms
expand	Expand polynomials and elementary functions
factor	Factorization
horner	Nested polynomial representation
numden	Numerator and denominator
simple	Search for shortest form
simplify	Simplification
subexpr	Rewrite in terms of subexprssions

Table 2.34

Basic operations	
ccode	C code representation of a symbolic expression
conj	Complex conjugate
findsym	Determine symbolic variables
fortran	Fortran representation of a symbolic expression
imag	Imaginary part of a complex number
latex	LaTeX representation of a symbolic expression
pretty	Pretty prints a symbolic expression
real	Real part of an imaginary number
sym	Create symbolic object
syms	Shortcut for creating multiple symbolic objects

Table 2.35

Arithmetic operations	
+	Addition
–	Subtraction
*	Multiplication
.*	Array multiplication
/	Right division
./	Array right division
\	Left division
.\	Array left division
^	Matrix or scalar raised to a power
.^	Array raised to a power
‘	Complex conjugate transpose
.‘	Real transpose

Table 2.36

Symbolic expressions	
horner(S)	Transposes **S** into its Horner, or nested, representation.
numden(S)	Returns two symbolic expressions that represent, respectively, the numerator expression and the denominator expression for the rational representation of **S**.
numeric(S)	Converts **S** to a numeric form (**S** must not contain any symbolic variables).
poly2sym(c)	Converts a polynomial coefficient vector **c** to a symbolic polynomial.
pretty(S)	Prints **S** in an output form that resembles typeset mathematics.
sym2poly(S)	Converts **S** to a polynomial coefficient vector. *
symadd(A,B)	Performs a symbolic addition, **A + B**.
symdiv(A,B)	Performs a symbolic division, **A / B.**
symmul(A,B)	Performs a symbolic multiplication, **A * B.**
sympow(S,p)	Performs a symbolic power, **S^p**.
symsub(A,B)	Performs a symbolic subtraction, **A – B.**

2.19.2 Solution to Differential Equations

Symbolic math functions can be used to solve a single equation, a system of equations, and differential equations. For example:

solve(f) : Solves a symbolic equation **f** for its symbolic variable. If **f** is a symbolic expression, this function solves the equation **f** = 0 for its symbolic variable.

solve(f1, …, fn): Solves the system of equations represented by **f1, …, fn.**

The symbolic function for solving ordinary differential equation is **dsolve** as shown below:

dsolve('equation', 'condition'): Symbolically solves the ordinary differential equation specified by **'equation'**. The optional argument **'condition'** specifies a boundary or initial condition.

The symbolic equation uses the letter **D** to denote differentiation with respect to the independent variable. A **D** followed by a digit denotes repeated differentiation. Thus, **Dy** represents dy/dx, and **D2y** represents d^2y/dx^2. For example, given the ordinary second order differential equation;

$$\frac{d^2x}{dt^2}+5\frac{dx}{dt}+3x=7$$

with the initial conditions $x(0)=0$ and $\dot{x}(0)=1$.

The MATLAB statement that determine the symbolic solution for the above differential equation is the following:

```
x = dsolve('D2x = -5*Dx - 3*x + 7', 'x(0) = 0', 'Dx(0) =1')
```

The symbolic functions are summarized in Table 2.37.

Table 2.37 Solution of equations

compose	Functional composition
dsolve	Solution of differential equations
finverse	Functional inverse
solve	Solution of algebraic equations

2.19.3 Calculus

There are four forms by which the symbolic derivative of a symbolic expression is obtained in MATLAB. They are:

diff(f) : Returns the derivative of the expression **f** with respect to the default independent variable.

diff(f, 't') : Returns the derivative of the expression **f** with respect to the variable **t**.

diff(f, n) : Returns the *n*th derivative of the expression **f** with respect to the default independent variable.

diff(f, 't', n) : Returns the *n*th derivative of the expression **f** with respect to the variable **t**.

The various forms that are used in MATLAB to find the integral of a symbolic expression **f** are given below and summarized in Table 2.38.

int(f) : Returns the integral of the expression **f** with respect to the default independent variable.

int(f, 't') : Returns the integral of the expression **f** with respect to the variable **t**.

int(f, a, b) : Returns the integral of the expression **f** with respect to the default independent variable evaluated over the interval **[a,b]**, where **a** and **b** are numeric expressions.

int(f,'t', a, b) : Returns the integral of the expression **f** with respect to the variable **t** evaluated over the interval **[a,b]**, where **a** and **b** are numeric expressions.

int(f, 'm', 'n') Returns the integral of the expression **f** with respect to the default independent variable evaluated over the interval **[m,n]**, where *m* and *n* are numeric expressions.

The other symbolic functions for pedagogical and graphical applications, conversions, integral transforms, and linear algebra are summarized in Tables 2.38 to 2.42.

Table 2.38

Calculus	
diff	Differentiate
int	Integrate
jacobian	Jacobian matrix
limit	Limit of an expression
symsum	Summation of series
taylor	Taylor series expansion

Table 2.39

Pedagogical and graphical applications	
ezcontour	Contour plotter
ezcontourf	Filled contour plotter
ezmesh	Mesh plotter
ezmeshc	Combined mesh and contour plotter
ezplot	Function plotter
ezplot	Easy-to-use function plotter
ezplot3	Three-dimensional curve plotter
ezpolar	Polar coordinate plotter
ezsurf	Surface plotter
ezsurfc	Combined surface and contour plotter
funtool	Function calculator
rsums	Riemann sums
taylortool	Taylor series calculator

Table 2.40

Conversions	
char	Convert sym object to string
double	Convert symbolic matrix to double
poly2sym	Function calculator
sym2poly	Symbolic polynomial to coefficient vector

Table 2.41

Integral transforms	
fourier	Fourier transform
ifourier	Inverse Fourier transform
ilaplace	Inverse Laplace transform
iztrans	Inverse Z-transform
laplace	Laplace transform
ztrans	Z-transform

Table 2.42

Linear algebra	
colspace	Basis for column space
det	Determinant
diag	Create or extract diagonals
eig	Eigenvalues and eigenvectors
expm	Matrix exponential
inv	Matrix inverse
jordan	Jordan canonical form
null	Basis for null space
poly	Characteristic polynomial
rank	Matrix rank
rref	Reduced row echelon form
svd	Singular value decomposition
tril	Lower triangle
triu	Upper triangle

2.20 THE LAPLACE TRANSFORMS

The Laplace transformation method is an operational method that can be used to find the transforms of time functions, the inverse Laplace transformation using the partial-fraction expansion of $B(s)/A(s)$, where $A(s)$ and $B(s)$ are polynomials in s. In this Chapter, we present the computational methods with MATLAB to obtain the partial-fraction expansion of $B(s)/A(s)$ and the zeros and poles of $B(s)/A(s)$.

MATLAB can be used to obtain the partial-fraction expansion of the ratio of two polynomials, $B(s)/A(s)$ as follows:

$$\frac{B(s)}{A(s)} = \frac{\text{num}}{\text{den}} = \frac{b(1)s^n + b(2)s^{n-1} + \cdots + b(n)}{a(1)s^n + a(2)s^{n-1} + \cdots + a(n)}$$

where $a(1) \neq 0$ and num and den are row vectors. The coefficients of the numerator and denominator of $B(s)/A(s)$ are specified by the num and den vectors.

Hence $\quad \text{num} = [b(1) \quad b(2) \quad \ldots \quad b(n)]$

$\quad\quad\quad \text{den} = [a(1) \quad a(2) \quad \ldots \quad a(n)]$

The MATLAB command

r, p, k = residue(num, den)

is used to determine the residues, poles, and direct terms of a partial-fraction expansion of the ratio of two polynomials $B(s)$ and $A(s)$ is then given by

$$\frac{B(s)}{A(s)} = k(s) + \frac{r(1)}{s - p(1)} + \frac{r(2)}{s - p(2)} + \cdots + \frac{r(n)}{s - p(n)}$$

The MATLAB command **[num, den] = residue(r, p, k)** where r, p, k are the output from MATLAB converts the partial fraction expansion back to the polynomial ratio $B(s)/A(s)$.

The command **printsys (num,den,'s')** prints the num/den in terms of the ratio of polynomials in s.

The command **ilaplace** will find the inverse Laplace transform of a Laplace function.
The MATLAB command **[z,p,k] = tf2zp(num, den)** is used to find the zeros, poles, and gain K of $B(s)/A(s)$. If the zeros, poles, and gain K are given, the following MATLAB command can be used to find the original num/den:

[num, den] = zp2tf (z,p,k)

2.21 ADDITIONAL EXAMPLES

Some more examples depicting the requirement of MATLAB are presented in this section.

Example E2.7: Consider the function

$$H(s) = \frac{n(s)}{d(s)}$$

where $n(s) = s^4 + 6s^3 + 5s^2 + 4s + 3$

$d(s) = s^5 + 7s^4 + 6s^3 + 5s^2 + 4s + 7$

(*a*) Find $n(-10)$, $n(-5)$, $n(-3)$ and $n(-1)$

(*b*) Find $d(-10)$, $d(-5)$, $d(-3)$ and $d(-1)$

(c) Find $H(-10)$, $H(-5)$, $H(-3)$ and $H(-1)$

Solution:

(*a*)
```
>> n=[1 6 5 4 3]; % n=s^4+6s^3+5s^2+4s+3
>> d=[1 7 6 5 4 7]; % d=s^5+7s^4+6s^3+5s^2+4s+7
>> n2=polyval(n,[-10])
n2 = 4463
>> nn10=polyval(n,[-10])
nn10 = 4463
>> nn5=polyval(n,[-5])
nn5 = -17
>> nn3=polyval(n,[-3])
nn3 = -45
>> nn1=polyval(n,[-1])
nn1 = -1
```

(*b*)
```
>> dn10=polyval(d,[-10])
dn10 = -35533
>> dn5=polyval(d,[-5])
dn5 = 612
>> dn3=polyval(d,[-3])
dn3 = 202
>> dn1=polyval(d,[-1])
dn1 = 8
```

(*c*)
```
>> Hn10=nn10/dn10
Hn10 = -0.1256
>> Hn5=nn5/dn5
Hn5 = -0.0278
>> Hn3=nn3/dn3
Hn3 = -0.2228
>> Hn1=nn1/dn1
Hn1 = -0.1250
```

Example E2.8: Generate a plot of

$$y(x) = e^{-0.7x} \sin \omega x$$

where $\omega = 15$ rad/s, and $0 \le x \le 15$. Use the colon notation to generate the x vector in increments of 0.1.

Solution:

```
>> x = [0 : 0.1 : 15];
>> w = 15;
>> y = exp(-0.7*x).*sin(w*x);
>> plot(x,y)
>> title('y(x) = e^-^0^.^7^xsin \omegax')
>> xlabel('x')
>> ylabel('y')
```

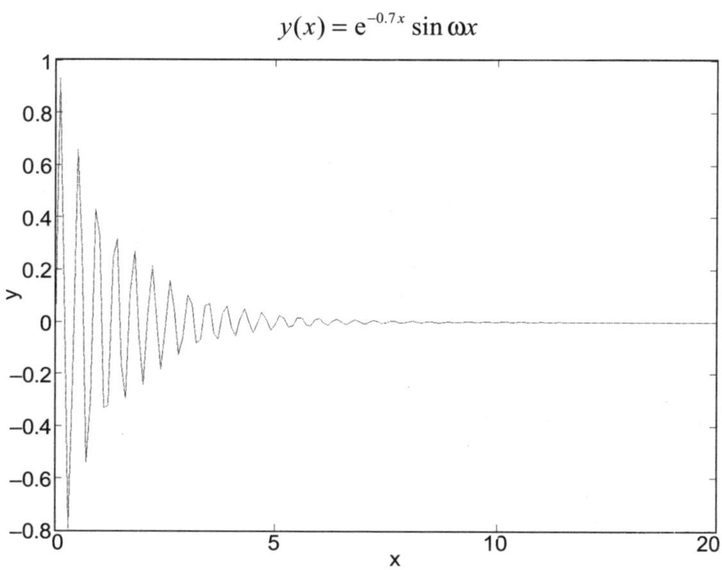

Fig. E2.8

Example E2.9: Generate a plot of

$$y(x) = e^{-0.6x} \cos \omega x$$

where $\omega = 10$ rad/s, and $0 \le x \le 15$. Use the colon notation to generate the x vector in increments of 0.05.

Solution:

```
>> x = [0 : 0.1 : 15];
>> w = 10;
>> y = exp(-0.6*x).*cos(w*x);
>> plot(x, y)
>> title('y(x) = e^-^0^.^6^xcos \omegax')
>> xlabel('x')
>> ylabel('y')
```

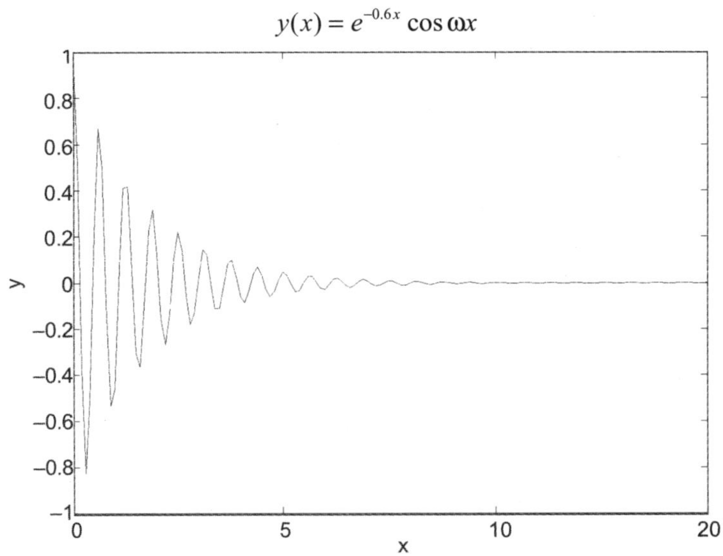

Fig. E2.9

Example E2.10: Using the functions for plotting x-y data given in Table 2.29 plot the following functions:

(*a*) $r^2 = 5 \cos 3t;\ 0 \le t \le 2\pi$

(*b*) $r^2 = 5 \cos 3t;\ 0 \le t \le 2\pi$

$x = r \cos t,\ y = r \sin t$

(*c*) $y_1 = e^{-2x} \cos x;\ 0 \le t \le 20$

$y_2 = e^{2x}$

(*d*) $y = \dfrac{\cos(x)}{x};\ -5 \le x \le 5\pi$

(*e*) $f = e^{-3t/5} \cos t;\ 0 \le t \le 2\pi$

(*f*) $z = -\dfrac{1}{3} x^2 + 2xy + y^2;\ |x| \le 7,\ |y| \le 7$

Solution:

(*a*)
```
t = linspace(0, 2*pi, 200);
r = sqrt(abs(5*cos(3*t)));
polar(t, r)
```

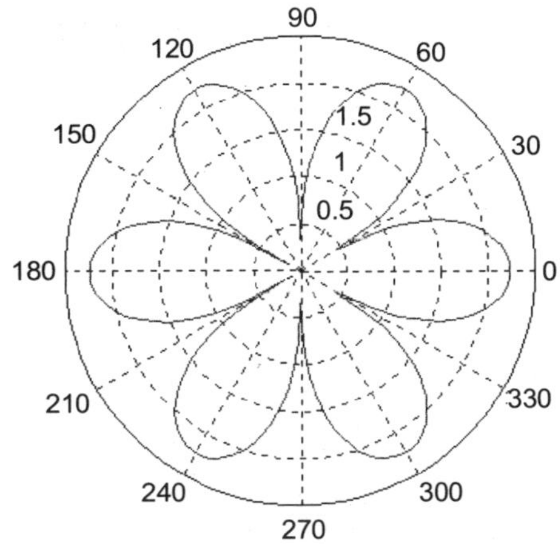

Fig. E2.10(a)

(*b*)
```
t = linspace(0, 2*pi, 200);
r = sqrt(abs(5*cos(3*t)));
x = r.*cos(t);
y = r.*sin(t);
fill(x, y, 'k'),
axis('square')
```

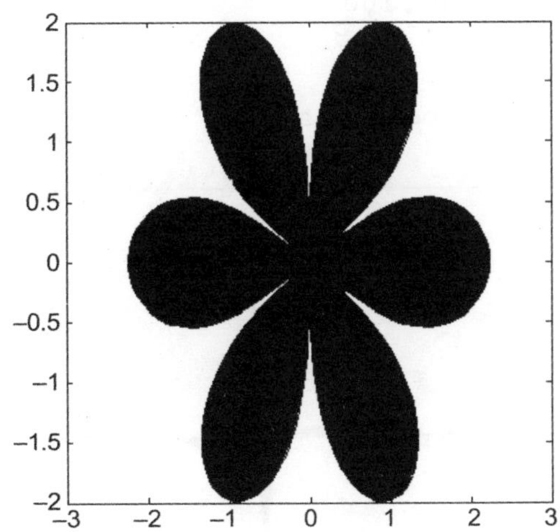

Fig. E2.10(b)

(*c*)
```
x = 1:0.1:20;
y1 = exp(-2*x).*cos(x);
y2 = exp(2*x);
Ax = plotyy(x,y1,x,y2);
hy1= get(Ax(1), 'ylabel');
hy2= get(Ax(2), 'ylabel');
set(hy1, 'string', 'exp(-2x).cos(x)')
set(hy2, 'string', 'exp(-2x)');
```

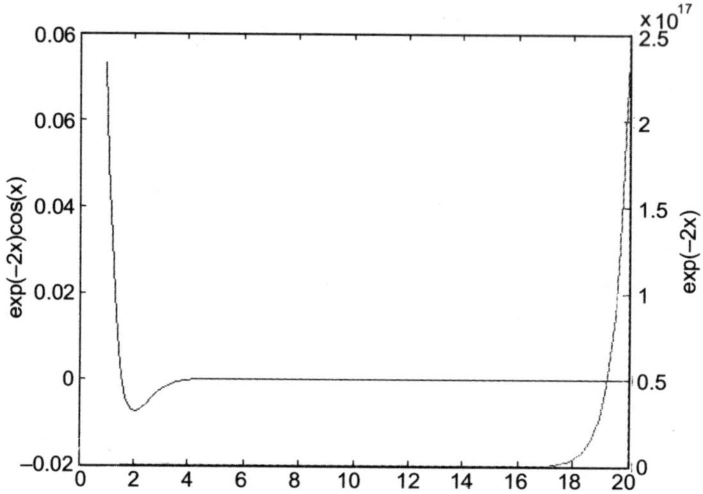

Fig. E2.10(c)

(*d*)
```
x = linspace(-5*pi, 5*pi,100);
y = cos(x)./x;
area(x, y);
xlabel('x (rad)'), ylabel('cos(x)/x')
hold on
```

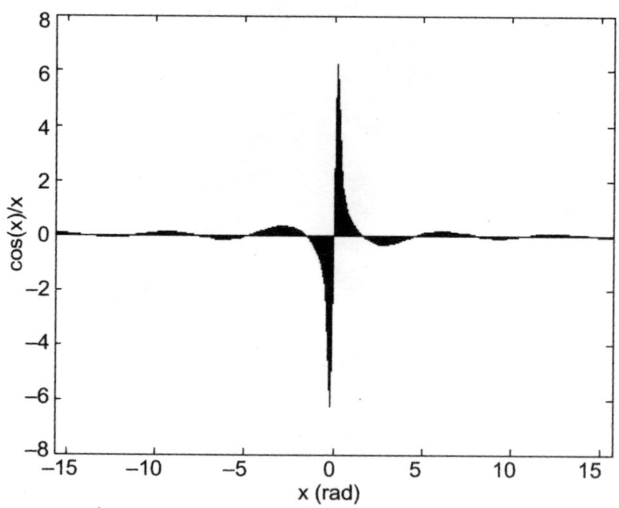

Fig. E2.10(d)

(*e*)
```
t = linspace(0,2*pi,200);
f = exp(-0.6*t).*sin(t);
stem(t, f)
```

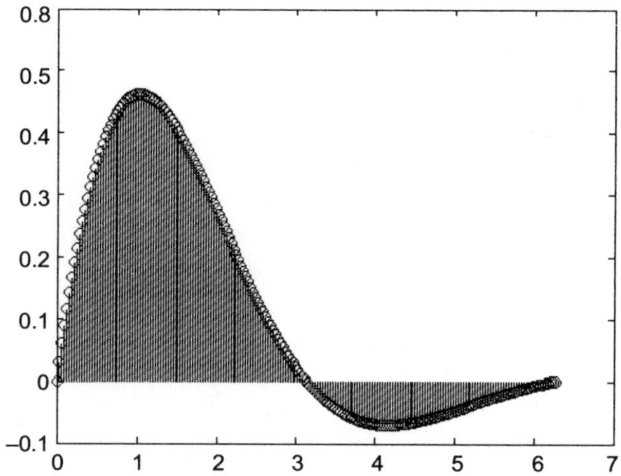

Fig. E2.10(e)

(*f*)
```
r= -7:0.2:7;
[X,Y]=meshgrid(r,r);
Z=-0.333*X.^2+2*X.*Y+Y.^2;
cs=contour(X,Y,Z);
label(cs)
```

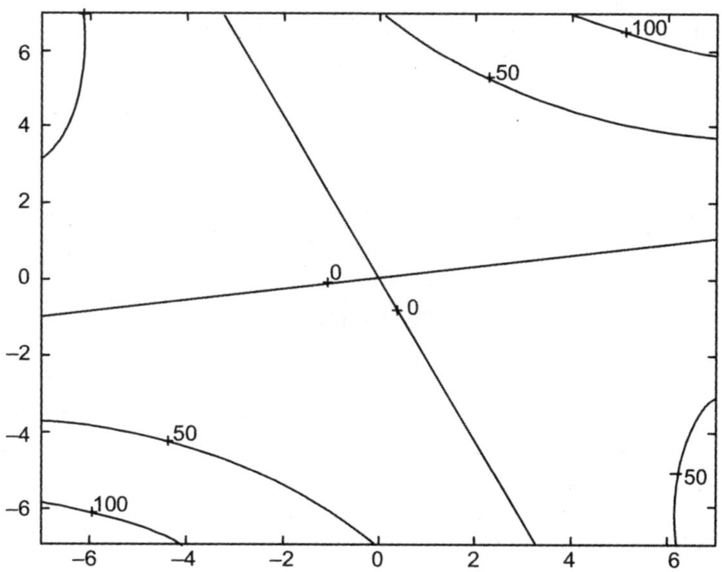

Fig. E2.10(f)

Example E2.11: Use the functions listed in Table 2.30 for plotting 3-*D* data for the following:

(*a*) $z = \cos x \cos y \; e^{-\frac{\sqrt{x^2+y^2}}{5}}$

$|x| \leq= 7, |y| \leq 7$

(*b*) Discrete data plots with stems

$x = t, y = t\cos(t)$

$z = e^{t/5} - 2; 0 \leq t \leq 5\pi$

(*c*) A cylinder generated by

$r = \sin(5\pi z) + 3$

$0 \leq z \leq 1; 0 \leq \theta \leq 2\pi$

Solution:

(*a*)
```
u=-7:0.2:7;x
[X,Y]=meshgrid(u,u);
Z=cos(X).*cos(Y).*exp(-sqrt(X.^2+Y.^2)/5);
surf(X,Y,Z)
```

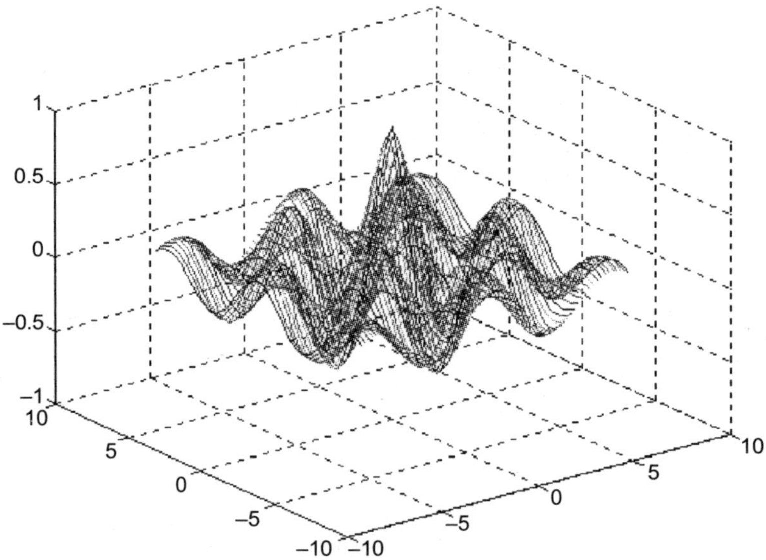

Fig. E2.11(a)

(*b*)
```
t=linspace(0,5*pi,200);
x=t;y=t.*cos(t);
z=exp(t/5)-2;
stem3(x,y,z,'filled');
xlabel('t'),ylabel('tcos(t)'), zlabel('e^t/5-1')
```

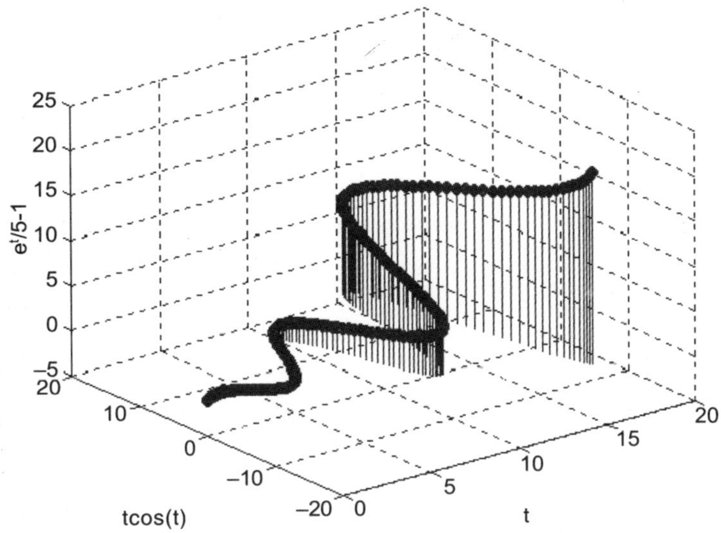

Fig. E2.11(b)

(*c*)
```
z=[0:0.2:1]';
r=sin(5*pi*z)+3;
cylinder(r)
```

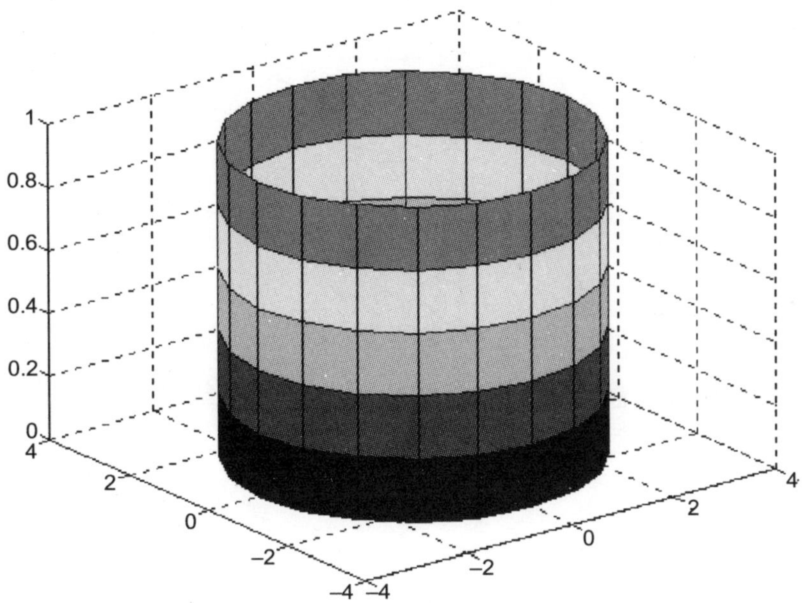

Fig. E2.11(c)

Example E2.12: Obtain the plot of the points for $0 \leq t \leq 6\pi$ when the coordinates x,y,z are given as a function of the parameter t as follows:

$$x = \sqrt{t}\sin(3t)$$
$$y = \sqrt{t}\cos(3t)$$
$$z = 0.8t$$

Solution:

```
% Line plots
>> t=[0:0.1:6*pi];
>> x=sqrt(t).*sin(3*t);
>> y=sqrt(t).*cos(3*t);
>> z=0.8*t;
>> plot3(x,y,z, 'k', 'linewidth',1)
>> grid on
>> xlabel('x'); ylabel('y'); zlabel('z')
```

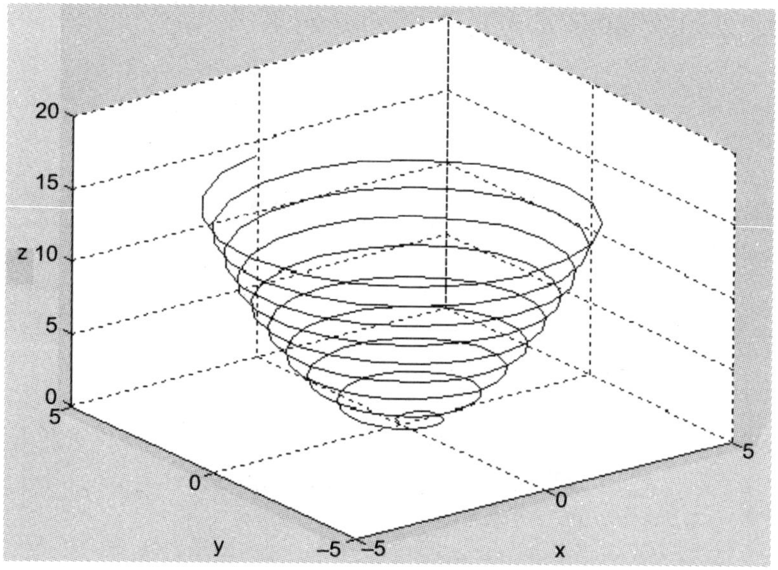

Fig. E2.12

Example E2.13: Obtain the mesh and surface plots for the function $z = \dfrac{2xy^2}{x^2 + y^2}$ over the domain $-2 \leq x \leq 6$ and $2 \leq y \leq 8$.

Solution:

```
% Mesh and surface plots
   x=-2:0.1:6;
>> y=2:0.1:8;
>> [x,y]=meshgrid(x,y);
>> z=2*x.*y.^2./(x.^2+y.^2);
>> mesh(x,y,z)
```

```
>> xlabel('x'); ylabel('y'); zlabel('z')
>> surf(x,y,z)
>> xlabel('x'); ylabel('y'); zlabel('z')
```

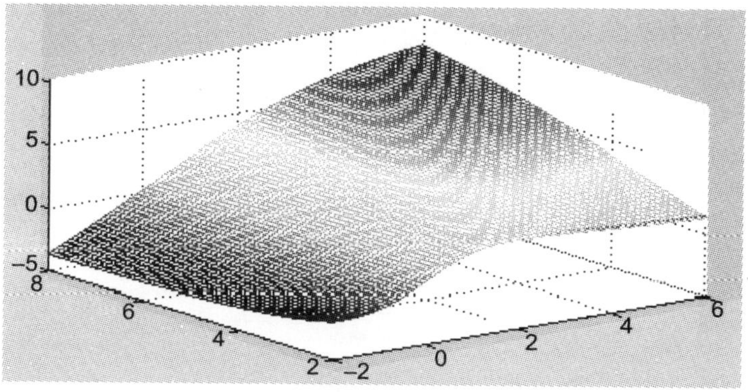

Fig. E2.13 (a)

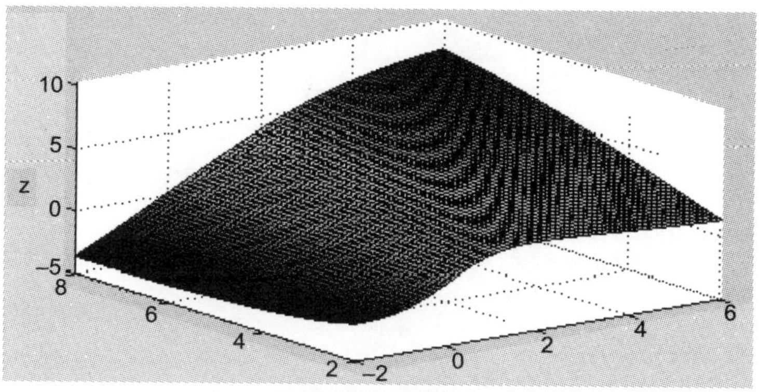

Fig. E2.13 (b)

Example E2.14: Plot the function $z = 2^{-1.5\sqrt{x^2+y^2}}\sin(x)\cos(0.5y)$ over the domain $-4 \le x \le 4$ and $-4 \le y \le 4$ using Table 2.30.

(*a*) Mesh plot
(*b*) Surface plot
(*c*) Mesh curtain plot
(*d*) Mesh and contour plot
(*e*) Surface and contour plot

Solution:

(*a*) % Mesh Plot

```
>> x=-4:0.25:4;
>> y=-4:0.25:4;
>> [x,y]=meshgrid(x,y);
>> z=2.^(-1.5*sqrt(x.^2 + y.^2)).*cos(0.5*y).*sin(x);
>> mesh(x,y,z)
>> xlabel('x');ylabel('y')
>> zlabel('z')
```

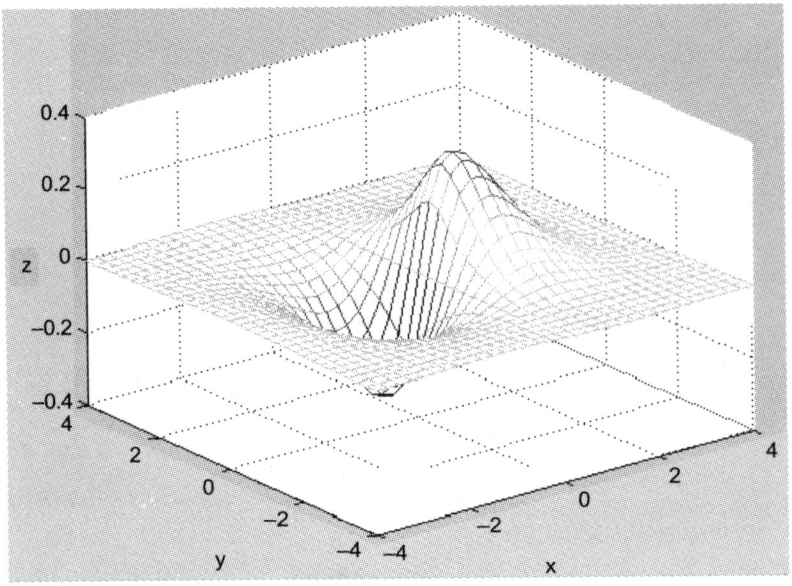

Fig. E2.14 (a)

(b) % Surface Plot

```
>> x=-4:0.25:4;
>> y=-4:0.25:4;
>> [x,y]=meshgrid(x,y);
>> z=2.0.^(-1.5*sqrt(x.^2+y.^2)).*cos(0.5*y).*sin(x);
>> surf(x,y,z)
>> xlabel('x'); ylabel('y')
>> zlabel('z')
```

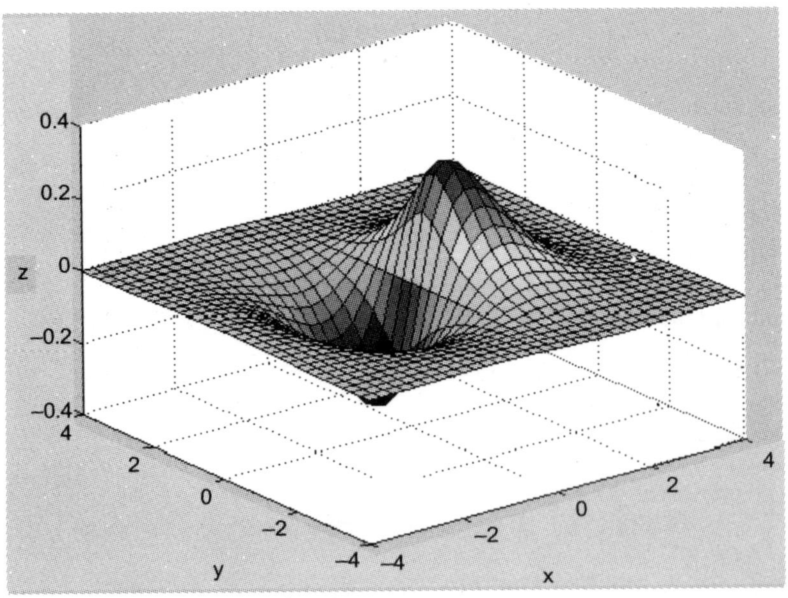

Fig. E2.14 (b)

(*c*) % Mesh Curtain Plot

```
>> x=-4.0:0.25:4;
>> y=-4.0:0.25:4;
>> [x,y]=meshgrid(x,y);
>> z=2.0.^(-1.5*sqrt(x.^2+y.^2)).*cos(0.5*y).*sin(x);
>> meshz(x,y,z)
>> xlabel('x');ylabel('y')
>> zlabel('z')
```

(*d*) % Mesh and Contour Plot

```
>> x=-4.0:0.25:4;
>> y=-4.0:0.25:4;
>> [x,y]=meshgrid(x,y);
>> z=2.0.^(-1.5*sqrt(x.^2+y.^2)).*cos(0.5*y).*sin(x);
>> meshc(x,y,z)
>> xlabel('x');ylabel('y')
>> zlabel('z')
```

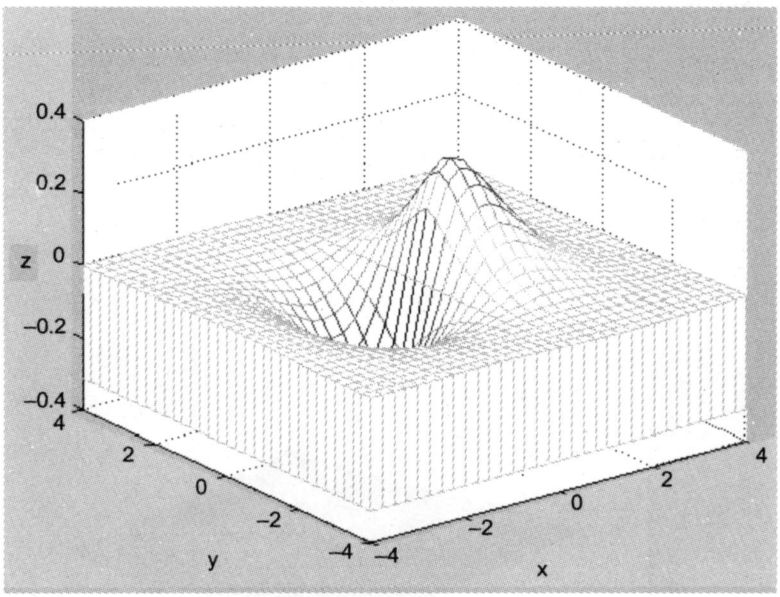

Fig. E2.14 (c)

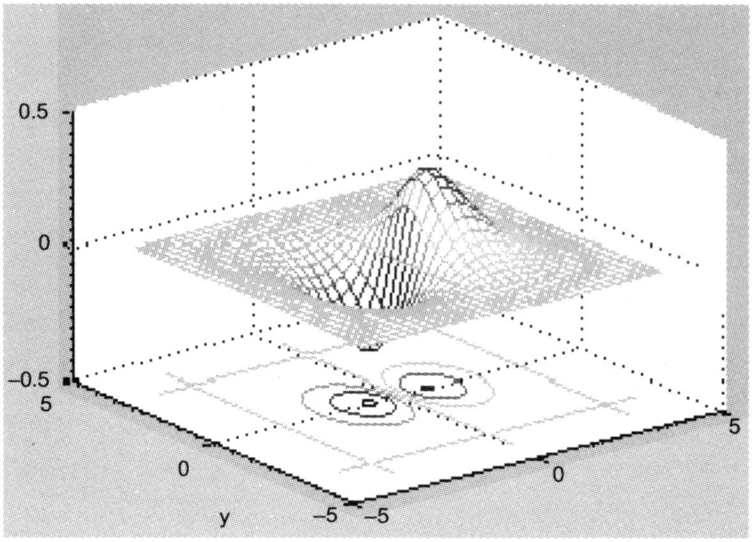

Fig. E2.14 (d)

(*e*) % Surface and Contour Plot

```
>> x=-4.0:0.25:4;
>> y=-4.0:0.25:4;
>> [x, y] =meshgrid(x, y);
>> z=2.0. ^ (-1.5*sqrt (x. ^2+y. ^2)).*cos (0.5*y).*sin(x);
>> surfc(x, y, z)
>> xlabel('x');ylabel('y')
>> zlabel('z')
```

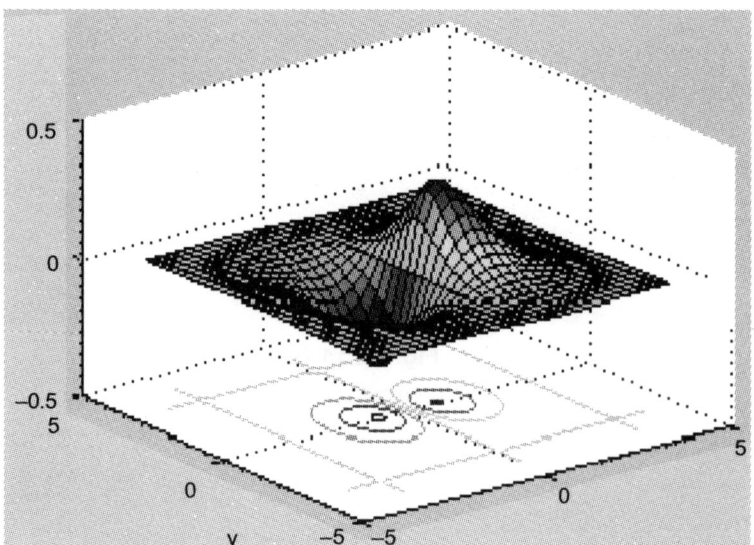

Fig. E2.14(e)

Example E2.15: Plot the function $z = 2^{-1.5\sqrt{x^2+y^2}} \sin(x)\cos(0.5y)$ over the domain $-4 \le x \le 4$ and $-4 \le y \le 4$ and using Table 2.30

(*a*) Surface plot with lighting
(*b*) Waterfall plot
(*c*) 3-D contour plot
(*d*) 2-D contour plot

Solution:

(*a*) % Surface Plot with lighting

```
>> x=-4.0:0.25:4;
>> y=-4.0:0.25:4;
>> [x,y]=meshgrid(x,y);
>> z=2.0.^(-1.5*sqrt(x.^2+y.^2)).*cos(0.5*y).*sin(x);
>> surfl(x,y,z)
>> xlabel('x');ylabel('y')
>> zlabel('z')
```

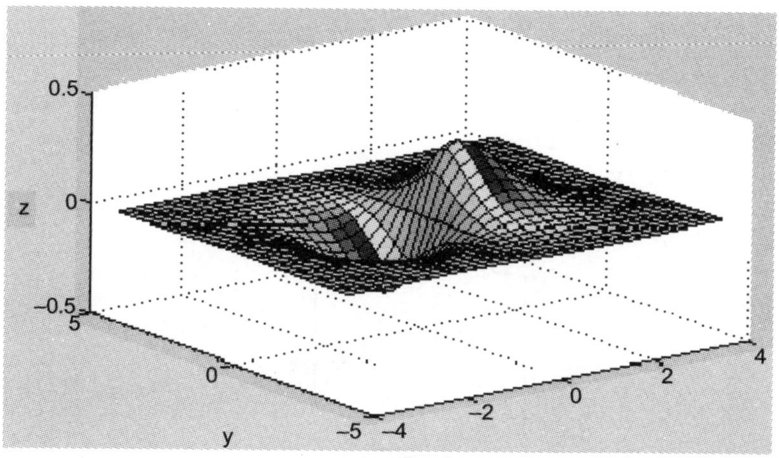

Fig. E2.15 (a)

(*b*) % Waterfall Plot

```
>> x=-4.0:0.25:4;
>> y=-4.0:0.25:4;
>> [x,y]=meshgrid(x,y);
>> z=2.0.^(-1.5*sqrt(x.^2+y.^2)).*cos(0.5*y).*sin(x);
>> waterfall(x,y,z)
>> xlabel('x');ylabel('y')
>> zlabel('z')
```

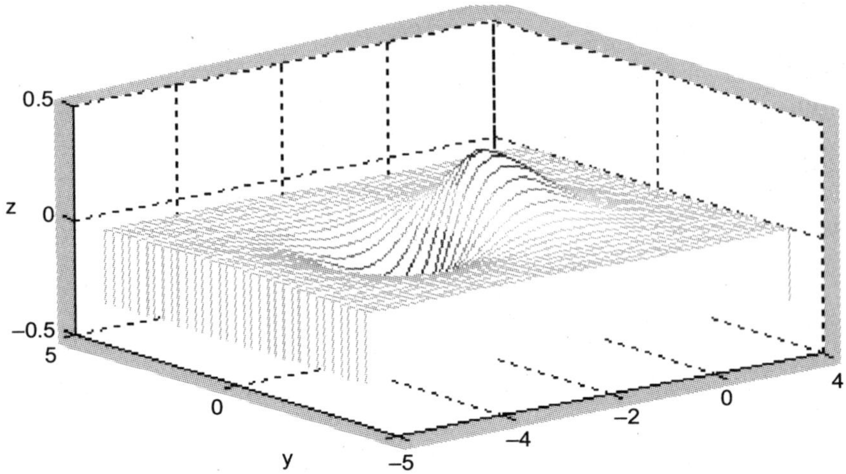

Fig. E2.15 (b)

(c) % 3-D Contour Plot

```
>> x=-4.0:0.25:4;
>> y=-4.0:0.25:4;
>> [x,y]=meshgrid(x,y);
>> z=2.0.^(-1.5*sqrt(x.^2+y.^2)).*cos(0.5*y).*sin(x);
>> contour3(x,y,z,15)
>> xlabel('x');ylabel('y')
>> zlabel('z')
```

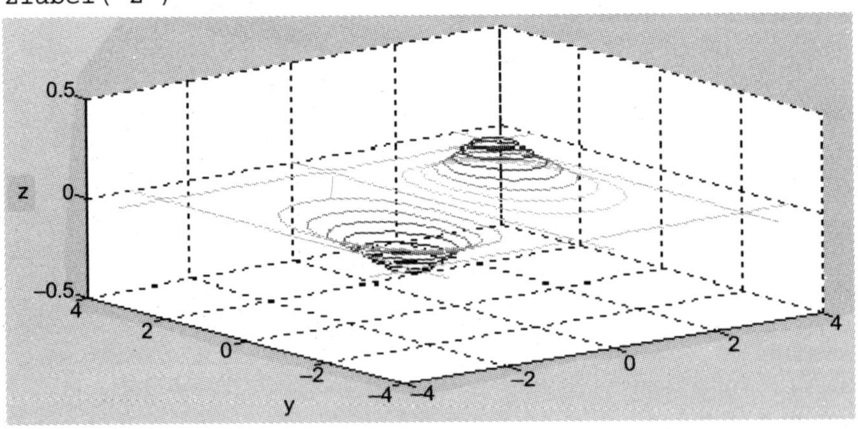

Fig. E2.15 (c)

(d) % 2-D Contour Plot

```
>> x=-4.0:0.25:4;
>> y=-4.0:0.25:4;
>> [x,y]=meshgrid(x,y);
```

```
>> z=2.0.^(-1.5*sqrt(x.^2+y.^2)).*cos(0.5*y).*sin(x);
>> contour(x,y,z,15)
>> xlabel('x');ylabel('y')
>> zlabel('z')
```

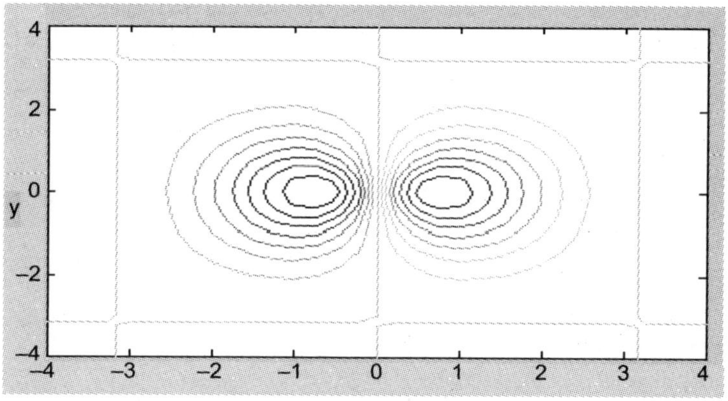

Fig. E2.15 (d)

Example E2.16: Using the functions given in Table 2.29 for plotting *x-y* data, plot the following functions:

(*a*) $f(t) = t\cos t;$ $\quad 0 \le t \le 10\pi$

(*b*) $x = e^{-2t}, y = t;$ $\quad 0 \le t \le 2\pi$

(*c*) $x = t, y = e^{2t};$ $\quad 0 \le t \le 2\pi$

(*d*) $x = e^{t}, y = 50 + e^{t}; 0 \le t \le 2\pi$

$r^2 = 3\sin 7t$

(*e*) $y = r\sin t \quad ;$ $\quad 0 \le t \le 2\pi$

$r^2 = 3\sin 4t$

(*f*) $y = r\sin t \quad ;$ $\quad 0 \le t \le 2\pi$

(*g*) $y = t\sin t;$ $\quad 0 \le t \le 5\pi$

Solution:

(*a*) % Use of plot command

```
>> fplot('x.*cos(x)', [0,10*pi])
```

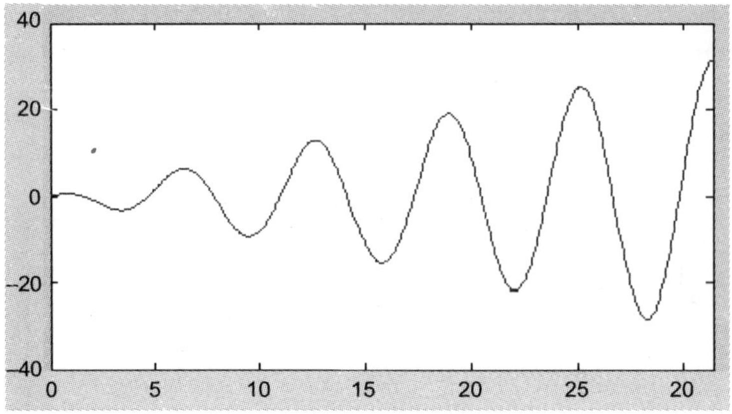

Fig. E2.16 (a)

(*b*) % Semilog *x* command

```
>> t=linspace(0,2*pi,200);
>> x=exp(-2*t); y=t;
>> semilog x(x,y),grid
```

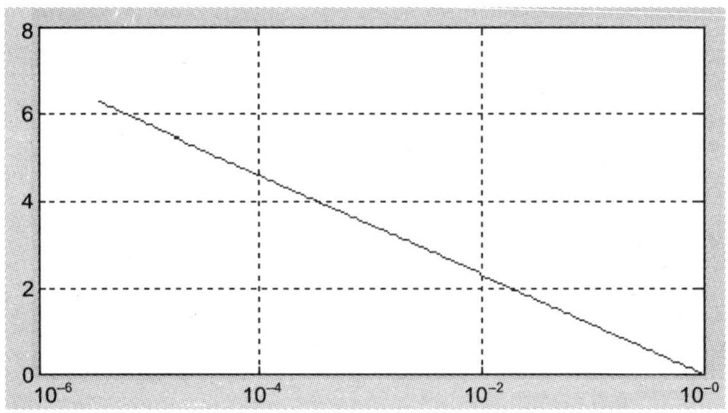

Fig. E2.16 (b)

(*c*) % Semilog *y* command

```
t=linspace(0,2*pi,200);
>> semilog y(t, exp(-2*t)),grid
```

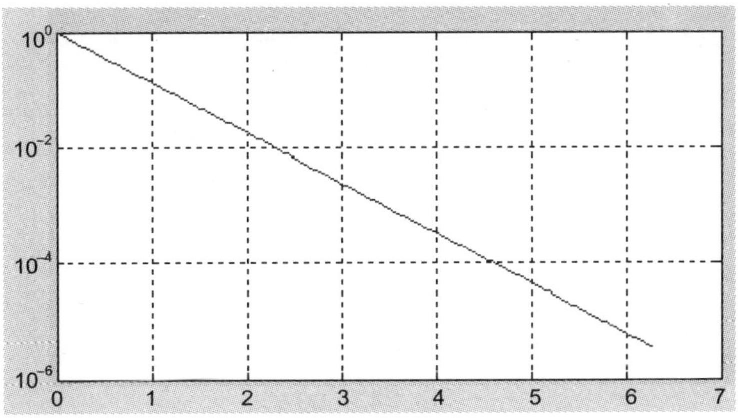

Fig. E2.16 (c)

(*d*) % Use of loglog command

```
>> t=linspace(0,2*pi,200);
>> x=exp(t);
>> y=50+exp(t);
>> loglog(x,y), grid
```

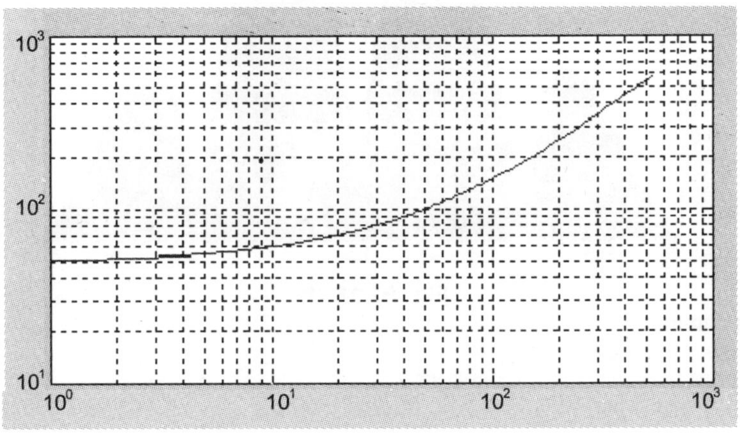

Fig. E2.16 (d)

(*e*) %Use of stairs command

```
>> t=linspace(0,2*pi,200);
>> r=sqrt(abs(3*sin(7*t)));
>> y=r.*sin(t);
>> stairs(t, y)
>> axis([0 pi 0 inf]);
```

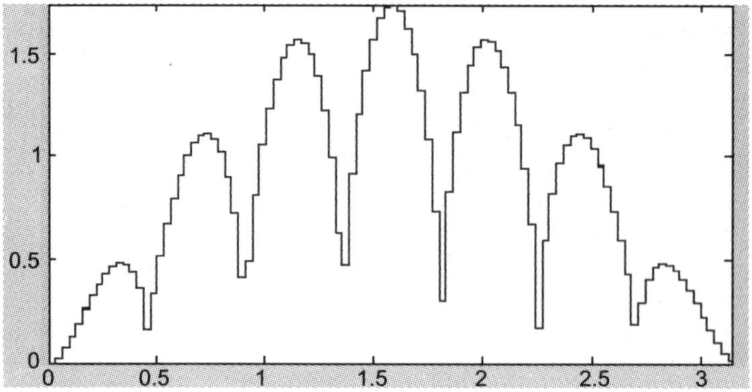

Fig. E2.16 (e)

(*f*) % Use of bar command

```
>> t=linspace(0,2*pi,200);
>> r=sqrt(abs(3*sin(4*t)));
>> y=r.*sin(t);
>> bar(t,y)
>> axis([0 pi 0 inf]);
```

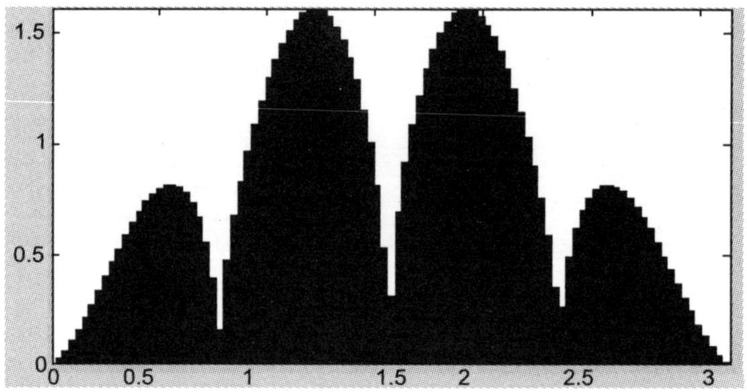

Fig. E2.16 (f)

(*g*) % use of comet command

```
>> q=linspace(0,5*pi,200);
>> y=q.*sin(q);
>> comet(q,y)
```

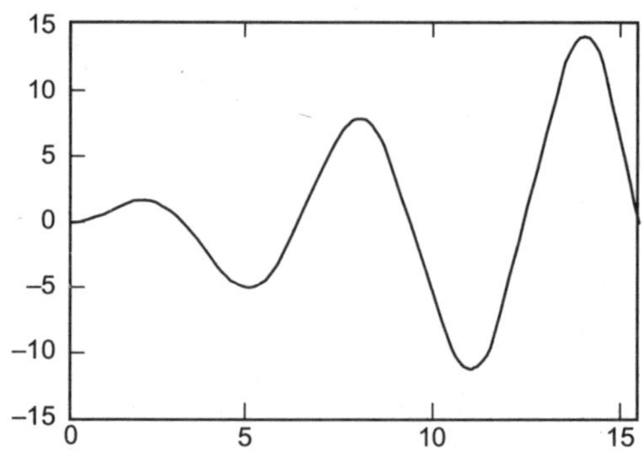

Fig. E2.16 (g)

Example E2.17: Consider the two matrices

$$A = \begin{bmatrix} 3 & 2\pi \\ 5j & 10+\sqrt{2}j \end{bmatrix} \quad \text{and} \quad B = \begin{bmatrix} 7j & -15j \\ 2\pi & 18 \end{bmatrix}$$

Using MATLAB , determine the following:

(*a*) $A+B$
(*b*) AB
(*c*) A^2
(*d*) A^T
(*e*) B^{-1}
(*f*) $B^T A^T$
(*g*) $A^2 + B^2 - AB$

Solution:

```
>> A = [3 2*pi;5j 10 + sqrt(2)*j];
>> B = [7j -15j;2*pi 18];
```

(*a*)
```
A + B
ans =
    3.0000 + 7.0000i  6.2832  -15.0000i
    6.2832 + 5.0000i  28.0000 + 1.4142i
```

(*b*)
```
>> A * B
ans =
    1.0e+002 *
    0.3948 + 0.2100i  1.1310 - 0.4500i
    0.2783 + 0.0889i  2.5500 + 0.2546i
```

(*c*)
```
>> A^2
ans =
     9.0000 + 31.4159i 81.6814 +  8.8858i
    -7.0711 + 65.0000i 98.0000 + 59.7002i
```

(*d*)
```
>> inv(A)
ans =
    0.1597 + 0.1917i  -0.1150 - 0.1042i
    0.0829 - 0.0916i   0.0549 + 0.0498i
```

(*e*)
```
>> B^-1
ans =
    0 -0.0817i 0.0681
    0 +0.0285i 0.0318
```

(*f*)
```
>> inv(B) * inv(A)
ans =
     0.0213 - 0.0193i  - 0.0048 + 0.0128i
   - 0.0028 + 0.0016i    0.0047 - 0.0017i
```

(*g*)
```
>> (A^2 + B^2) - (A * B)
ans =
    1.0e + 002 *
   -0.7948 - 0.8383i  0.7358 - 2.1611i
    0.7819 +1.0010i  1.6700 - 0.6000i
```

Example E2.18: Find the inverse of the following matrices using MATLAB:

$$(a)\begin{bmatrix}3 & 2 & 0\\ 2 & -1 & 7\\ 5 & 4 & 9\end{bmatrix} \qquad (b)\begin{bmatrix}-4 & 2 & 5\\ 7 & -1 & 6\\ 2 & 3 & 7\end{bmatrix} \qquad (c)\begin{bmatrix}-1 & 2 & -5\\ 4 & 3 & 7\\ 7 & -6 & 1\end{bmatrix}$$

Solution:

```
>> clear % Clears the workspace
>> A = [3 2 0; 2 -1 7; 5 4 9]; % Spaces separate matrix columns -
  semicolons separate matrix rows
>> B = [-4 2 5; 7 -1 6; 2 3 7]; % Spaces separate matrix columns -
semicolons separate matrix rows
>> C = [-1 2 -5; 4 3 7; 7 -6 1]; % Spaces separate matrix columns -
semicolons separate matrix rows
>> inv(A); % Finds the inverse of the selected matrix
>> inv(B); % Finds the inverse of the selected matrix
>> inv(C) % Finds the inverse of the selected matrix
% Inverse of A

ans =

    0.4805  0.2338 -0.1818

   -0.2208 -0.3506  0.2727

    0.1688  0.0260  0.0909

% Inverse of B

ans =

   -0.1773   0.0071   0.1206

   -0.2624  -0.2695   0.4184

    0.1631   0.1135   0.0709

% Inverse of C

ans =
    0.1667   0.1037   0.1074

    0.1667   0.1259  -0.0481

   -0.1667   0.0296  -0.0407
```

Example E2.19: Determine the eigenvalues and eigenvectors of matrix **A** using MATLAB

$$(a)\qquad A=\begin{bmatrix}4 & -1 & 5\\ 2 & 1 & 3\\ 6 & -7 & 9\end{bmatrix} \quad \text{and} \quad (b)\ A=\begin{bmatrix}3 & 5 & 7\\ 2 & 4 & 8\\ 5 & 6 & 10\end{bmatrix}$$

Solution:

(*a*)
```
A = [4 -1 5;2 1 3;6 -7 9]
A =
    4  -1   5
    2   1   3
    6  -7   9
%The eigenvalues of A
format short e
eig(A)
ans =
      1.0000e + 001
      5.8579e - 001
      3.4142e + 000
%The eigenvectors of A
[Q,d] =eig(A)
Q =
    -5.5709e - 001 - 8.2886e - 001 - 7.3925e - 001
    -3.7139e - 001 - 3.9659e - 002 - 6.7174e - 001
    -7.4278e - 001  5.5805e - 001 - 4.7739e - 002
d =
    1.0000e + 001     0    0
    0 5.8579e-001    0
    0                    0 3.4142e + 000
```

(*b*)
```
A =
    3     5    7
    2     4    8
    5     6   10
%The eigenvalues of A
format short e
eig(A)
ans =
      1.7686e + 001
     -3.4295e - 001 +1.0066e + 000i
     -3.4295e - 001 -1.0066e + 000i
%The eigenvectors of A
[Q,d] =eig(A)
Q =
```

```
    Column 1
            5.0537e - 001
            4.8932e - 001
            7.1075e - 001

    Column 2
            -2.0715e - 001 - 5.2772e - 001i
             7.1769e - 001
            -3.3783e - 001 +2.2223e - 001i

    Column 3
            - 2.0715e - 001 + 5.2772e - 001i
              7.1769e - 001
            - 3.3783e - 001- 2.2223e - 001i

d =
    Column 1
            1.7686e + 001
            0
            0

    Column 2
            0
            -3.4295e - 001 +1.0066e + 000i
            0

    Column 3
            0
            0
            -3.4295e - 001 -1.0066e + 000i
```

Example E2.20: Determine the eigenvalues and eigenvectors of **AB** using MATLAB.

$$\mathbf{A} = \begin{bmatrix} 3 & 0 & 2 & 1 \\ 1 & 2 & 5 & 4 \\ 7 & -1 & 2 & 6 \\ 1 & -2 & 3 & 4 \end{bmatrix}; \quad \mathbf{B} = \begin{bmatrix} 1 & 3 & 5 & 7 \\ 2 & -1 & -2 & 4 \\ 3 & 2 & 1 & 1 \\ 4 & 1 & 0 & 6 \end{bmatrix}$$

Solution:

% MATLAB Program

% The matrix "*a*" = *A*B*

```
>> A = [ 3 0 2 1; 1 2 5 4; 7 -1 2 6; 1 -2 3 4 ];
>> B = [ 1 3 5 7; 2 -1 -2 4; 3 2 1 1; 4 1 0 6 ];
>> a = A*B
```

```
a =
   13 14 17 29
   36 15  6 44
   35 32 39 83
   22 15 12 26
>> eig (a)

ans =
   98.5461
    2.2964
   -1.3095
   -6.5329
```

The eigenvectors are:

```
>> [Q, d] = eig (a)
Q =
  - 0.3263  -0.2845    0.3908   0.3413
  - 0.3619   0.7387   -0.7816  -0.9215
  - 0.8168  -0.6026    0.4769   0.0962
  - 0.3089   0.1016   -0.0950   0.1586

d =
     98.5461  0           0          0
     0        2.2964      0          0
     0        0          -1.3095     0
     0        0           0         -6.5329
```

Example E2.21: Solve the following set of equations using MATLAB:

(*a*) $x_1 + 2x_2 + 3x_3 + 5x_4 = 21$
$-2x_1 + 5x_2 + 7x_3 - 9x_4 = 18$
$5x_1 + 7x_2 + 2x_3 - 5x_4 = 25$
$-x_1 + 3x_2 - 7x_3 + 7x_4 = 30$

(*b*) $x_1 + 2x_2 + 3x_3 + 4x_4 = 8$
$2x_1 - 2x_2 - x_3 - x_4 = -3$
$x_1 - 3x_2 + 4x_3 - 4x_4 = 8$
$2x_1 + 2x_2 - 3x_3 + 4x_4 = -2$

Solution:

(*a*)
```
>> A = [1 2 3 5;-2 5 7 -9; 5 7 2 -5; -1 -3 -7 7];
>> B = [21; 18; 25; 30];
>> S = A\B
S =
  -8.9896
  14.1285
  -5.4438
   3.6128
```

% Therefore $x_1 = -8.9896$, $x_2 = 14.1285$, $x_3 = -5.4438$, $x_4 = 3.6128$.

(*b*)
```
>> A = [1 2 3 4; 2 -2 -1 1; 1 -3 4 -4; 2 2 -3 4];
>> B = [8;-3; 8;-2];
>> S =A\B
S =
   2.0000
   2.0000
   2.0000
  -1.0000
```

% Therefore $x_1 = 2.0000$, $x_2 = 2.0000$, $x_3 = 2.0000$, $x_4 = -1.0000$.

Example E2.22: Use *diff* command for symbolic differentiation of the following functions:

(*a*) $S_1 = e^{x^8}$

(*b*) $S_2 = 3\,x^3\,e^{x^5}$

(*c*) $S_3 = 5x^3 - 7x^2 + 3x + 6$

Solution:

(*a*)
```
>> syms x
>> S1= exp (x^8);
>> diff (S1)
ans =
      8*x^7*exp(x^8)
```

(*b*)
```
>> S2=3*x^3*exp(x^5);
>> diff (S2)
ans =
      9*x^2*exp(x^5) +15*x^7*exp(x^5)
```

(*c*)
```
>> S3=5*x^3-7*x^2+3*x+6;
>> diff (S3)
ans =
      15*x^2-14*x + 3
```

Example E2.23: Use MATLAB's symbolic commands to find the values of the following integrals:

(*a*) $\int_{0.2}^{0.7} |x|\,dx$ (*b*) $\int_0^{\pi} (\cos y + 7y^2)dy$

(*c*) $\sqrt{x}$ (*d*) $7x^5 - 6x^4 + 11x^3 + 4x^2 + 8x + 9$

(*e*) $\cos a$

Solution:

(*a*)
```
>>syms x, y, a, b
>> S1= abs (x)
>> int (S1, 0.2, 0.7)
ans =
      9/40
```

(*b*)
```
>> S2=cos (y) +7*y^2
>> int (S2, 0, pi)
ans =
      7/3*pi^3
```

(*c*)
```
>> S3=sqrt (x)
>> int (S3)
ans =
      2/3*x^ (3/2)
      >> int (S3,'a', 'b')
ans =
      2/3*b^ (3/2)-2/3*a^ (3/2)
      >> int (S3, 0.4, 0.7)
ans =
      7/150*70^ (1/2)-4/75*10^ (1/2)
```

(*d*)
```
>> S4 = 7*x^5-6*x^4+11*x^3+4*x^2+8*x-9
>> int (S4)
ans =
      7/6*x^6-6/5*x^5+11/4*x^4+4/3*x^3+4*x^2-9*x
```

(*e*)
```
>> S5=cos (a)
>> int (S5)
ans =
      sin (a)
```

Example E2.24: Obtain the general solution of the following first order differential equations:

(*a*) $\frac{dy}{dt} = 5t - 6y$ (*b*) $\frac{d^2y}{dt^2} + 3\frac{dy}{dt} + y = 0$

(*c*) $\frac{ds}{dt} = Ax^3$ (*d*) $\frac{ds}{dA} = Ax^3$

Solution:

(*a*)
```
>> solve ('Dy=5*t-6*y')
ans =
      5/6*t-5/36+exp (-6*t)*C1
```

(*b*)
```
>> dsolve ('D2y +3*Dy + y = 0')
ans =
      C1*exp (1/2*(5^ (1/2)-3)*t) + C2*exp (-1/2*(5^ (1/2) +3)*t)
```

(*c*)
```
>> dsolve ('Ds =A*x^3','x')
ans =
      1/4*A*x^4 + C1
```

(*d*)
```
>> dsolve ('Ds=A*x^3', 'A')
ans =
      1/2*A^2*x^3 + C1
```

Example E2.25: Determine the solution of the following differential equations that satisfies the given initial conditions.

(*a*) $\frac{dy}{dx} = -7x^2$; $y(1) = 0.7$

(*b*) $\frac{dy}{dx} = 5x\cos^2 y$; $y(0) = \pi/4$

(*c*) $\frac{dy}{dx} = -y + e^{3x}$; $y(0) = 2$

(*d*) $\frac{dy}{dt} = +5y = 35$; $y(0) = 4$

Solution:

(*a*)
```
>> dsolve ('Dy = -7*x^2','y (1) =0.7')
ans =
      -7*x^2*t + 7*x^2 + 7/10
```

(*b*)
```
>> dsolve ('Dy=5*x*cos (y) ^2','y (0) =pi/4')
ans =
      atan (5*t*x + 1)
```

(*c*)
```
>> dsolve ('Dy = -y + exp (3*x)', 'y (0) =2')
ans =
       exp (3*x) + exp (- t)*(-exp (3*x) +2)
```

(*d*)
```
>> dsolve ('Dy + 5*y =35', 'y (0) =4')
ans =
      7 - 3*exp (-5*t)
```

Example E2.26: Given the differential equation

$$\frac{d^2x}{dt^2} + 7\frac{dx}{dt} + 5x = 8u(t) \; ; t \geq 0$$

Using MATLAB program, find

(*a*) $x(t)$ when all the initial conditions are zero.

(*b*) $x(t)$ when $x(0) = 1$ and $\dot{x} = 2$.

Solution:

(*a*) x (t) when all the initial conditions are zero
```
>> x = dsolve ('D2x = -7*Dx - 5*x +8', 'x (0) = 0')
x =
    8/5+ (-8/5-C2)*exp (1/2*(-7+29^ (1/2))*t) + C2*exp (-1/2*
     (7+29^ (1/2))*t)
```

(*b*) x (t) when x (0) = 1 and $\dot{x}$ (0) = 2

```
>> x = dsolve ('D2x = -7*Dx - 5*x +8', 'x (0) = 1', 'Dx (0) = 2')
x =
   8/5+ (-3/10-1/290*29^ (1/2))*exp (1/2*(-7+29^ (1/2))*t)-1
/290*(-1+3*29^ (1/2))*29^ (1/2)*exp (-1/2*(7+29^ (1/2))*t)
```

Example E2.27: Given the differential equation

$$\frac{d^2x}{dt^2}+12\frac{dx}{dt}+15x=35\ ;\ t\geq 0$$

Using MATLAB program, find

(*a*) $x(t)$ when all the initial conditions are zero.

(*b*) $x(t)$ when $x(0) = 0$ and $\dot{x}(0) = 1$.

Solution:

(*a*) $x(t)$ when all the initial conditions are zero

```
>> x = dsolve ('D2x = -12*Dx - 15*x +35', 'x (0) = 0')
x =
  7/3+ (-7/3-C2)*exp ((-6+21^ (1/2))*t) +C2*exp (-(6+21^ (1/2))*t)
```

(*b*) $x(t)$ when $x(0) = 0$ and $\dot{x}(0) = 1$.

```
>> x = dsolve ('D2x = -12*Dx - 15*x + 35', 'x (0) = 0', 'Dx (0) = 1')
x =
   7/3+ (-7/6-13/42*21^ (1/2))*exp ((-6+21^ (1/2))*t)-1/126*
   (-39+7*21^ (1/2))*21^ (1/2)*exp (-(6+21^ (1/2))*t0)
```

Example E2.28 : Find the inverse of the following matrix using MATLAB.

$$A=\begin{bmatrix} s & 2 & 0 \\ 2 & s & -3 \\ 3 & 0 & 1 \end{bmatrix}$$

Solution:

```
>> A = [s 2 0; 2 s -3; 3 0 1];
>> inv (A)
ans =
    [s/(s^2-22),            -2/(s^2-22),  -6/(s^2-22)]
    [-11/(s^2-22),          s/(s^2-22),   3*s/(s^2-22)]
    [-3*s/(s^2-22),         6/(s^2-22),   (s^2-4)/(s^2-22)]
```

Example E2.29: Expand the following function $F(s)$ into partial fractions using MATLAB. Determine the inverse Laplace transform of $F(s) = \dfrac{1}{s^4+5s^3+7s^2}$.

The MATLAB program for determining the partial fraction expansion is given below:

Solution:

```
>> b = [0 0 0 0 1];
>> a = [1 5 7 0 0];
>> [r, p, k] = residue (b, a)
r =
        0.0510 - 0.0648i
        0.0510 + 0.0648i
    -0.1020
        0.1429
p =
     -2.5000 + 0.8660i
     -2.5000 - 0.8660i
        0
        0
k = [ ]
```

% From the above MATLAB output, we have the following expression:

$$F(s) = \frac{r_1}{s - p_1} + \frac{r_2}{s - p_2} + \frac{r_3}{s - p_3} + \frac{r_4}{s - p_4}$$

$$F(s) = \frac{0.0510 - 0.0648i}{s - (-2.5000 + 0.8660i)} + \frac{(0.0510 + 0.0648i)}{s - (-2.5000 - 0.8660i)} + \frac{-0.1020}{s - 0} + \frac{0.1429}{s - 0}$$

% Note that the row vector *k* is zero implies that there is no constant term in this example problem.
% The MATLAB program for determining the inverse Laplace transform of *F*(*s*) is given below:

```
>> syms s
>> f = 1/(s^4 + 5*s^3 + 7*s^2);
>> ilaplace ( f )
ans =
 1/7*t-5/49+5/49*exp (-)*cos (1/2*3^ (1/2)*t) +11/147*exp (-5/2*t)*3^
  (1/2)*sin(1/2*3^(1/2)*t)
```

Example E2.30: Expand the following function *F*(*s*) into partial fractions using MATLAB. Determine the inverse Laplace transform of *F*(*s*).

$$F(s) = \frac{5s^2 + 3s + 6}{s^4 + 3s^3 + 7s^2 + 9s + 12}$$

Solution:
The MATLAB program for determining the partial fraction expansion is given below:

```
>> b = [0 0 5 3 6];
>> a = [1 3 7 9 12];
>> [r, p, k] = residue(b, a)
r =
     -0.5357 -1.0394i
     -0.5357 +1.0394i
      0.5357 -0.1856i
      0.5357 +0.1856i
```

```
p =
     -1.5000 + 1.3229i
     -1.5000 - 1.3229i
     -0.0000 + 1.7321i
     -0.0000 - 1.7321i
k = [ ]
```

% From the above MATLAB output, we have the following expression:

$$F(s) = \frac{r_1}{s - p_1} + \frac{r_2}{s - p_2} + \frac{r_3}{s - p_3} + \frac{r_4}{s - p_4}$$

$$F(s) = \frac{-0.5357 - 1.0394i}{s - (-1.500 + 1.3229i)} + \frac{(-0.5357 + 1.0394i)}{s - (-1.5000 - 1.3229i)}$$

$$+ \frac{0.5357 - 0.1856i}{s - (-0 + 1.7321i)} + \frac{-0.5357 + 0.1856i}{s - (-0 - 1.7321i)}$$

% Note that the row vector k is zero implies that there is no constant term in this example problem.
% The MATLAB program for determining the inverse Laplace transform of $F(s)$ is given below:

```
>> syms s
>> f = (5*s^2 + 3*s +6)/(s^4 + 3*s^3 + 7*s^2 + 9*s +12);
>> ilaplace(f )
ans =
     11/14*exp(-3/2*t)*7^(1/2)*sin(1/2*7^(1/2)*t) -15/14*exp(-3/
      2*t)*cos(1/2*7^(1/2)*t) +3/14*3^ (1/2)*sin(3^(1/2)*t)+15/
      14*cos(3^(1/2)*t)
```

Example E2.31: For the following function $F(s)$:

$$F(s) = \frac{s^4 + 3s^3 + 5s^2 + 7s + 25}{s^4 + 5s^3 + 20s^2 + 40s + 45}$$

Using MATLAB, find the partial fraction expansion of $F(s)$. Also, find the inverse Laplace transformation of $F(s)$.

Solution:

$$F(s) = \frac{s^4 + 3s^3 + 5s^2 + 7s + 25}{s^4 + 5s^3 + 20s^2 + 40s + 45}$$

The partial fraction expansion of $F(s)$ using MATLAB program is given as follows:

```
num = [ 1  3  5  7  25];
den = [1  5  20  40  45];
[r, p, k] = residue(num, den)

r =
    -1.3849 +1.2313i
    -1.3849 -1.2313i
     0.3849 -0.4702i
     0.3849 +0.4702i
```

```
p =
    -0.8554 +3.0054i
    -0.8554 -3.0054i
    -1.6446 +1.3799i
    -1.6446 -1.3799i
k =
    1
```

From the MATLAB output, the partial fraction expansion of $F(s)$ can be written as follows:

$$F(s) = \frac{r_1}{(s-p_1)} + \frac{r_2}{(s-p_2)} + \frac{r_3}{(s-p_3)} + \frac{r_4}{(s-p_4)} + k$$

$$F(s) = \frac{(-1.3849 + j1.2313)}{(s + 0.8554 - j3.005} + \frac{(-1.3849 - j1.2313)}{(s + 0.8554 + j3.005)}$$

$$+ \frac{(0.3849 - j0.4702)}{(s + 1.6446 - j1.3799)} + \frac{(0.3849 + j0.4702)}{(s + 1.6446 + j1.3779)} + 1$$

Example E2.32: Obtain the partial fraction expansion of the following function using MATLAB:

$$F(s) = \frac{8(s+1)(s+3)}{(s+2)(s+4)(s+6)^2}$$

Solution:

$$F(s) = \frac{8(s+1)(s+3)}{(s+2)(s+4)(s+6)^2} = \frac{(8s+8)(s+3)}{(s^2+6s+8)(s^2+12s+36)}$$

The partial fraction expansion of $F(s)$ using MATLAB program is given as follows:

```
EDU>> num=conv([8 8], [1 3]);
EDU>> den=conv([1 6 8], [1 12 36]);
EDU>> [r, p, k]= residue(num, den)
r =
    3.2500
   15.0000
   -3.0000
   -0.2500

p =
   -6.0000
   -6.0000
   -4.0000
   -2.0000

k = [ ]
```

From the above MATLAB result, we have the following expansion:

$$F(s) = \frac{r_1}{(s-p_1)} + \frac{r_2}{(s-p_2)} + \frac{r_3}{(s-p_3)} + \frac{r_4}{(s-p_4)} + k$$

$$F(s) = \frac{3.25}{(s+6)} + \frac{15}{(s-15)} + \frac{-3}{(s+3)} + \frac{-0.25}{(s+0.25)} + 0$$

It should be noted here that the row vector k is zero, because the degree of the numerator is lower than that of the denominator.

$$F(s) = 3.25e^{-6t} + 15e^{15t} - 3e^{-3t} - 0.25e^{-0.25t}$$

Example E2.33: Find the Laplace transform of the following function using MATLAB.

(*a*) $f(t) = 7t^3\cos(5t + 60°)$
(*b*) $f(t) = -7t\,e^{-5t}$
(*c*) $f(t) = -3\cos 5t$
(*d*) $f(t) = t\sin 7t$
(*e*) $f(t) = 5\,e^{-2t}\cos 5t$
(*f*) $f(t) = 3\sin(5t + 45°)$
(*g*) $f(t) = 5\,e^{-3t}\cos(t - 45°)$

Solution:

% MATLAB Program

(*a*)
```
>> syms t % tell MATLAB that "t" is a symbol.
>> f = 7 * t^3*cos(5*t + (pi/3)); % define the function.
>> laplace( f )

ans =
   -84/(s^2+25)^3*s^2+21/(s^2+25)^2+336*(1/2*s-5/2*3^(1/2))/
    s^2+25)^4*s^3-168*(1/2*s-5/2*3^(1/2))/(s^2+25)^3*s

>> pretty(laplace( f )) % the pretty function prints symbolic output
% in a format that resembles typeset mathematics.
```

$$\frac{-84s^2}{(s^2+25)^3} + \frac{21}{(s^2+25)^2} + \frac{336\left(\frac{1}{2}s - \frac{5}{2}(3)^{1/2}\right)s^3}{(s^2+25)^4} - \frac{168\left(\frac{1}{2}s - \frac{5}{2}(3)^{1/2}\right)s}{(s^2+25)^3}$$

(*b*)
```
>>syms t x
>>f = -7*t*exp(-5*t);
>> laplace(f, x)
ans =
   -7/(x + 5)^2
```

(*c*)
```
>>syms t x
>>f = -3*cos(5*t);
>> laplace(f,x)
ans =
   -3*x/(x^2 + 25)
```

(*d*)
```
>>syms t x
>>f = t*sin(7*t);
>> laplace( f, x)
ans =
     1/(x^2 + 49)*sin(2*atan(7/x))
```

(*e*)
```
>>syms t x
>>f = 5*exp(-2*t)*cos(5*t);
>> laplace(f, x)
ans =
     5*(x + 2)/((x + 2)^2 +25)
```

(*f*)
```
>>syms t x
>>f = 3*sin(5*t + (pi/4));
>> laplace( f, x)
ans =
     3*(1/2*x*2^(1/2) + 5/2*2^(1/2))/(x^2 + 25)
```

(*g*)
```
>>syms t x
>>f = 5*exp(-3*t)*cos(t - (pi/4));
>> laplace( f, x)
ans =
     5*(1/2*(x + 3)*2^(1/2)+1/2*2^(1/2))/((x + 3)^2 + 1)
```

Example E2.34: Generate partial fraction expansion of the following function:

$$F(s) = \frac{10^5(s+7)(s+13)}{s(s+25)(s+55)(s^2+7s+75)(s^2+7s+45)}$$

Solution:

```
Generate the partial fraction expansion of the following function:
numg=poly[-7 -13];
numg=poly([-7 -13]);
deng=poly([0 -25 -55 roots([1 7 75])' roots([1 7 45])']);
[numg,deng]=zp2tf(numg',deng',1e5);
Gtf=(numg,deng);
Gtf=tf(numg,deng);
G=zpk(Gtf);
[r,p,k]=residue(numg,deng)
r =
   1.0e - 017*
   0.0000
 - 0.0014
   0.0254
 - 0.1871
   0.1621
 - 0.0001
   0.0000
   0.0011
```

```
p =
  1.0e + 006 *
  4.6406
  1.4250
  0.3029
  0.0336
  0.0027
  0.0001
  0.0000
       0
k = [ ]
```

Example E2.35: Determine the inverse Laplace transform of the following functions using MATLAB.

$$(a) \quad F(s) = \frac{s}{s(s+2)(s+6)} \qquad (b) \quad F(s) = \frac{1}{s^2(s+5)}$$

$$(c) \quad F(s) = \frac{3s+1}{(s^2+2s+9)} \qquad (d) \quad F(s) = \frac{s-25}{s(s^2+3s+20)}$$

Solution:

(*a*)

```
>> syms s
>> f = s/(s*((s + 2)*(s + 6)));
>> ilaplace( f )
ans =
    1/2*exp(-4*t)*sinh(2*t)
```

(*b*)

```
>> syms s
>> f = 1/((s^2)*(s + 5));
>> ilaplace(f )
ans =
    1/3*t - 2/9*exp(-3/2*t)*sinh(3/2*t)
```

(*c*)

```
>>syms s
>> f = (3*s + 1)/(s^2 + 2*s + 9);
>> ilaplace( f )
ans =
    3*exp(-t)*cos(2*2^(1/2)*t)-1/2*2^(1/2)*exp(-t)*sin(2*2^(1/2)*t)
```

(*d*)

```
>>syms s
>> f = (s -25)/(s*(s^2 + 3*s +25));
>> ilaplace( f )
ans =
    5/4*exp(-3/2*t)*cos(1/2*71^(1/2)*t)+23/284*71^(1/2)*exp(-3/2*t)*
        sin(1/2*71^(1/2)*t)-5/4
```

Example E2.36: Find the inverse Laplace transform of the following function using MATLAB.

$$G(s) = \frac{(s^2 + 9s + 7)(s + 7)}{(s + 2)(s + 3)(s^2 + 12s + 150)}$$

Solution:

```
% MATLAB Program
>> syms s     % tell MATLAB that "s" is a symbol.
>>G = (s^2 + 9*s +7)*(s + 7)/[(s + 2)*(s + 3)*(s^2 + 12*s + 150)]; % define
the function.
>>pretty(G) % the pretty function prints symbolic output
% in a format that resembles typeset mathematics.
```

$$\frac{(s + 9s + 7)(s + 7)}{(s + 2)(s + 3)(s + 12s + 150)}$$

```
>> g = ilaplace(G); % inverse Laplace transform
>>pretty(g)
```

$$-7/26\exp(-2t) + \frac{44}{123}\exp(-36) + \frac{2915}{3198}\exp(-6t)\cos(114t)^{1/2}$$

$$+\frac{889}{20254}\exp(-6(t))114^{1/2}\ \sin(114(t))^{1/2}$$

2.22 SUMMARY

In this chapter the interactive environment for numeric computation, data analysis, and graphics using MATLAB was presented. Arithmetic operations, display formats, elementary built-in functions, arrays, scalars, vectors or matrices, operations with arrays including dot product, array multiplication, array division, inverse and transpose of a matrix, determinants, element by element operations, eigenvalues and eigenvectors, random number generating functions, polynomials, system of linear equation, script files, programming in MATLAB, the commands used for printing information and generating 2-D and 3-D plots, input/output in MATLAB was presented with illustrative examples. MATLAB's functions for symbolic mathematics were introduced. These functions are useful in performing symbolic operations and developing closed-form expressions for solutions to linear algebraic equations, ordinary differential equations and systems of equations. Symbolic mathematics for determining analytical expressions for the derivative and integral of an expression was also presented. Some more details of MATLAB for engineering applications are given in Chapter-4.

CHAPTER 3

MATLAB TUTORIAL FOR MECHANICS

3.1 INTRODUCTION

MATLAB has an excellent collection of commands and functions that are useful for solving engineering mechanics problems. The problems presented in this chapter are basic and are normally presented in introductory mechanics courses. The application of MATLAB to the problems in the analysis and design of engineering mechanics is presented in this chapter with a number of illustrative examples. These examples cover different topics of Mechanics. Each solution begins with a problem formulation followed by a corresponding MATLAB code with simple explanations wherever possible. The background required for solving Mechanics problems is the knowledge of differentiation, integration, trigonometry along with basic physical laws. Table 3.1 summarizes some of the important mathematical relations often used for computations in statics and dynamics.

Table 3.1 Some mathematical expressions

S.No.	Mathematical entity
1.	If $ax^2 + bx + c = 0$, then $x = \dfrac{-b \pm \sqrt{b^2 - 4ac}}{2a}$ is root.
2.	If $\sin\theta = A$ and $\cos\theta = B$ then (*i*) $A^2 + B^2 = 1$, (*ii*) $\sin 2\theta = 2AB$, (*iii*) $\cos 2\theta = B^2 - A^2$ (*iv*) $\tan\theta = \dfrac{A}{B}$, (*v*) $\sin(\theta \pm \phi) = \sin\theta\cos\phi \pm \cos\theta\sin\phi$ In a triangle: (*a*) **sine rule** $a/\sin(\theta) = b/\sin(\phi)$, where a is side opposite to angle ϕ and b is side opposite to angle θ. (*b*) **Cosine rule**: $c^2 = a^2 + b^2 - 2ab\cos(\theta)$, where θ is angle opposite to side c.
3.	Important derivatives: (*i*) $\dfrac{d}{dx}(u^n) = nu^{n-1}\dfrac{du}{dx}$ (*ii*) $\dfrac{d}{dx}(uv) = u\dfrac{dv}{dx} + v\dfrac{du}{dx}$ (*iii*) $\dfrac{d}{dx}\left(\dfrac{u}{v}\right) = \dfrac{v\dfrac{du}{dx} - u\dfrac{dv}{dx}}{v^2}$ (*iv*) $\dfrac{d}{dx}(\sin\theta) = \cos\theta\dfrac{d\theta}{dx}$ (*v*) $\dfrac{d}{dx}(\cos\theta) = -\sin\theta\dfrac{d\theta}{dx}$ (*vi*) $\dfrac{d}{dx}(\tan\theta) = \sec^2\theta\dfrac{d\theta}{dx}$

4.	Important integrals (here c is a constant) (*i*) $\int x^p dx = \frac{x^{p+1}}{p+1} + c, p \neq -1$ (*ii*) $\int \frac{dx}{p+qx} = \frac{1}{q}\log_e(p+qx) + c$ (*iii*) $\int \frac{dx}{p+qx^2} = \frac{1}{2\sqrt{-pq}}\log_e \frac{p+x\sqrt{-pq}}{p-x\sqrt{-pq}} + c,$ where $pq < 0$ (*iv*) $\int \sin x\, dx = -\cos x + c,$ (*v*) $\int \cos x\, dx = \sin x + c,$ (*vi*) $\int e^{ax}\, dx = \frac{1}{a}e^{ax} + c$

In computational approach the first step is to understand theory necessary to solve the problem. Second step is to develop an algorithm that outlines how the problem input can be used along with the theory to generate the required output. In final step, this algorithm must be translated into correct MATLAB statements. It must be noted that MATLAB solutions need not be unique. In all the examples solved in this chapter, a script is written in the form of '*.m*' files and the program is executed by typing the m file name with extension at the command prompt. It is emphasized to develop each program with different commands to learn MATLAB with perfection.

3.2 EXAMPLE PROBLEMS AND SOLUTIONS

3.2.1 Statics

Problems in statics are presented on the following common topics: forces on a particle, rigid-bodies and equivalent forces, equilibrium of rigid bodies, truss analysis, beams, friction and distributed forces (centroids).

ES 3.1: Write a MATLAB program to determine the magnitude and direction of the resultant of 3-coplanar forces applied at point A in Fig.ES3.1. Use the following values:

$$F_1 = 20 \text{ kN},\ F_2 = 40 \text{ kN},\ F_3 = 200 \text{ kN},\ \alpha_1 = 40°,\ \alpha_2 = 25° \text{ and } \alpha_3 = 58°.$$

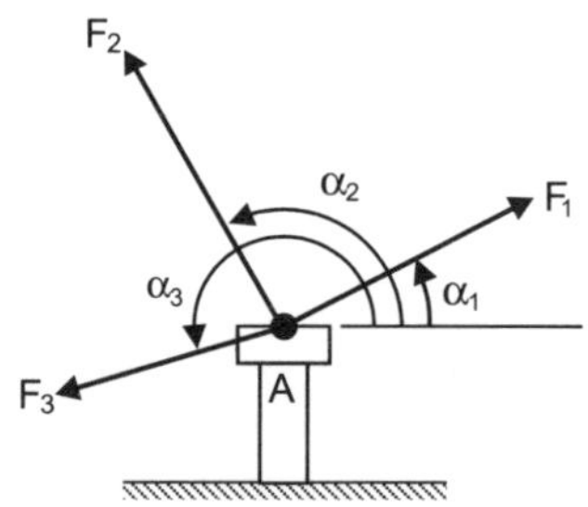

Fig. ES3.1

Solution:

We know that for a coplanar force system

$$R_x = \sum_{i=1}^{n} F_i \cos\alpha_i, \qquad R_y = \sum_{i=1}^{n} F_i \sin\alpha_i \qquad ...(1)$$

Therefore $$R = \sqrt{R_x^2 + R_y^2} \qquad ...(2)$$

and $$\alpha_R = \tan^{-1}\frac{R_y}{R_x} \qquad ...(3)$$

Let α_R^* be the value defined by Eq. (3) and such that $-90^\circ \le \alpha_R^* \le 90^\circ$. Then, we have

If $R_x \ge 0$ and $R_y \ge 0 : \alpha_R = \alpha_R^*$ ...(4)

If $R_x \ge 0$ and $R_y < 0 : \alpha_R = 360^\circ + \alpha_R^*$ ...(5)

If $R_x < 0 : \alpha_R = 180^\circ + \alpha_R^*$ ...(6)

MATLAB Solution:

```
%%%%%%%%%%%%%%%%%%%%%%%%%%%%
n=3; % Number of forces
alpha= [40 25 58];
alpha1=alpha*pi/180;
force= [20 40 200];
sumx=0;
sumy=0;
for i=1: n
  sumx=sumx + force (i)*cos (alpha1 (i));
  sumy=sumy + force (i)*sin (alpha1(i));
end
 r=sqrt (sumx^2+sumy^2);
 alphar = atan2 (sumy, sumx);
 alphar=alphar*180/pi;
 if alphar < 0
 alphar = alphar + 360;
end
fprintf ('The resultant R is %4.2f kN\n', r);
fprintf ('The angle between the resultant and x axis is % 4.2f degrees\n', alphar);
%%%%%%%%%%%%%%%%%%%%%%%%%%%%
```

Output is given below:

The resultant *R* is 254.11 kN.

The angle between the resultant and *x* axis is 51.68 degrees.

ES 3.2: Figure ES3.2 shows two forces, one 500 N and the other *P* applied by cables on each side of the obstruction *A* in order to remove the spike. Write a MATLAB program to determine

(*a*) the magnitude of *P* necessary to such that the resultant *T* is directed along the spike

(*b*) the magnitude of *T*

(*c*) plot *P* and *T* as a function of *d*. (Range of *d* between 1 and 20 mm).

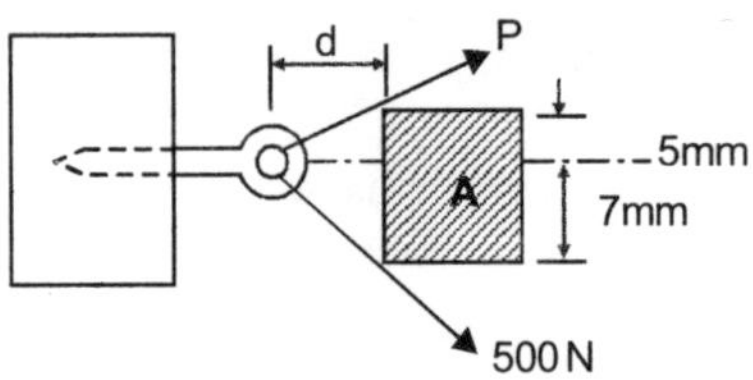

Fig. ES3.2

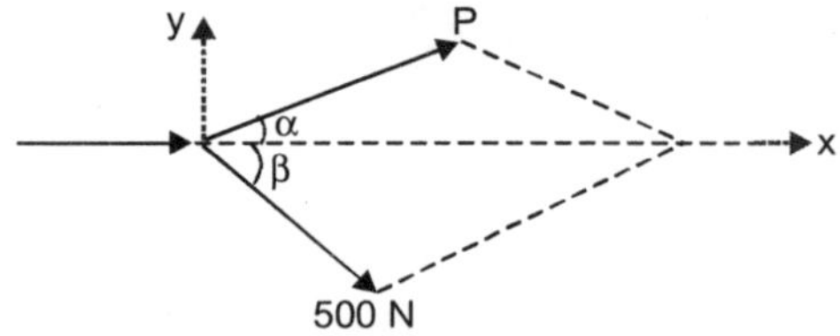

Fig. ES3.2 (a)

Solution: Free body diagram of the system is shown in Fig. ES3.2 (*a*).

From equilibrium equations, resolving the forces along positive x and y directions:

$$T = \Sigma F_x = P\cos\alpha + 500\cos\beta$$

$$0 = \Sigma F_y = P\sin\alpha - 500\sin\beta$$

Solving the above two equations, we obtain

$$P = \frac{500\sin\beta}{\sin\alpha}$$

and $$T = 500\,(\sin\beta\cot\alpha + \cos\beta)$$

From the geometry

$$\alpha = \tan^{-1}(5/d) \text{ and } \beta = \tan^{-1}(7/d)$$

Complete MATLAB program is given below:

```
% Range of d
d = 1:1:20;
% Define alpha
alpha = atan (5./d);
%Define beta
beta = atan(7./d);
% Compute force P
P = 500*sin(beta)./sin(alpha);
% Define force T
T=500*(sin (beta).*cot (alpha) +cos (beta));
plot (d, P,'-*', d, T,'-p')
xlabel ('d (mm)');
ylabel ('Force (N)');
legend('Force P', 'Net force T');
grid on;
```

Figure ES3.2 (*b*) shows the plot of P and T as a function of d.

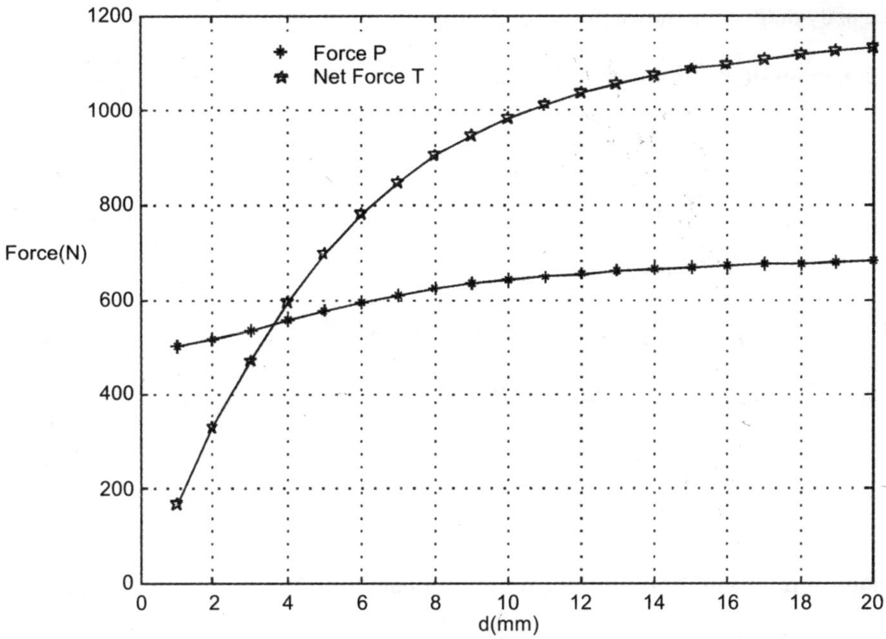

Fig. ES3.2(b)

ES 3.3: Figure ES3.3 shows the two cables *MO* and *NO* tied together at *O* and the loadings are also shown. The magnitude of *F* is 150 N.

(*a*) Derive the expressions relating the tension in each cable as a function of α.

(*b*) Write a MATLAB program to plot the tension in each cable for $0° \leq \alpha \leq 90°$.

(*c*) Determine the smallest value of α for which both cables are in tension.

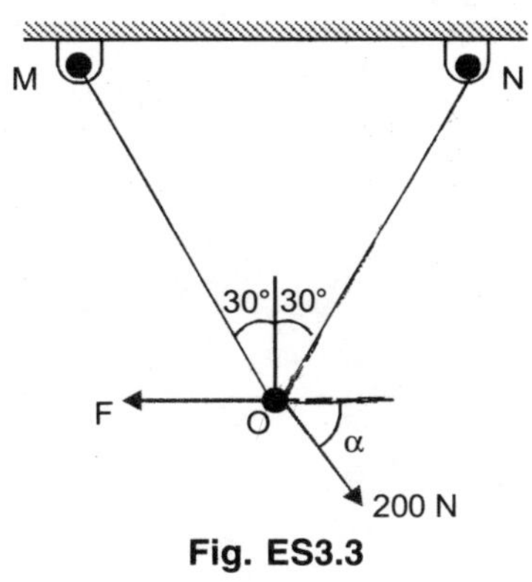

Fig. ES3.3

Solution: Free body diagram is shown in Fig. ES3.3 (*a*).

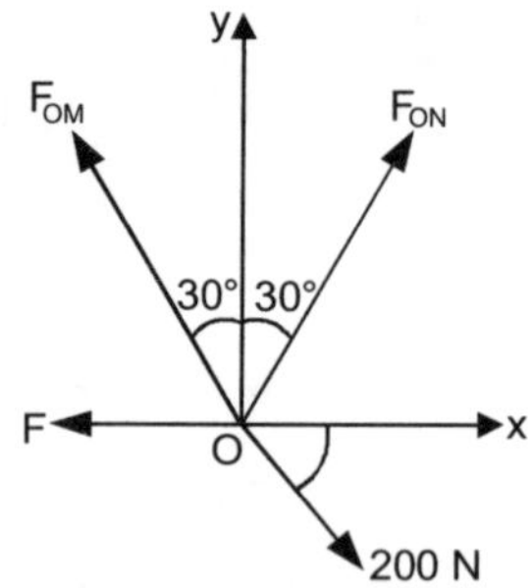

Fig. ES3.3 (a)

Applying the equations of equilibrium:

(*a*) $\Sigma F_x = 0 \Rightarrow (F_{ON} - F_{OM}) \sin 30° - F + 200 \cos \alpha = 0$

or $(F_{ON} - F_{OM}) = 2F - 400 \cos \alpha$

Substituting $F = 150$ N

$(F_{ON} - F_{OM}) = 300 - 400 \cos \alpha$ ...(1)

$\Sigma F_y = 0 \Rightarrow (F_{OM} + F_{ON}) \cos 30° - 200 \sin \alpha = 0$

or $(F_{OM} + F_{ON}) = 200 \times \dfrac{2}{\sqrt{3}} \cdot \sin \alpha = 230.94 \sin \alpha$ ...(2)

Solving (1) and (2),

$F_{ON} = 150 - 200 \cos \alpha + 115.47 \sin \alpha$

$F_{OM} = 115.47 \sin \alpha + 200 \cos \alpha - 150$

For finding range of α, at which tensions are positive equate $F_{ON} = 0$ and $F_{OM} = 0$ and find α.

(*b*) MATLAB Program:

```
Alpha= [0:2:90];
alpha=Alpha*pi/180;
fon =150-200*cos (alpha) +115.47*sin (alpha);
fom=-150+115.47*sin (alpha) +200*cos (alpha);
ton=abs (fon);
tom=abs (fom);
[ton_min, i]=min (ton);
[tom_min, j]=min (tom);
Ang1_min=Alpha (i);
Ang2_min=Alpha (j);
plot (Alpha, fon, Alpha, fom);
legend ('Fon', 'Fom', 2)
xlabel ('Alpha (degrees)')
ylabel ('Cable tension (N)')
grid on
fprintf ('(c)\n')
fprintf ('Smallest value of Alpha for which the tensions are positive is from %g to
%g degrees\n', Ang1_min, Ang2_min)
```

The output is shown in Fig. ES3.3 (*b*).

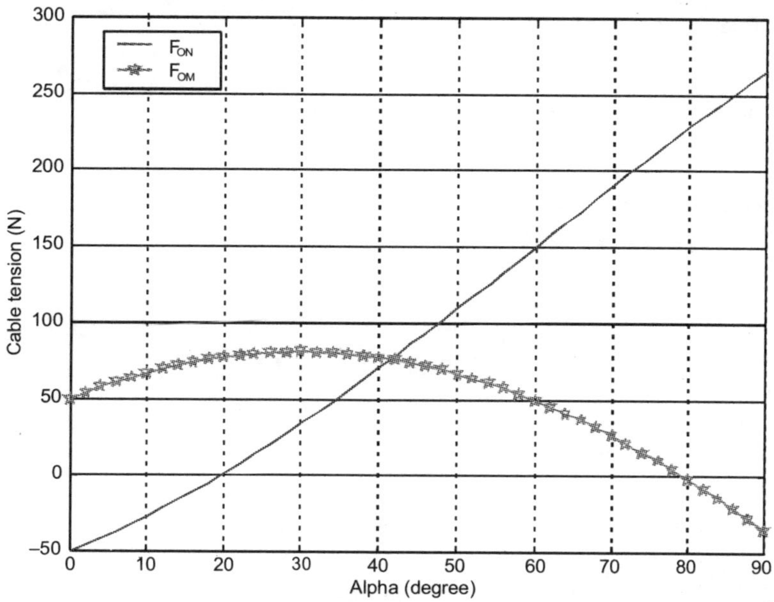

Fig. ES3.3 (b)

(*c*) Smallest value of alpha for which the tensions are positive is from 20 to 80 degrees.

ES 3.4: Figure ES3.4 shows a weight *W* hung from the end of a horizontal pole of negligible weight.

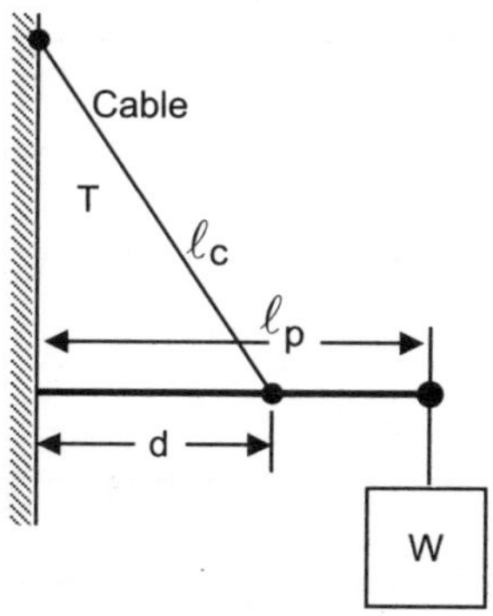

Fig. ES3.4

The pole is attached to the wall by a pivot and is supported by a cable attached to the wall at a higher point. The tension *T*, in the cable is given by

$$T = \frac{W \ell_c \ell_p}{d\sqrt{\ell_p^2 - d^2}}$$

where T = tension in the cable, W = weight of the object, ℓ_c = length of the cable, ℓ_p = length of the pole and d = distance along the pole at which the cable is attached.

Write a MATLAB program to (*a*) determine the distance (*d*), at which the cable can be attached to the pole in order to minimize the tension in the cable, (*b*) plot the tension in the cable as a function of *d*.

Given: $W = 250$ N, $\ell_p = 50$ cm, $\ell_c = 40$ cm.

Solution: The free-body diagram of the system is shown in Fig. ES3.4 (*a*).

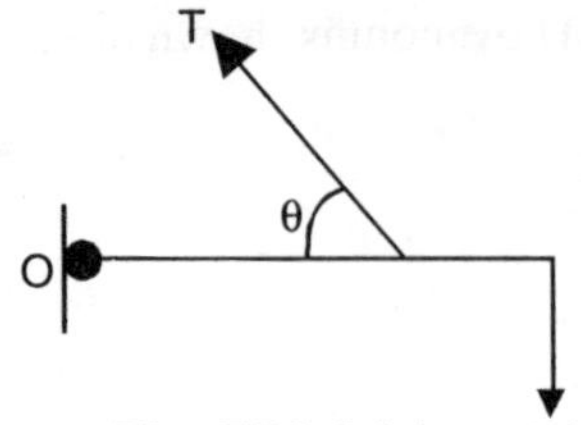

Fig. ES3.4 (a)

MATLAB program is given below:

```
%%%%%%%%%%%%%%%%%%%%%%%%%%%%%%%%%%%%%
lc=0.40;
lp=0.50;
W=250;
d=0.05:0.05:lp;
% Calculate tension
T=W *lc *lp./(d.*sqrt(lp^2-d.^2));
plot(d*100,T,'-p');
xlabel('Distance d  in cm));
ylabel('Tension in string in N');
grid on;
[Tmin, I]=min(T);
fprintf('Minimum tension is %g N at %g cm',Tmin,d(I)*100)
%%%%%%%%%%%%%%%%%%%%%%%%%%%%%%%%%%%%%%%
```

The output is given in Fig. ES3.4(*b*).

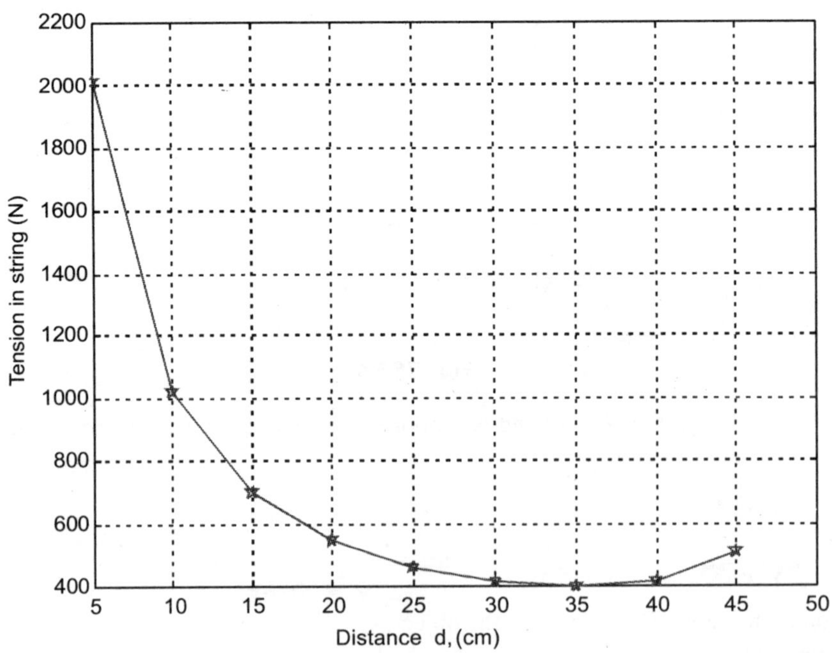

Fig. ES3.4 (b)

At the command prompt following output is obtained:

>> Minimum tension is 400.08 N at 35 cm.

ES 3.5: Figure ES3.5 shows the location of the center of gravity of a 5000 N truck for the unloaded condition. The location of the added load W_L is at a distance of x inches behind the rear axle. Write a MATLAB program and plot W_L as a function of x for x ranging from 0 to 60 mm.

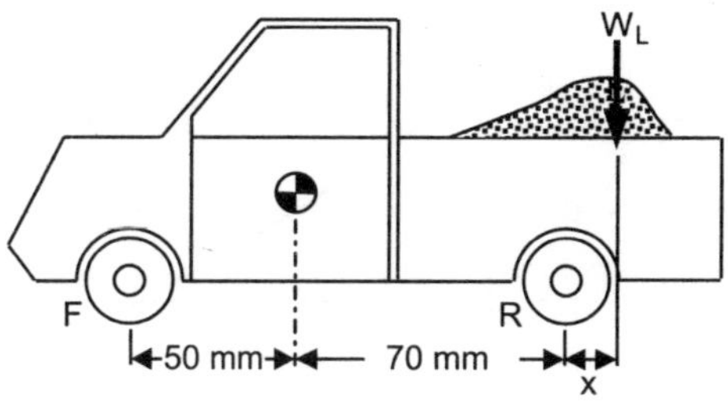

Fig. ES3.5

Solution: Free body diagram of the system is shown in Fig. ES3.5(*a*).

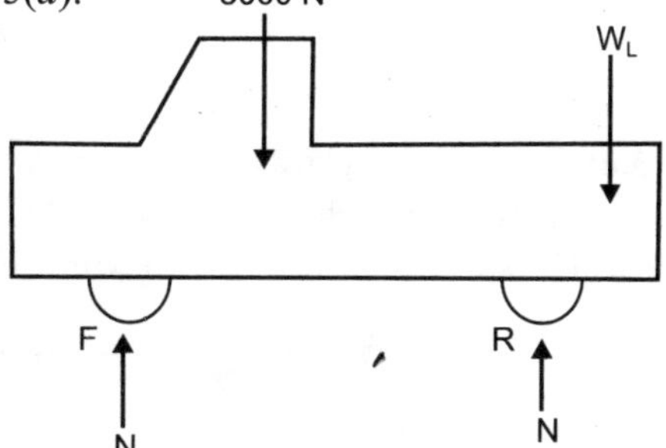

Fig. ES3.5 (a)

The equilibrium equations can be written as
Moment about rear wheel axle:

$$\Sigma M_R = 0 = 5000(70) - N(120) - W_L x = 0$$

$$\Sigma F_y = 0 = N + N - 5000 - W_L = 0$$

Solving the above equations for W_L, we obtain

$$W_L = \frac{5000}{(60+x)}$$

The plot of W_L as a function of x is shown in Fig.ES3.5 (*b*) from the following program.

MATLAB Solution:

```
% Define the range of x for the plot
x = 0:0.05:60;
% Define Wl
Wl=50000. /(60.+x);
plot(x, Wl)
% Labels
xlabel ('x (mm)')
ylabel ('Load weight (N)')
grid on;
```

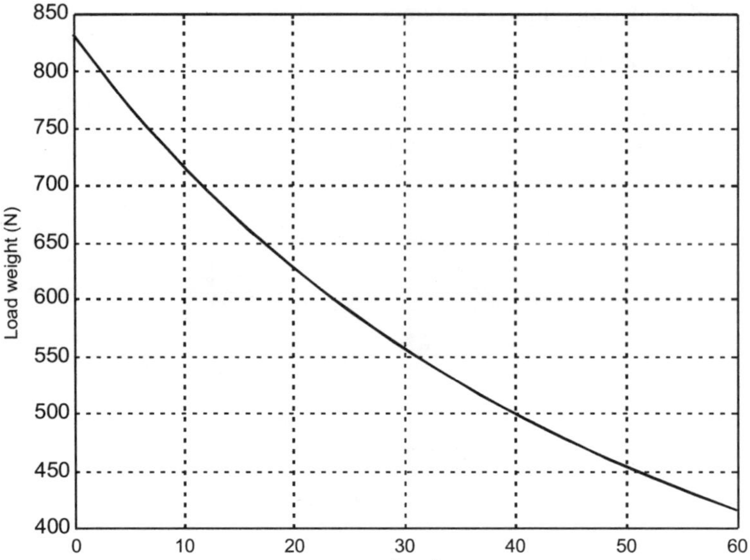

Fig. ES3.5 (b)

ES 3.6: Figure ES3.6 shows a container of weight W suspended from ring C to which cable CB of length 6 m and spring AC are attached. Write a MATLAB program to determine the tension in the cable when $W = 150$ N. Given: The spring constant as 120 N/m, and its unstreched length as 4 m.

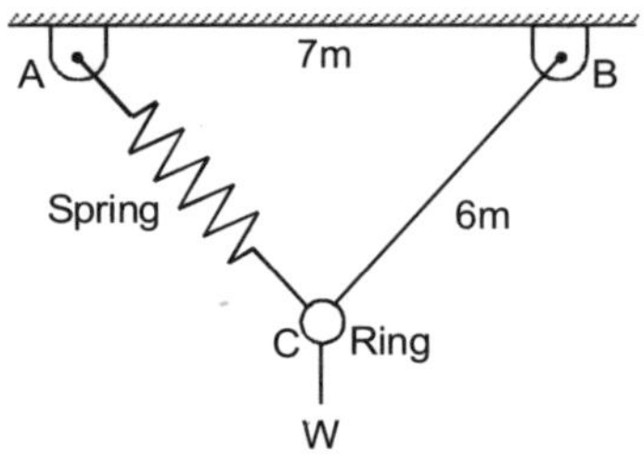

Fig. ES3.6

Solution: Free body diagram of the system is shown in Fig. ES3.6 (*a*).

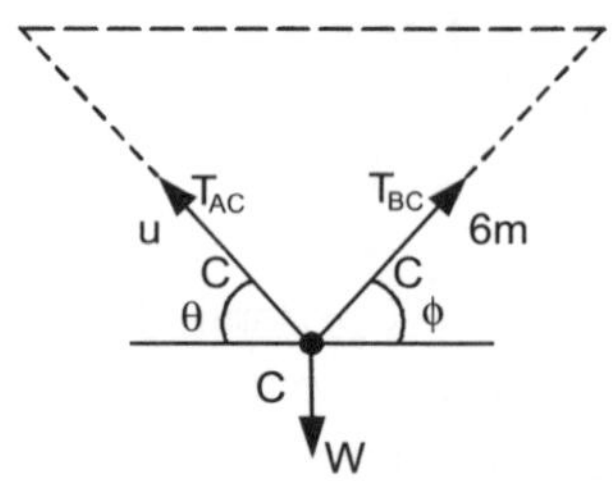

Fig. ES3.6 (a)

$$\Sigma F_x = 0 \Rightarrow T_{BC} \cdot \cos\phi - T_{AC}\cos\theta = 0 \quad ...(1)$$

$$\Sigma F_y = 0 \Rightarrow T_{BC} \cdot \sin\phi + T_{AC}\sin\theta - W = 0 \quad ...(2)$$

Geometry: $$u^2 = 6^2 + 7^2 - 2(6)(7)\cos\phi = 78 - 84\cos\phi \quad ...(3)$$

$$\frac{u}{\sin\phi} = \frac{6}{\sin\theta} \quad ...(4)$$

Stiffness: $$T_{AC} = k(u - u_0) = 120(u - 4) \quad ...(5)$$

Replacing T_{AC} in terms of u, the problem reduces to find four unknowns, T_{BC}, u, θ and ϕ.
Calling them respectively as x_1, x_2, x_3, x_4 following MATLAB program is developed to solve the equations.

MATLAB Program:

```
eq1='x1*cos(x4)-120*(x2-4)*cos(x3)=0'
eq2='x1*sin(x4)+120*(x2-4)*sin(x3)-150=0'
eq3='x2^2-78+84*cos(x4)=0'
eq4='x2-6*sin(x4)/sin(x3)=0';
[x1,x2,x3,x4]=solve(eq1,eq2,eq3,eq4)
```

This gives output as set of solutions. Choose by proper reasoning following values are obtained. Output is $x_1 = T_{BC} = 60.509$ N and $x_2 = u = 4.921$ m, $x_3 = \tan^{-1}(0.9365)$ and $x_4 = \tan^{-1}(0.7681)$.

ES 3.7: Figure ES3.7 shows the members *CJ* and *CF* of the loaded truss cross which are not connected to members *BI* and *DG*. Determine the values of a for which the truss cannot be in equilibrium. Write a MATLAB program to plot the forces in members *BC*, *JC*, *IC*, and *IG* as a function of α.

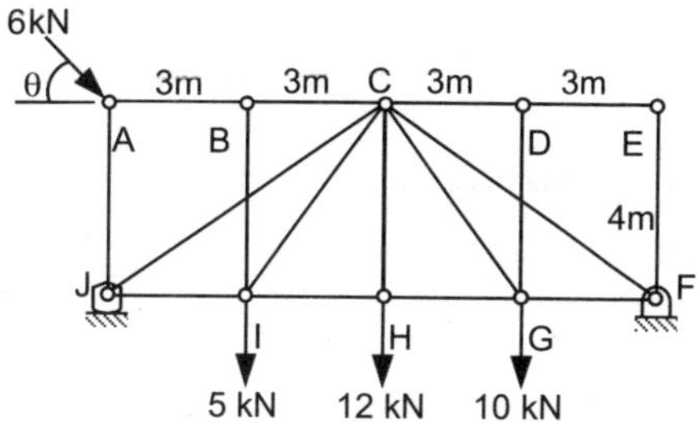

Fig. ES3.7

Solution: Free body diagram is shown in Fig. ES3.7 (*a*).

First we determine the reaction force at *J* from a free-body diagram for the entire truss.

$$\circlearrowleft \Sigma M_F = 0 = 6\sin\theta(12) - 6\cos\theta(4) + 4(9) + 10(6) + 8(3) - J_y(12)$$

$$J_y = 10 + 6\sin\theta - 2\cos\theta$$

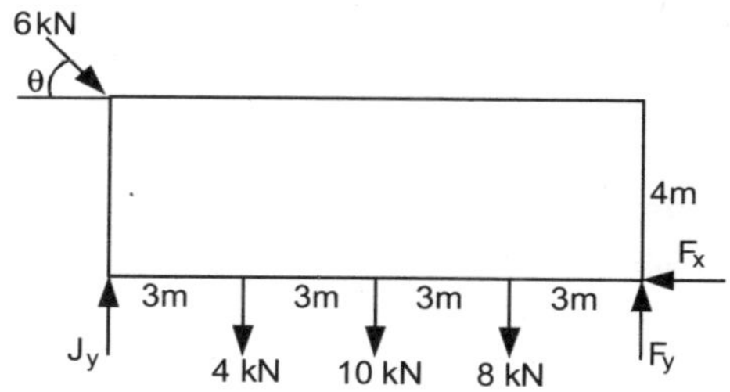

Fig. ES3.7 (a)

Note that the rocker can only exert an upward force at *A*. Thus, to be in equilibrium, J_y must be positive. Since sin θ and cos θ vary between plus and minus one, the above equation indicates that J_y will be positive for all θ. Thus, the truss will be in equilibrium for all values of θ.

To obtain the required forces we now consider free-body diagrams for joints *A*, *J* and *I*. Note that each member is assumed to be in tension. Thus, positive answers will imply tension and negative answers compression. Also note the order in which the joints are analyzed. In each case there are only two unknown forces.

Joint A:

$$\Sigma F_x = 0 = AB + 6\cos\theta,\ \Sigma F_y = 0 = -6\sin\theta - AJ$$

$$AJ = -6\sin\theta,\ AB = -6\cos\theta$$

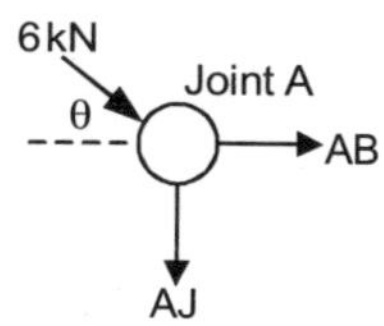

Note that *AB* = *BC* since *BI* is a zero-force member.

Joint J:

$$\Sigma F_x = 0 = IJ + \frac{3}{\sqrt{13}} JC,\ \Sigma F_y = 0 = AJ + J_y + \frac{2}{\sqrt{13}} JC$$

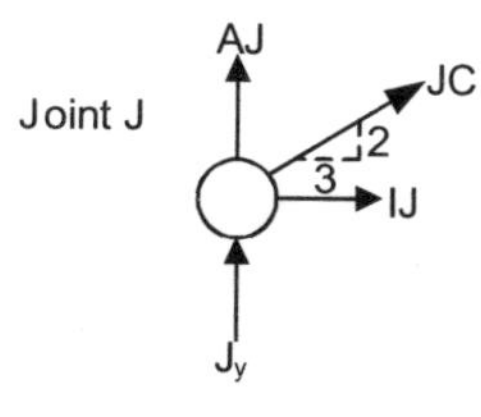

$$JC = -\frac{\sqrt{13}}{2}(AJ + J_y) = -\frac{\sqrt{13}}{2}(10 - 2\cos\theta)$$

$$IJ = -\frac{3}{\sqrt{13}} JC = 15 - 3\cos\theta$$

Joint I:

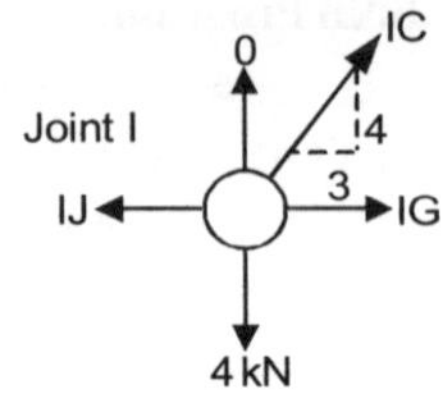

$$\Sigma F_x = 0 = GI - IJ + \frac{3}{5} IC, \; \Sigma F_y = 0 = \frac{4}{5} IC - 4$$

$$IC = 5 \text{ kN}; \; IG = IJ - \frac{3}{5} 5 = 12 - 3 \cos \theta$$

Note in the above that substitutions have been made in order to express each force explicitly in terms of θ. This has been done primarily for completeness. The program below shows that the explicit substitution is not necessary. The computer will substitute automatically provided each force is expressed in terms of functions that have been previously defined.

MATLAB Program:

```
th=0:0.05:2*pi;
Jy=10+6*sin (th)-2*cos (th);
AJ=-6*sin (th);
BC=-6*cos (th);
JC=sqrt (13)/2*(AJ + Jy);
IJ=-3/sqrt (13)*JC;
IC=5;
IG=IJ-3;
plot(th,BC,th,JC,th,IC,th,IG)
legend ('BC','JC','IC','IG')
xlabel ('Theta (rads)')
ylabel ('Force (kN)')
```

Output is shown in Fig. ES3.7(*b*).

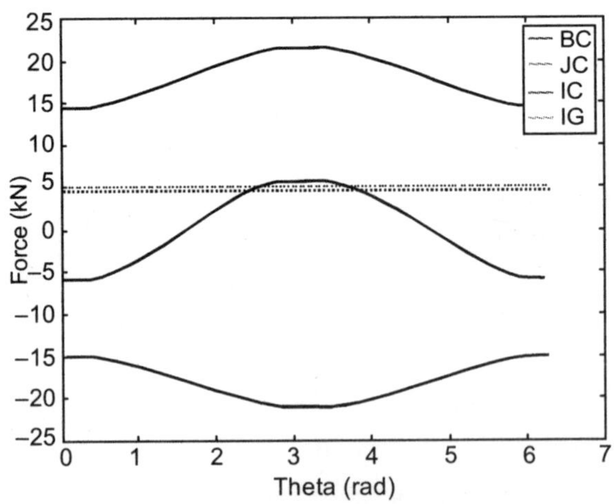

Fig. ES3.7 (b)

ES 3.8: In Fig.ES3.8 rod *CB* is held by a cord *AC* which has a tension *T*.

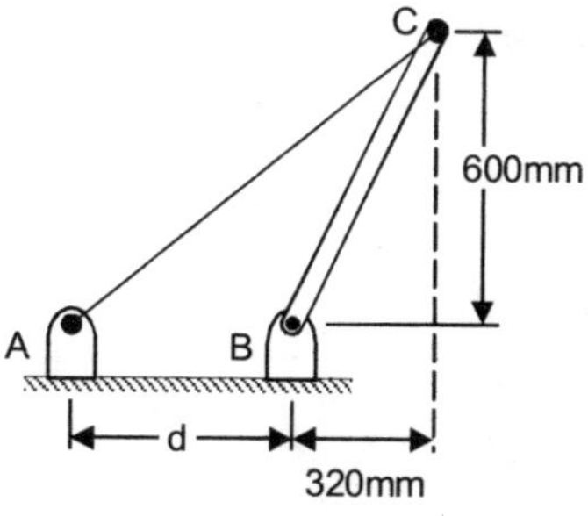

Fig. ES3.8

Write a MATLAB program to determine,

(*a*) the moment about *B* of the force exerted by the cord at point *C* as a function of the tension *T* and the distance *d*.

(*b*) plot the moment about *B* for 300 mm ≤ *d* ≤ 1000 mm when (*i*) *T* = 60 N, (*ii*) *T* = 80 N, (*iii*) *T* = 110 N.

Solution: Free body diagram is shown in Fig. ES3.8 (*a*).

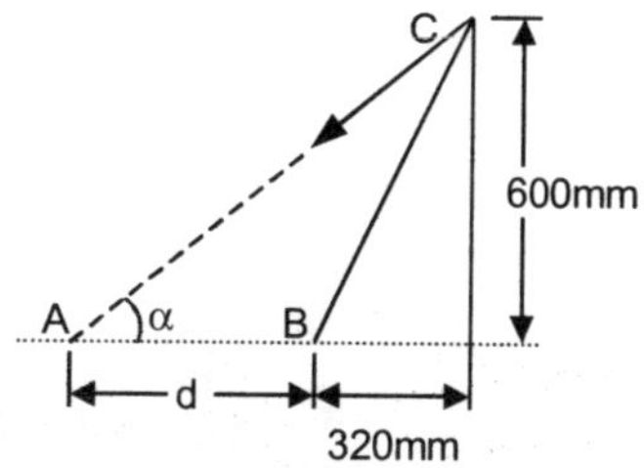

Fig. ES3.8 (a)

The length of *AC* is from the figure,

$$AC = \sqrt{(600)^2 + (320+c)^2}$$

For the angle α we have

$$\cos\alpha = \frac{320+c}{AC}$$

$$\sin\alpha = \frac{600}{AC}$$

The tension **T** is given by

$$\mathbf{T} = -T\cos\alpha\boldsymbol{i} - T\sin\boldsymbol{j}$$

and the position vector from point *B* as

$$\mathbf{r} = 320\boldsymbol{i} + 600\boldsymbol{j}$$

The moment about point *B* is

$$\mathbf{M}_B = \mathbf{r}\times\mathbf{T} = (320\boldsymbol{i} + 600\boldsymbol{j}) \times (-\cos\alpha\boldsymbol{i} - \sin\alpha\boldsymbol{j})\,T$$

$$= T(600\cos\alpha - 320\sin\alpha)\,\mathbf{k}$$

The magnitude of the moment is

$$M_B = \frac{600Tc}{\sqrt{c^2 + 640c + 462400}}$$

$$M_B = \frac{T}{AC}[600(320+c) - 320(600)]$$

$$M_B = \frac{600Tc}{\sqrt{(600)^2 + (320+c)^2}}$$

or $$M_B = \frac{600Tc}{\sqrt{c^2 + 640c + 4624000}}$$

MATLAB Program:

```
c=[320:1:960];
Mb1=600*60*c./(sqrt(c.^2+640*c+462400));
Mb2=600*80*c./(sqrt(c.^2+640*c+462400));
Mb3=600*110*c./(sqrt(c.^2+640*c+462400));
plot(c,Mb1,c,Mb2,c,Mb3)
legend('T=60 N','T=80 N','T=110 N',2)
xlabel('c(mm)')
ylabel('Moment(N·mm)')
grid on
```

Output is shown in Fig. ES3.8(*b*).

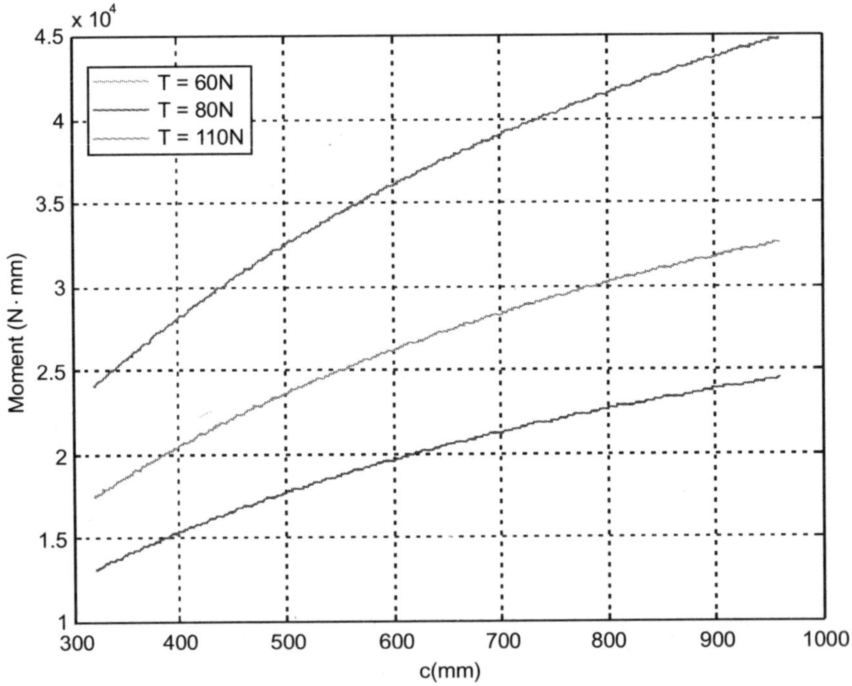

Fig. ES3.8 (b)

ES 3.9: Figure ES3.9 shows a frame in which the structural members support the 5 kN load. The load may be applied at any angle α ($-90°$ to $+90°$). The pins at A and B need to be designed to support the maximum force transmitted to them. Write a MATLAB program to plot the forces at A and B as a function of α and find their maximum values and corresponding angles α.

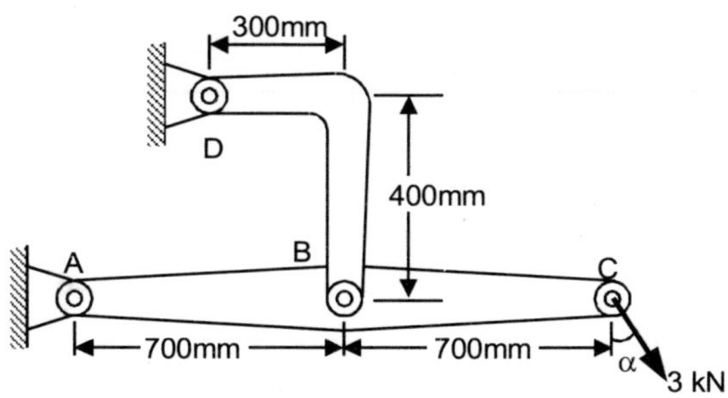

Fig. ES3.9

Solution: Free body diagram of *ABC* is shown in Fig. ES3.9(*a*).

Note that member *BD* is a two-force member, thus the direction of the force **B** is from *B* to *D*. The equilibrium equations are,

$$\Sigma M_A = \frac{4}{5}B\,(0.6) - 5\cos\alpha(1.4) = 0$$

$$\Sigma F_x = A_x - \frac{3}{5}B + 5\sin\alpha = 0$$

$$\Sigma F_y = A_y + \frac{4}{5}B - 5\cos\alpha = 0$$

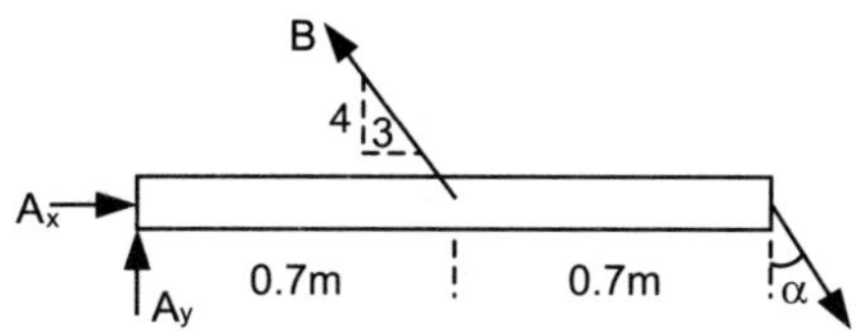

Fig. ES3.9 (a)

Solving these equations yields,

$$B = 12.5\cos\alpha,\quad A_x = 0.6\ B - 5\cos\alpha \quad \text{and} \quad A_y = 5\cos\alpha - 0.8B$$

$$A = \sqrt{A_x^2 + A_y^2}$$

Substitution and simplification yields

$$A = \sqrt{81.25\cos^2\alpha + 25\sin^2\alpha - 75\cos\alpha\sin\alpha}$$

Maximum value of *A* is obtained from MATLAB program while maximum value of *B* is 12.5 N at $\alpha = 0$.

The maximum value of *A* and the corresponding angle α will be found in the MATLAB program.

MATLAB Program:

```
al=-pi/2:0.01:pi/2;
B=12.5*cos(al);
Ax=0.6*B-5*sin(al);
Ay=5*cos(al)-0.8*B;
A=sqrt(Ax.^2+Ay.^2);
[Amax,K]=max(A);
plot(al,B,al,A)
legend('A','B')
xlabel('Alpha (rads)')
ylabel('Force (kN)')
fprintf('Maximum value of A=%f and corresponding angle =%f\n',Amax,al(K)*180/pi);
```

Output is shown in Fig. ES3.9 (*b*).

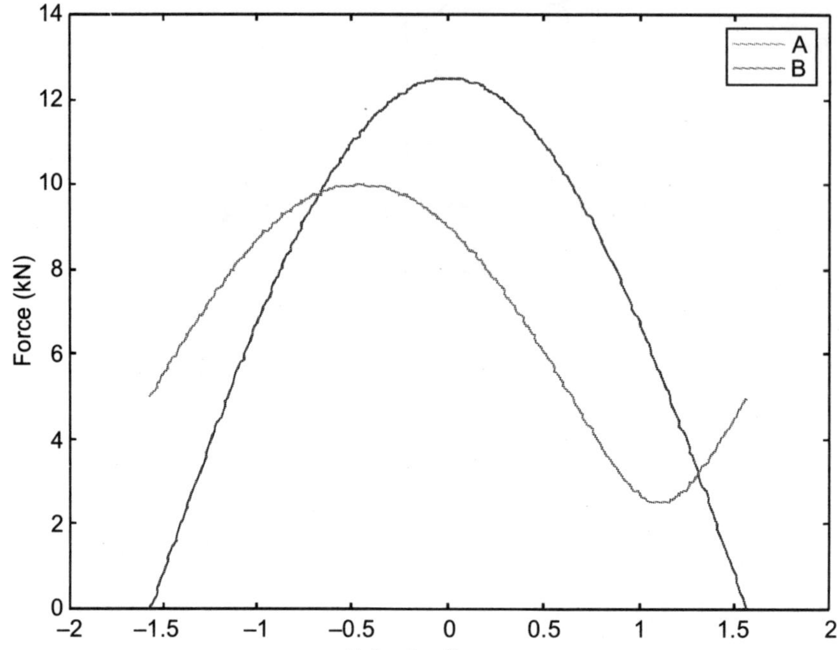

Fig. ES3.9 (b)

ES 3.10: Figure ES3.10 shows a uniform quarter-circular member of mass *m* lying in the vertical plane and hinged at *A* and supported against the vertical wall by a small roller at *B*.

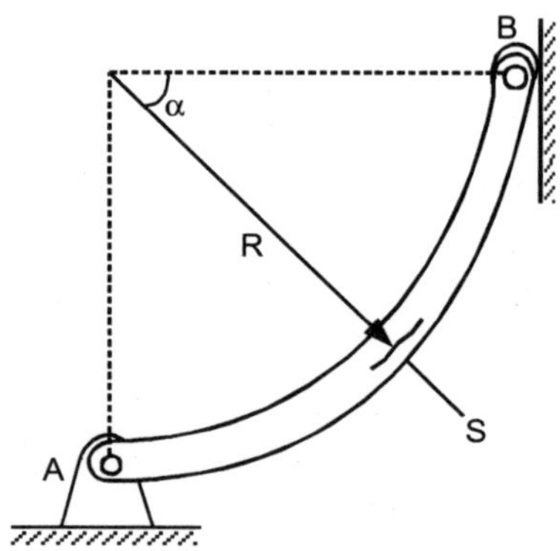

Fig. ES3.10

(*a*) Derive expressions for the shear force *V*, the compressive force *C*, and the bending moment *M* due to the weight of the member.

(*b*) Write a MATLAB program to plot the non-dimensional forces and moment (V/mg, C/mg and M/mgR) as a function of the section orientation angle α.

(*c*) Determine the maximum bending moment and its location angle α.

Solution: First consider the free-body diagram for the entire member. The centroid for a quarter-circular arc is $2r/\pi$. (see the Fig. ES3.10(a)). All we need from this free-body diagram is the force B, which we can find by summing moments about A.

$$\circlearrowleft \Sigma M_A = 0 = Br - mg\left(\frac{2r}{\pi}\right);\ B = \frac{2mg}{k}$$

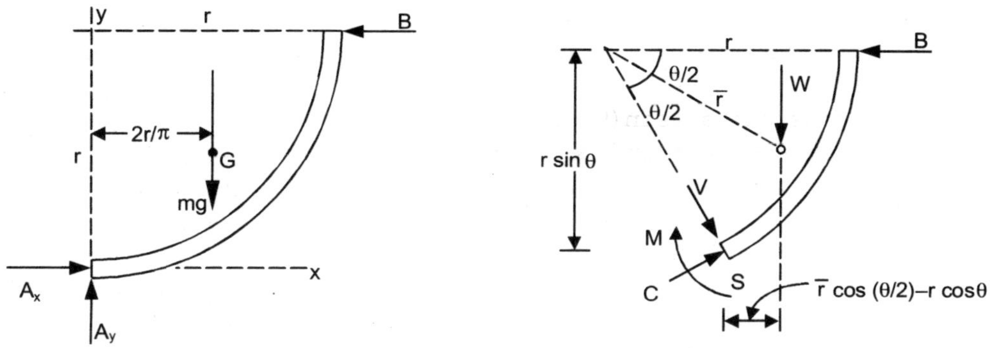

Fig. ES3.10(a)

Now we construct another free-body diagram by cutting through at an arbitrary angle θ as shown to the figure above. From Table $D/3$, the centroid $\bar{r}$ is,

$$\bar{r} = \frac{r\sin(\theta/2)}{\theta/2} = \frac{2r\sin(\theta/2)}{\theta}$$

The weight W of the section in the diagram will be mg (the total weight of the member) times the ratio of the length of the section $(r\theta)$ to the total length of the member $(r\pi/2)$. Thus,

$$W = \frac{r\theta}{r\pi/2}mg = \frac{2\theta}{\pi}mg$$

Now we sum moment about S. For sake of clarity, the moment arms for W and B are shown in the diagram.

$$\Sigma M_S = 0 = M + W(\bar{r}\cos(\theta/2) - r\cos\theta) - Br\sin\theta$$

Substitution and simplification yields,

$$M = \frac{2mgr}{\pi}(\theta\cos\theta + \sin\theta - 2\sin(\theta/2)) = \frac{2mgr}{\pi}\theta\cos\theta$$

Summing forces in the x and y directions gives the following two equations.

$$\Sigma F_x = 0 = C\sin\theta + V\cos\theta - B,\ \Sigma F_y = 0 = C\cos\theta - V\sin\theta - W$$

Solving these two equations followed by substitution and simplification yields.

$$C = \frac{2mg}{\pi}(\theta\cos\theta + \sin\theta);\quad V = \frac{2mg}{\pi}(\cos\theta - \theta\sin\theta)$$

In non-dimensional form we have,

$$\frac{M}{mgr} = \frac{2}{\pi}\theta\cos\theta;\ \frac{C}{mg} = \frac{2}{\pi}(\theta\cos\theta + \sin\theta);\ \frac{V}{mg} = \frac{2}{\pi}(\cos\theta - \theta\sin\theta)$$

The maximum moment and where it occurs will be found in the worksheet below. The results are,

$$M_{max} = 0.357\ mgr \text{ at } \theta = 0.860 \text{ radian } (49.3°).$$

Complete MATLAB program is given below:

MATLAB Program:

```
%%%%%%%%%%%%%%%%%%%Script #1%%%%%%%%%%%%%%%%%%%%%
%This script plots the non-dimensional forces C and V as % functions of theta.
theta=0:0.01:pi/2;
C=2/pi.*(theta.*cos(theta)+sin(theta));
V=2/pi.*(cos(theta)-theta.*sin(theta));
plot(theta*180/pi,C,theta*180/pi,V)
legend('C','V')
xlabel('Theta(deg)')
title('Non-dimensional Forces')
%%%%%%%%%%%%%%%%%%%%%%%%%%%%%%%%%%%%%%%%%%%
```

Output is shown in Fig.ES3.10(*b*).

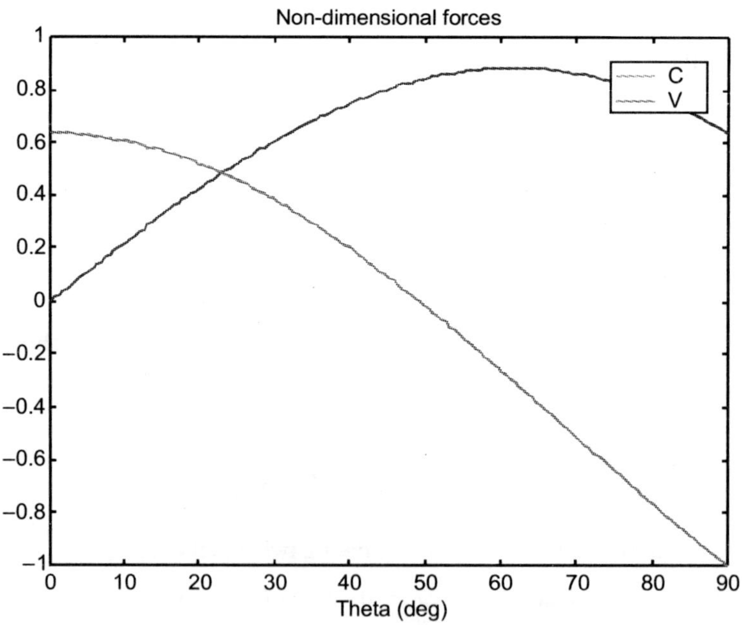

Fig. ES3.10 (b)

```
% This script plots the non-dimensional bending moment
% as a function of theta
theta = 0:0.01:pi/2;
M = 2/pi*theta.*cos (theta);
plot (theta*180/pi, M)
xlabel ('Theta (deg)')
Title('Non-dimensional bending moment')
```

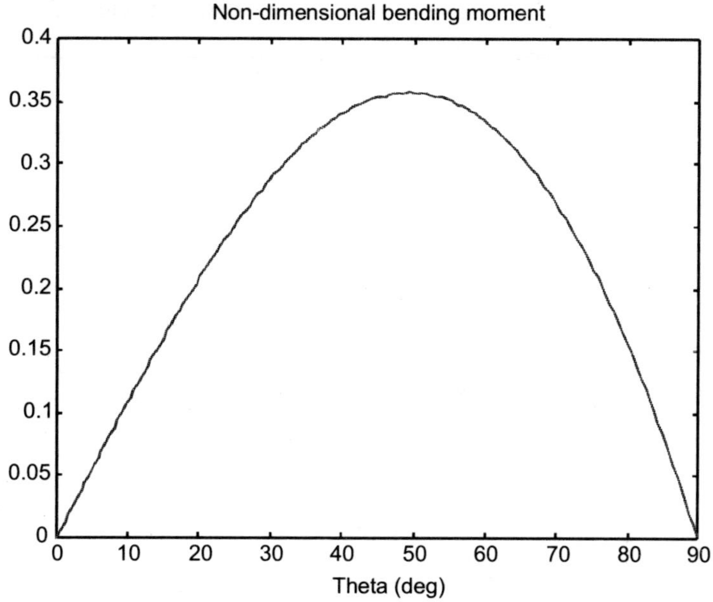

Fig. ES3.10(c)

```
% This program finds the maximum moment
% using MATLAB's max function
theta=0:0.01:pi/2;
M=2/pi*theta.*cos(theta);
[M_max, i]=max (M)
theta_max=theta (i)
```

```
Output is shown below:
M_max =
       0.3572
i =
     87
theta_max =
           0.8600
```

```
%This script plots y versus x using the values of T0
%found in the previous script

y = inline('T0./0.1177.*(cosh (0.1177.*x./T0)-1)');
x = -150:1:150;
y10 = 132.61./0.1177.*(cosh(0.1177.*x./132.61) -1);
y30 = 44.71./0.1177.*(cosh(0.1177.*x./44.71) -1);
y60 = 23.16./0.1177.*(cosh(0.1177.*x./23.16) -1);
plot (x, y10, x, y30, x, y60)
xlabel ('x (m)')
ylabel ('y (m)')
```

The output is shown in Fig. ES3.10 (*d*).

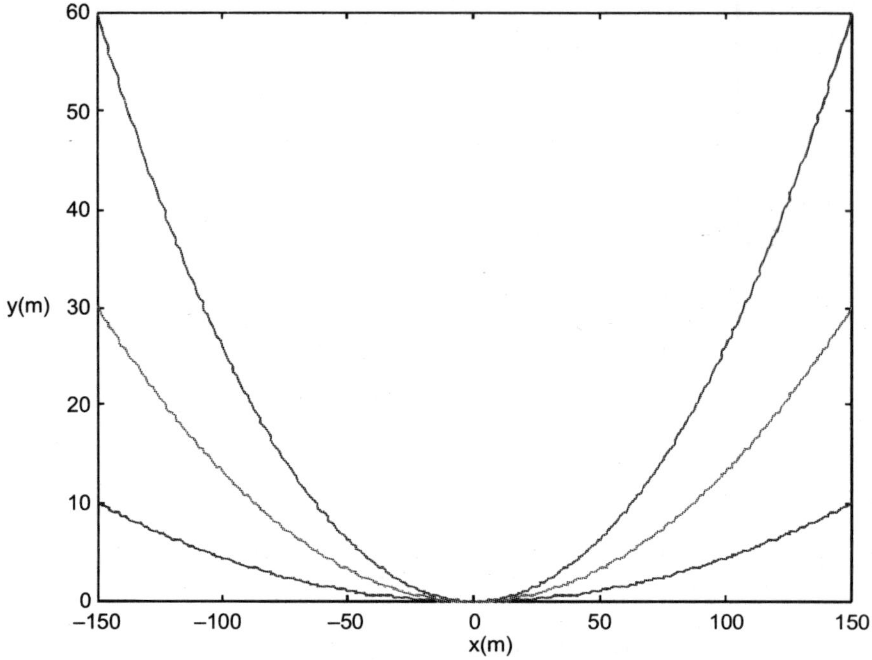

Fig. ES3.10(d)

ES 3.11: Figure ES3.11 shows a slender rod *AB* of weight *W* attached to blocks at *A* and *B* which move freely in the guides.

(*a*) Derive a relationship between α, *W*, *L* and *k* that need to be satisfied for the rod to be in equilibrium. The spring constant is *k*, *W* = 20 N and *L* = 50 mm.

(*b*) Write a MATLAB program to compute and plot *k* as a function of α for 15° ≤ α ≤ 40°.

(*c*) Determine the two values of α corresponding to equilibrium when *k* = 0.8 N/mm.

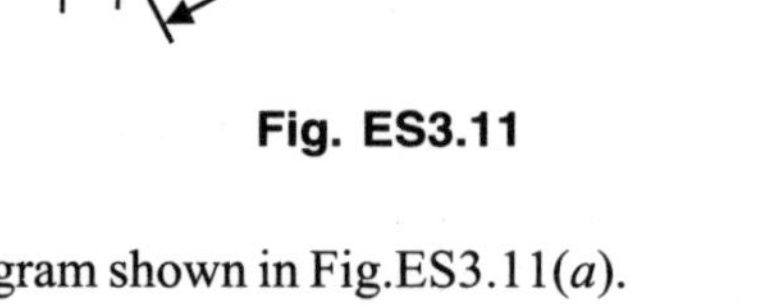

Fig. ES3.11

Solution: Refer to the free-body diagram shown in Fig.ES3.11(*a*).

If x is equal to the elongation of the spring, then

$$x = AC + BC - L$$
$$= L(\sin\alpha + \cos\alpha - 1)$$

The tension in the spring is given by

$$T = kx = kL(\sin\alpha + \cos\alpha - 1)$$

The moment about point D gives

$$\Sigma M_D = 0: \quad T(L\sin\alpha) - T(L\cos\alpha) + W\left(\frac{1}{2}L\cos\alpha\right) = 0$$

or,
$$T(\sin\alpha - \cos\alpha) + \frac{W}{2}\cos\alpha = 0$$

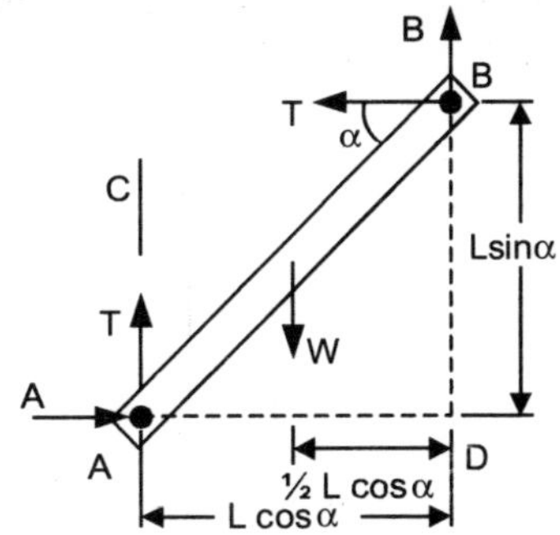

Fig. ES3.11(a)

Substituting for T gives

$$kL(\sin\alpha + \cos\alpha - 1)(\sin\alpha - \cos\alpha) + \frac{W}{2}\cos\alpha = 0$$

$$(\sin\alpha + \cos\alpha - 1)(\tan\alpha - 1) + \frac{W}{2}\cos\alpha = 0$$

and
$$\frac{W}{2kL} = (\sin\alpha + \cos\alpha - 1)(1 - \tan\alpha)$$

For $W = 20$ N and $L = 50$ mm

$$k = \frac{20}{100(\sin\alpha + \cos\alpha - 1)(1 - \tan\alpha)}$$

$$= \frac{1}{5(\sin\alpha + \cos\alpha - 1)(1 - \tan\alpha)}$$

MATLAB Program:

```
syms pt real % DEFINING pt and real as symbolic quantities
W=20;
L=50;
K=0.8;
rhs=W/(2*K*L);
Tha=[15:0.1:40];
t=Tha*pi/180;
k=1./(5*(sin(t)+cos(t)-1).*(1-tan(t)));
plot(Tha,k)
xlabel('Theta(deg)')
ylabel('Spring constant k (N/mm)')
grid on
eq1=(sin(pt)+cos(pt)-1)*(1-tan(pt))-rhs;
[pt]=solve(eq1);
Ang=real(double(pt));
Theta=Ang*180/pi;
fprintf('The calculated angles for Theta are:%8.2fdegrees\n',Theta)
```

Output is shown in Fig. ES 3.11 (*b*).

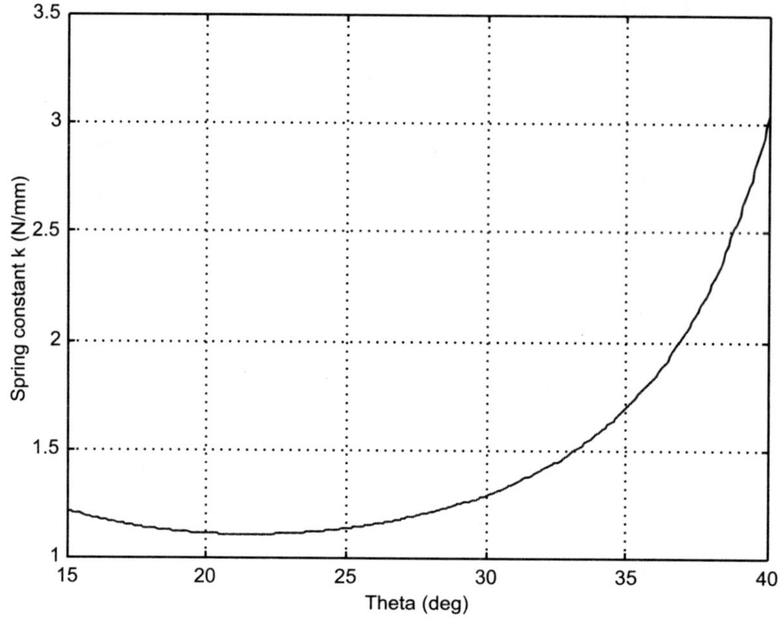

Fig. ES3.11(b)

The calculated angles for Theta are: –132.18 degrees
The calculated angles for Theta are: 21.09 degrees
The calculated angles for Theta are: 21.09 degrees

ES 3.12: Figure ES3.12 shows the loading on a semicircular member. The radius of the semicircular member is 25 mm. Write a MATLAB program to plot the internal forces, namely, the axial forces, shearing force and bending moment as functions of α for $0 < \alpha < 90°$.

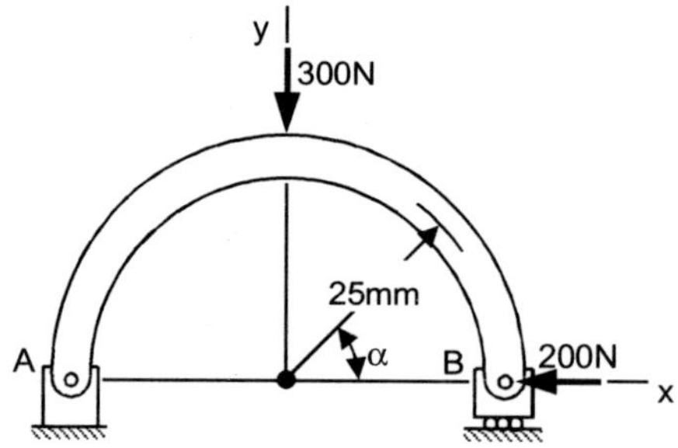

Fig. ES3.12

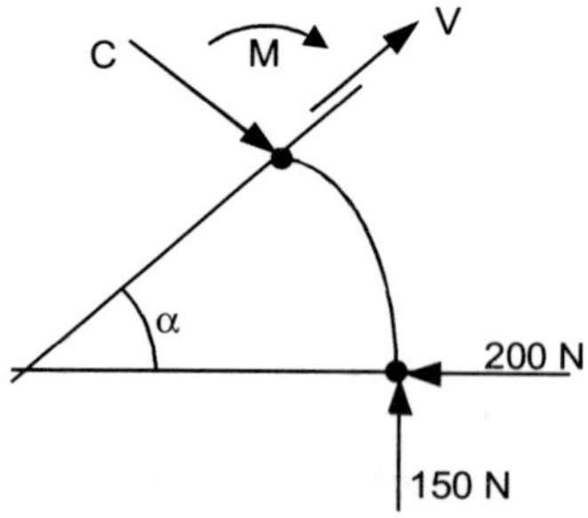

Fig. ES3.12(a)

Solution: Figure 3.12 (*a*) shows free-body diagram of the system.

For $0 \le \alpha \le 90°$:

Axial force $C = 200\sin\alpha + 150\cos\alpha$

Shear force $V = 200\cos\alpha - 150\sin\alpha$

Moment $M = -200r\sin\alpha + 150\,r\,(1 - \cos\alpha)$

Plot these values as a function of θ for $0° \le \theta \le 90°$ with $r = 25$ mm using following MATLAB function.

MATLAB Program:

```
r=25;
alpha = 0:5:90;
alprad = alpha*pi/180;
C=200.*sin(alprad) +150.*cos(alprad);
V=200.*cos(alprad) -150.*sin(alprad);
M=-200*r.*sin(alprad)+150*r*(1-cos(alprad));
subplot(3,1,1);
plot(alpha,C,'-p');
ylabel('Axial force (N)');grid on;
subplot(3,1,2);
plot(alpha,V, '-p');
ylabel('Shear force (N)');
grid on;
subplot(3,1,3);
plot(alpha,M,'-p');
ylabel('Moment(Nm)');
xlabel('Angle in degrees)');
grid on;
```

Output is shown in Fig. ES3.12 (*b*).

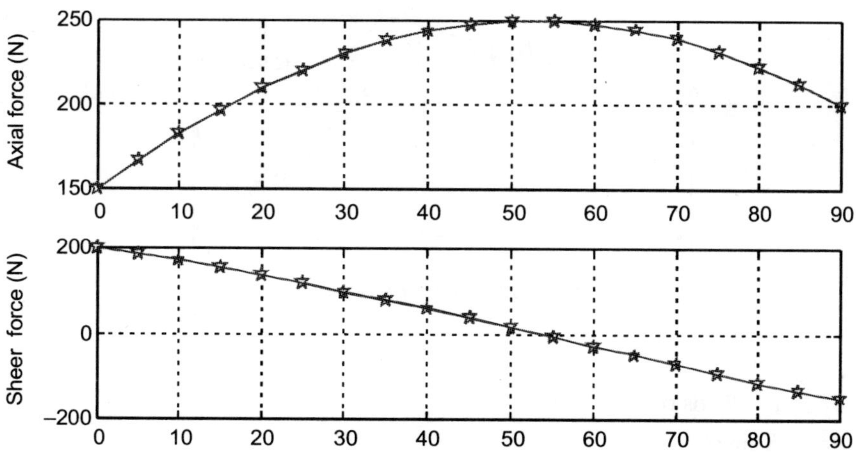

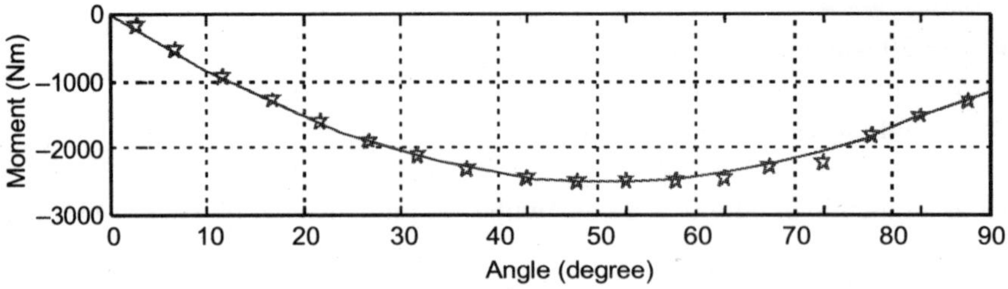

Fig. ES3.12(b)

ES3.13: In figure ES3.13, the spring is unstretched when $\alpha = 0$ and k is the spring constant. Write a MATLAB program to compute and plot the mass m corresponding to equilibrium as a function of α for values of α from 0° to 90°. Find the value of α corresponding to equilibrium m = 2.5 kg. Given R = 210 mm, d = 50 mm and k = 1.2 kN/m.

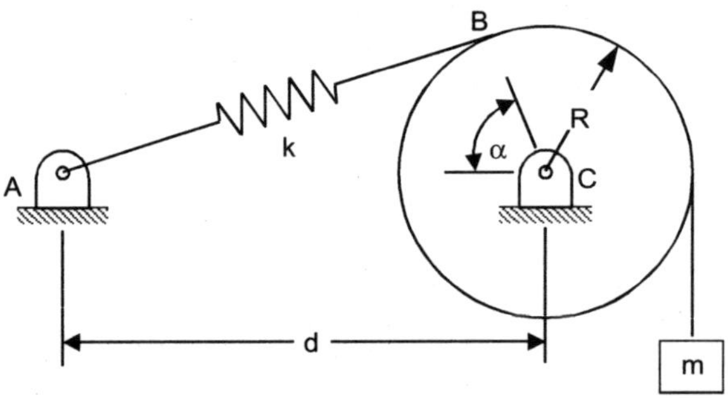

Fig. ES3.13

Solution: The free-body diagram is shown in Fig. ES3.13 (*a*).

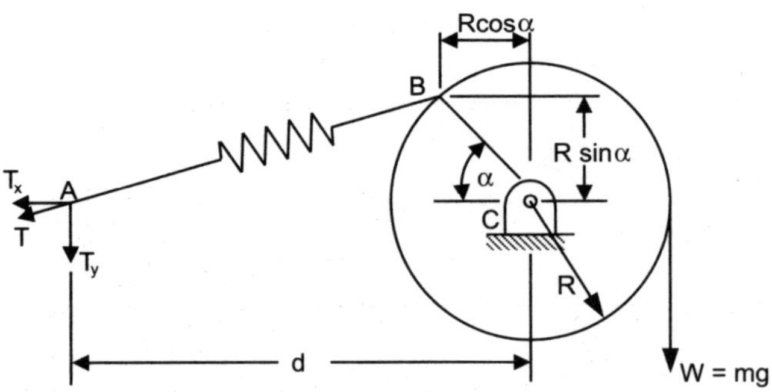

Fig. ES3.13 (a)

The length of spring AB is

$$AB_x = d - R\cos\alpha$$

$$AB_y = R\sin\alpha$$

$$AB = \sqrt{AB_x^2 + AB_y^2}$$

The tension in the spring is

$$T = k[AB - (d - R)]$$

The vertical component T_y

$$T_y = T\frac{AB_y}{AB}$$

Summing moments about point C gives

$$\Sigma M_C = 0: \qquad T_y d - mgR = 0;\ m = T_y\frac{d}{Rg}$$

For this MATLAB program is as follows:

```
% Input Data
syms d R al g k real
ABx=d-R*cos(al);
ABy=R*sin(al);
AB=sqrt(ABx^2+ABy^2);
T=k*(AB-(d-R));
Ty=T*ABy/AB;
m=Ty*d/(R*g);
u=Ty*d-2*R*g;
u=subs(u,{R,g,d,k},{0.21,9.81,0.5,1200});
m=subs(m,{R,g,d,k},{0.21,9.81,0.5,1200});
th=solve(u,al);
Angle=th*180/pi;
Alpha=double(Angle)
td=[0:5:90];
ang=td*pi/180;
yd=subs(m,al,ang);
%
plot(td,yd)
xlabel('Alpha(deg)')
ylabel('Mass m, (kg)')
grid on
```

Output is as follows along with Fig. ES3.13 (*b*).

```
Alpha =
    1.0e+002 *
  -0.0957+0.1850i
   1.7683
  -0.0957-0.1850i
   0.2271
```

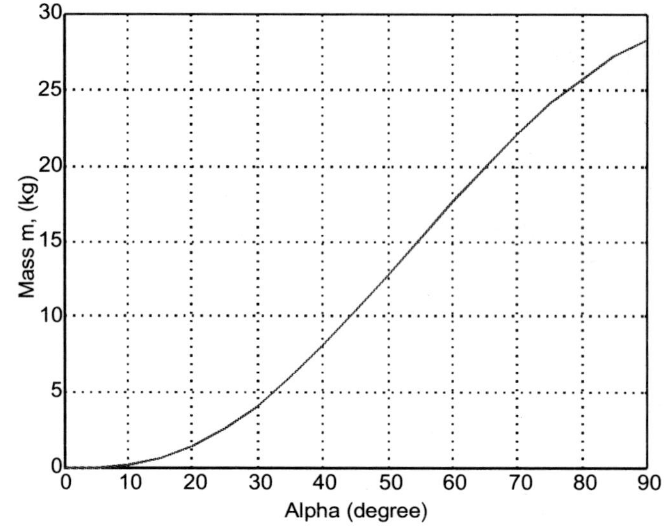

Fig. ES3.13 (b)

ES 3.14: Figure ES3.14 shows a cable loaded uniformly along the horizontal, which has a mass of 15 kg per meter of its own length and supports its own weight only.

(*a*) Determine the tension at mid length, the maximum tension and the total length of the cable for $H = 10, 20, 30, 40, 50, 60, 70$ and 80 meters.

(*b*) Write a MATLAB program to plot y as a function of x for the above values of H.

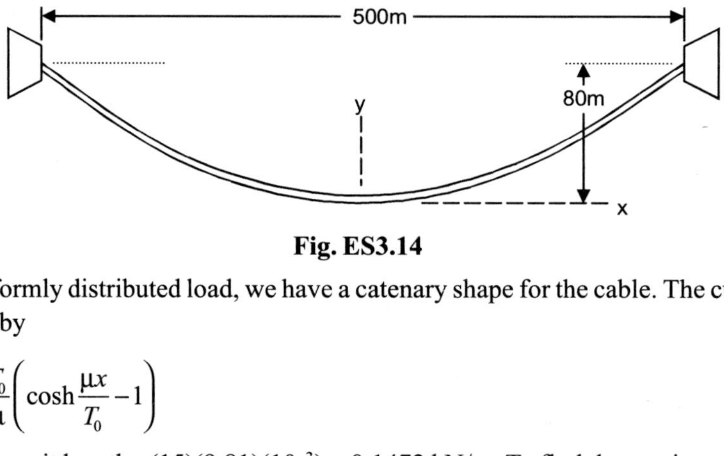

Fig. ES3.14

Solution: For a uniformly distributed load, we have a catenary shape for the cable. The curve assumed by the cable ($y(x)$) is given by

$$y = \frac{T_0}{\mu}\left(\cosh\frac{\mu x}{T_0} - 1\right)$$

where μ = weight per unit length= $(15)(9.81)(10^{-3}) = 0.1472$ kN/m. To find the tension at mid length (T_0) we substitute $x = 250$ m and $y = h$, giving

$$h = \frac{T_0}{\mu}\left(\cosh\frac{250\mu}{T_0} - 1\right)$$

To use the solve function in MATLAB, we first need to re-write the above in terms of a function f whose root (zero) provides the solution to the original equation.

$$f = h - \frac{T_0}{0.1472}\left(\cosh\frac{250\times0.1472}{T_0} - 1\right)$$

The roots of this equation will be obtained for the three specified values of h. With T_0 known we can easily find the maximum cable tension T_{max} (with $y = h$) and the total length of the cable ($L_c = 2s$) (with $x = 250$ m).

$$T_{max} = T_0 + \mu h$$

$$L_c = 2s = \frac{2T_0}{\mu}\sinh\left(\frac{250\mu}{T_0}\right)$$

MATLAB Program:

```
% This script finds the roots (T0) of function f
%(see problem formulation) for h= 10,30,and
% 60 meters. the maximum tension and total cable
%length is then found for the same values of h.

f='h-T0/0.1472*(cosh(0.1472*250/T0)-1)'
f10=subs(f,'h',10);
T0_10=solve(f10,'T0')
f30=subs(f,'h',30);
T0_30=solve(f30,'T0')
f60=subs(f,'h',60);
T0_60=solve(f60, 'T0')
% Now we find the maximum cable tension.
Tm=inline('T0+0.1472*h')
Tmax_10=Tm(T0_10,10)
Tmax_30=Tm(T0_30,30)
Tmax_60=Tm(T0_60,60)
% Next we determine the total length of the cable.
Lc=inline('2*T0/0.1472*(sinh(0.1472*250/T0))')
Lc_10=Lc(T0_10)
Lc_30=Lc(T0_30)
Lc_60=Lc(T0_60)
```

Output is:

```
f =
h-T0/0.1472*(cosh(0.1472*250/T0)-1)
T0_10 =
460.24512430858055194348798961110
T0_30 =
154.06375896035471416316009559463
```

```
TO_60 =
78.0958258286435987215366060353 24
Tm =
   Inline function:
    Tm(TO,h) = TO+0.1472*h
Tmax_10 =
461.7171243085805519434879896111 0
Tmax_30 =
158.4797589603547141631600955946 3
Tmax_60 =
86.9278258286435987215366060353 24
Lc =
    Inline function:
    Lc(TO) = 2*TO/0.1472*(sinh(0.1472*250/TO))
Lc_10 =
500.5329357124506077758692774382 2
Lc_30 =
504.7681758474918209213363228205 3
Lc_60 =
518.7102289478608014316048995918 3
```

The results are summarized as follows:

h(m)	T_0(kN)	T_{max} (kN)	L_c (m)
10	132.6	133.8	300.9
30	44.7	48.2	307.9
60	23.2	30.2	329.9

ES 3.15: For the system shown in Fig.ES3.15, obtain the force in each member of the truss as a function of d. Write a MATLAB program to plot the force in each member for $30 \le d \le 250$ mm.

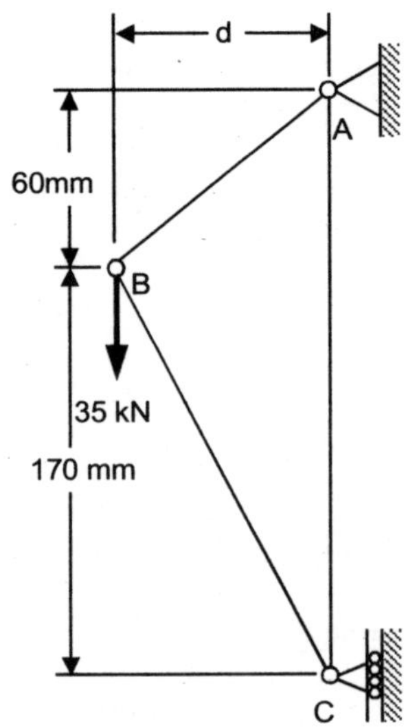

Fig. ES3.15

Solution: Free body diagram is given in Fig. ES 3.15(*a*).

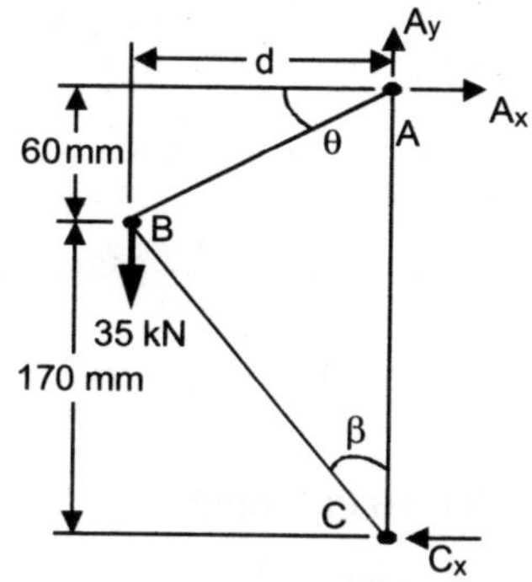

Fig. ES 3.15(a)

The length of *AB* and *BC* is given by

$$AB = \sqrt{d^2 + (60)^2}$$

$$BC = \sqrt{d^2 + (172)^2}$$

and the angles θ and β by

$$\sin\theta = \frac{60}{AB}, \quad \cos\theta = \frac{d}{AB}$$

$$\sin\beta = \frac{d}{BC}, \quad \cos\beta = \frac{170}{BC}$$

Reactions:

$\Sigma F_x = 0$: $\quad A_x - C_x = 0$

$\Sigma F_y = 0$: $\quad A_y - 35000 = 0, \quad A_y = 35000$

$\Sigma M_C = 0$: $\quad d(35000) - 230,\ A_x = 0$

Thus $\quad A_x = \dfrac{35000}{230} d = 152.17d = C_x$

Joint *A* and *C*: (Fig. ES3.15 (*b*))

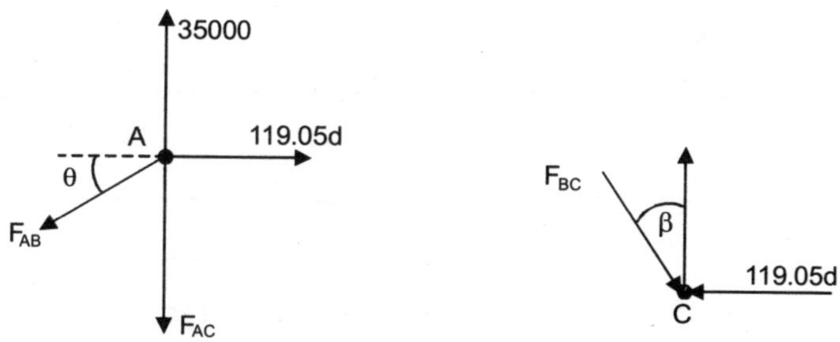

Fig. ES3.15 (b)

For *A*:

$\Sigma F_x = 0$: $\quad 0 = 152.17d - F_{AB}\cos\theta$

or $\quad F_{AB} = \dfrac{AB}{d}(152.17d)$

$$F_{AB} = 152.17AB \text{ N(Tension)}$$

$\Sigma F_y = 0$: $\quad 0 = 35000 - F_{AC} - F_{AB}\sin\theta$

or $\quad F_{AC} = 35000 - 152.17AB\left(\dfrac{60}{AB}\right)$

$F_{AC} = 25868.9$ N(Tension)

For C:

$\Sigma F_y = 0$: $\quad 0 = 25869 - F_{BC}\cos\beta$

or $\quad F_{BC} = \dfrac{BC}{170}25868.9$

$F_{BC} = 152.17BC$ N(Compression)

MATLAB Program:

```
d=30:10:250;
AB=sqrt(d.^2+60^2);
BC=sqrt(d.^2+170^2);
Fab=152.17*AB;
Fac=25868.9+0*d;
Fbc=-152.17*BC;
plot(d,Fab,'-p',d,Fac,'-*',d,Fbc,'-o')
xlabel('d(mm)')
ylabel('Member force(N)')
legend('FAB','FAC','FBC')
grid on
```

Output is shown in Fig. ES3.15 (*c*).

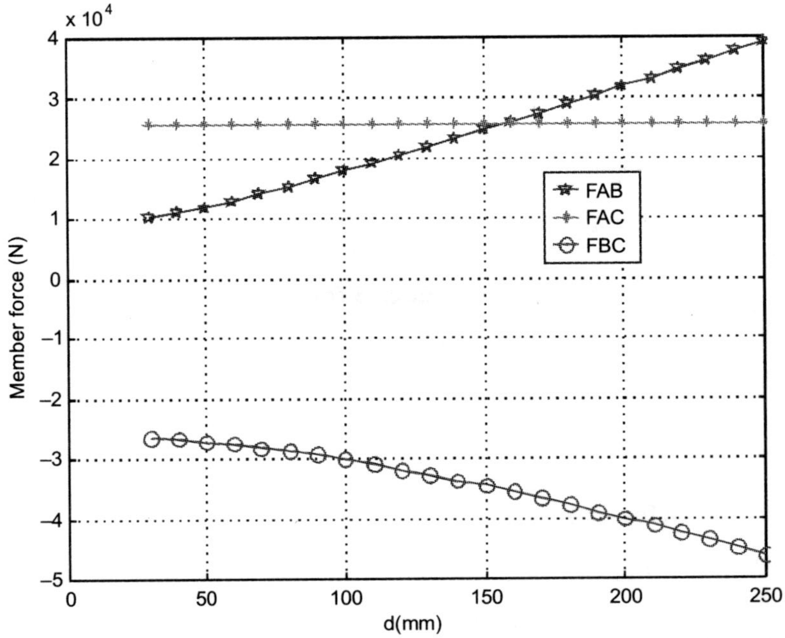

Fig. ES3.15 (c)

ES 3.16: Figure ES3.16 shows a curved beam member AB which is a parabola and the vertex is at A and $\ell = nH$. Write a MATLAB program to determine,

(*a*) the internal forces and bending moment at an arbitrary point C.

(*b*) plot the internal forces and bending moment as a function of x for $0.1\ell \le x \le 0.9\ell$.

Given: The magnitude of the vertical load $F = 1.5$ kN, $\ell = 500$ mm and $n = 2, 3, 4$.

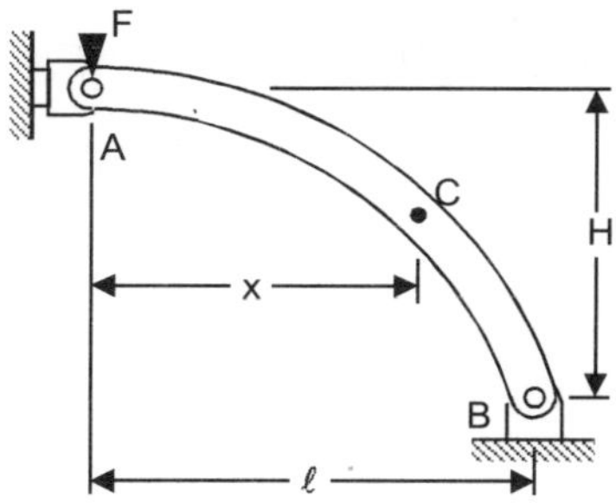

Fig. ES3.16

Solution: Free body diagram is given in Fig. ES 3.16(*a*).

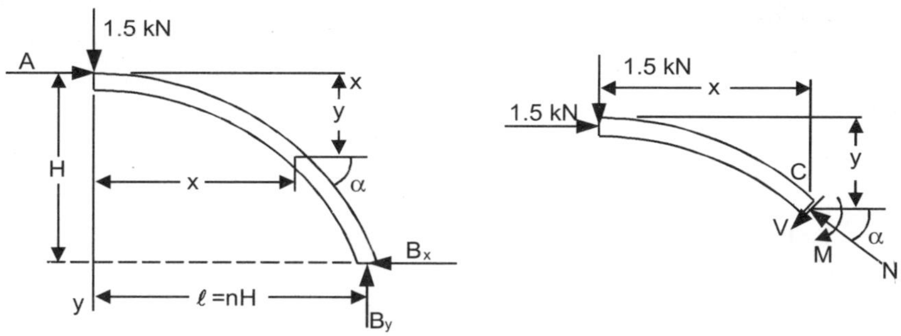

Fig. ES3.16 (a)

Summing forces and moments yields

$\Sigma F_y = 0$: $\quad 0 = B_y - 1.5$

or $\quad B_y = 1.5$ kN

$\Sigma M_B = 0$: $\quad 0 = 1.5(L) - A(H)$

or $\quad A = 1.2\dfrac{\ell}{H} = 1.5n$ kN

For a parabola $y = kx^2$, then we have

$$h = k\ell^2 = k(nH)^2$$

and $\quad k = \dfrac{1}{Hn^2}$

Arbitrary section *AJ*:
At an arbitrary point *C*, we have

$$\text{slope} = \frac{dy}{dx} = \frac{2x}{Hn^2}$$

and $$\tan\alpha = \frac{2x}{Hn^2} \qquad \text{or,} \quad \alpha = \tan^{-1}\left(\frac{2x}{Hn^2}\right)$$

At an arbitrary section *AC*:

$\Sigma M_J = 0$: $\quad 0 = 1.5(x) - 1.5n(y) - M$

or $$M = 1.5(x - ny)$$

$$= 1.5\left(x - \frac{nx^2}{Hn^2}\right)$$

$$M = 1.5x\left(1 - \frac{x}{Hn}\right)$$

Summing forces gives:

$\Sigma F_N = 0$: $\quad 0 = N - 1.5(\sin\alpha) - 1.5n(\cos\alpha)$

or $\quad N = 1.5(\sin\alpha - n\cos\alpha)$

$\Sigma F_V = 0$: $\quad 0 = -V - 1.5(\cos\alpha) + 1.5n(\sin\alpha)$

or $\quad V = 1.5(n\sin\alpha - \cos\alpha)$

With $\ell = 0.5\text{ m} = Hn$

$$M = 1.5x\left(1 - \frac{x}{0.5}\right)$$

where $\alpha = \tan^{-1}\left(\dfrac{2x}{\ell n}\right) = \tan^{-1}\left(\dfrac{x}{0.25n}\right)$.

MATLAB program for this problem is given below:

```
%%%%%%%%%%%%%%%%%%%%%%%%%%%%%%%%
syms n t real
L=0.5;
x1=0.1*L;
xf=0.9*L;
N= inline('1.5*(sin(atan(5*x/n))+n*cos(atan(5*x/n)))','x','n');
V= inline('1.5*(n*sin(atan(5*x/n))-cos(atan(5*x/n)))', 'x','n');
x=[x1:0.01:xf];
M=1.5*x.*(1-x./0.5);
figure(1)
plot(x,N(x,2),x,N(x,3),x,N(x,4));
xlabel('x(m)')
ylabel('Normal force(kN)')
legend('n=2','n=3','n=4',4)
```

```
grid on
figure(2)
plot(x,V(x,2),x,V(x,3),x,V(x,4));
xlabel('x(m)')
ylabel('Shear force(kN)')
legend('n=2','n=3','n=4',2)
grid on
figure(3)
plot(x,M)
xlabel('x(m)')
ylabel('Moment(kN·m)')
grid on
%%%%%%%%%%%%%%%%%%%%%%%%%%%%%%%%
```

Output of the program is shown in Figures ES3.16 (*b*) (*c*) and (*d*).

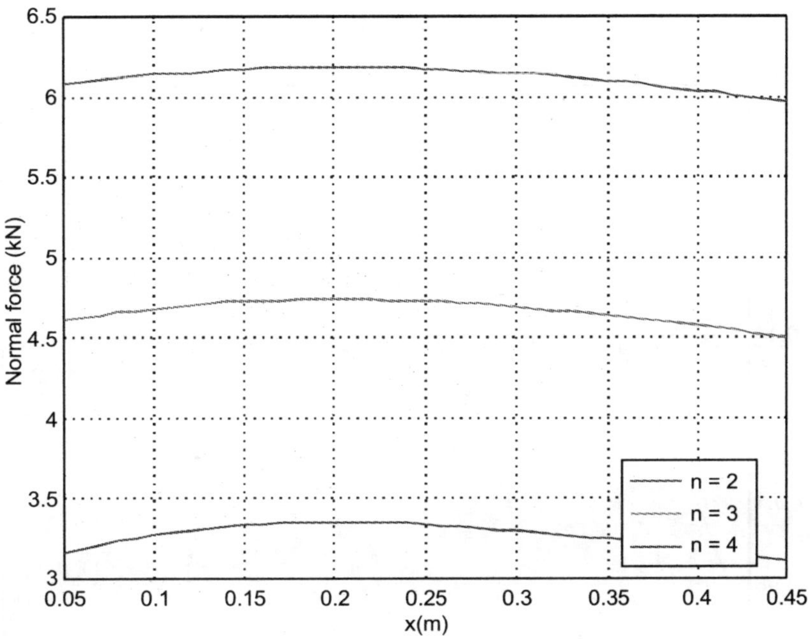

Fig. ES3.16(b)

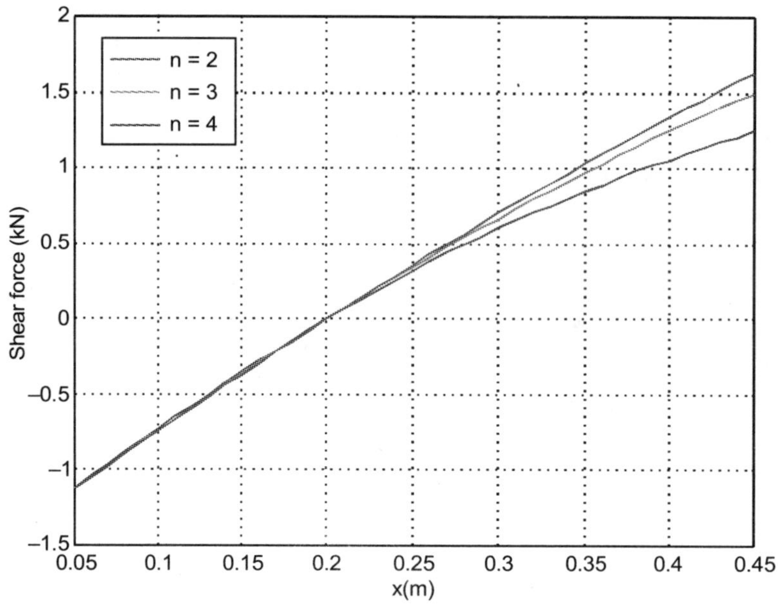

Fig. ES3.16(c)

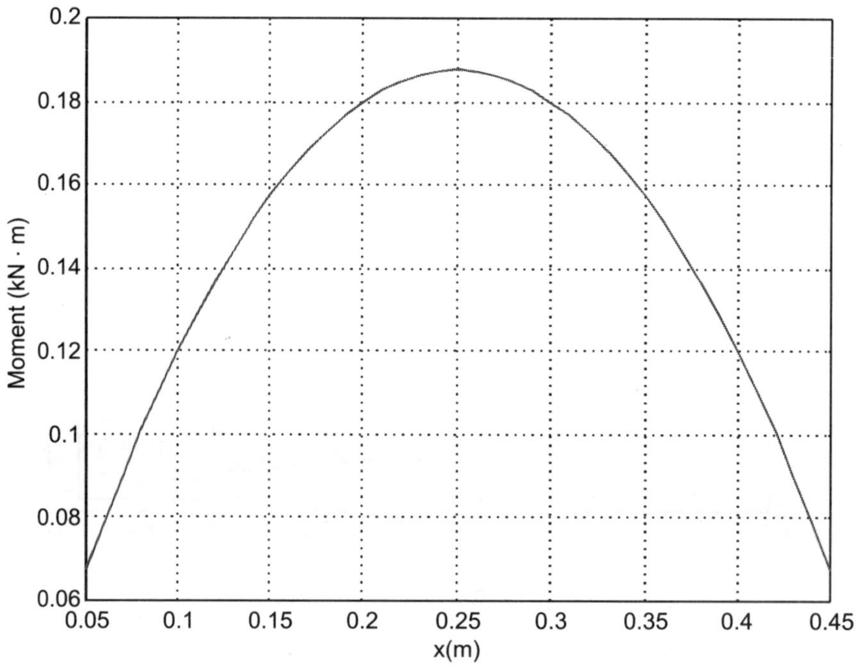

Fig. ES3.16(d)

ES 3.17: Write a MATLAB program to plot the shear and bending moment diagrams for the beam shown in Fig.ES3.17. The length of the beam $\ell = 4$ m and $w_0 = 20$ kN/m.

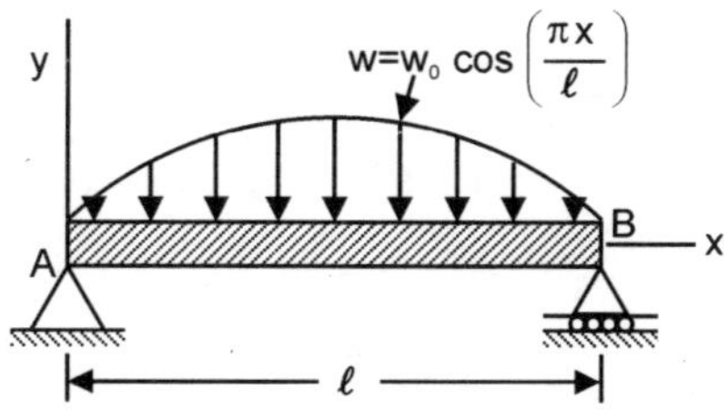

Fig. ES3.17

Solution: It is known that

$$\frac{dV}{dx} = -w \qquad ...(1)$$

$$\frac{dM}{dx} = V \qquad ...(2)$$

Integrating equations (1) and (2), we get

$$V = -\int w dx = \int w_0 \sin\frac{\pi x}{\ell} dx = w_0 \frac{L}{\pi}\cos\frac{\pi x}{\ell} + C_1$$

$$M = \int V dx = \int\left(w_0 \frac{\ell}{\pi}\cos\frac{\pi x}{\ell} + C_1\right)dx = w_0\left(\frac{\ell}{\pi}\right)^2 \sin\frac{\pi x}{\ell} + C_1 x + C_2$$

The boundary conditions are written as

At $x = 0$: $M = 0 = w_0\left(\frac{\ell}{\pi}\right)^2 \sin\frac{\pi(0)}{\ell} + C_1(0) + C_2$, which implies that $C_2 = 0$.

At $x = \ell$: $M = 0 = w_0\left(\frac{\ell}{\pi}\right)^2 \sin\frac{\pi\ell}{\ell} + C_1\ell$, which implies that $C_1 = 0$.

Hence, $V = w_0 \frac{\ell}{\pi}\cos\frac{\pi x}{\ell}$, $M = w_0\left(\frac{\ell}{\pi}\right)^2 \sin\frac{\pi x}{\ell}$

MATLAB Solution:

```
% Input w0 and l%%%%%%%%%%%%%%%%%%%%%%%%%%%
w0=20;
ℓ = 4;
x = [0:0.1:4];
v1 = w0*(ℓ/pi)*cos(pi*x/ℓ);
m1= w0*(ℓ/pi)^2*sin(pi*x/ℓ);
plot(x,v1)
xlabel('x(m)')
ylabel('Shear force (kN)')
grid on
plot(x,m1)
xlabel('x,(m)')
ylabel('Bending moment  kN·m)')
%%%%%%%%%%%%%%%%%%%%%%%%%%%%%%%
```

Output for this program is shown in Figs. ES3.17 (*a*) and (*b*).

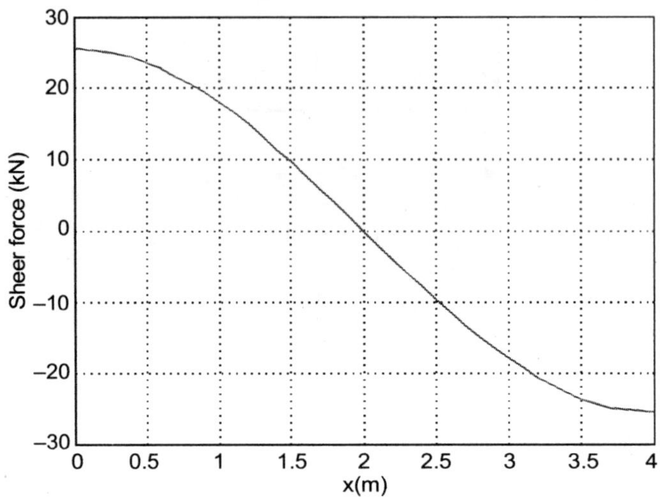

Fig. ES3.17(a)

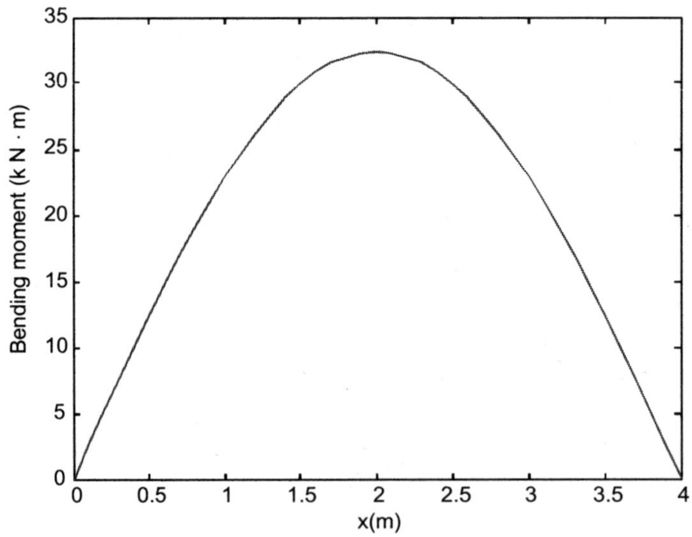

Fig. ES3.17(b)

ES 3.18: Figure ES3.18 shows a crank shaft mechanism where a couple M is applied to the crank AB to maintain the equilibrium of the system. The force applied to the system is F.

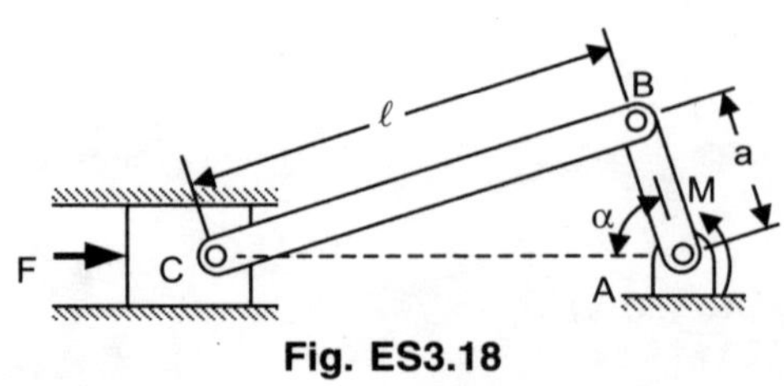

Fig. ES3.18

Write a MATLAB program to plot the ratio of *M*/*F* as a function of crank angle α from 0 to 180 degrees. Given $a = 50$ mm and $\ell = 150$ mm. Determine the value of crank angle α for which the ratio *M*/*F* is maximum and the corresponding value of *M*/*F*.

Solution: Free body diagram of the system is shown in Fig. ES3.18 (*a*).

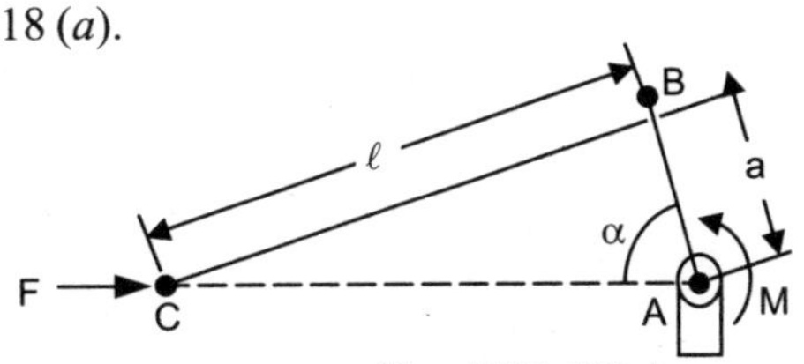

Fig. ES3.18(a)

Applying the law of cosines, we get

$$\ell^2 = (CA)^2 + a^2 - 2(CA)(a)\cos\alpha$$

(or) $$(CA)^2 - 2(CA)(a)\cos\alpha = \ell^2 - a^2$$

Solving the above quadratic equation, we get

$$CA = \frac{2a\cos\alpha + \left[(2a\cos\alpha)^2 + 4(\ell^2 - a^2)\right]^{1/2}}{2}$$

Differentiating gives

$$\delta(CA) = -a\sin\alpha\,\delta\alpha - \frac{4a^2\cos\alpha\sin\alpha\,\delta\alpha}{\left[(2a\cos\alpha)^2 + 4(\ell^2 - a^2)\right]}$$

Applying the principle of virtual work, we get

$$\delta U = -P\delta(CA) - M\,\delta\alpha = 0$$

or $$\frac{M}{P} = -\delta(CA) = a\sin\alpha + \frac{4a^2\cos\alpha\sin\alpha}{\left[(2a\cos\alpha)^2 + 4(\ell^2 - a^2)\right]}$$

MATLAB Solution:

```
% Program inputs: a and l%%%%%%%%%%%%%%%%%
a = 0.050;
ℓ = 0.150;
alpha = Ho:1:180;
alpha=alpha*pi/180;
den=sqrt(4*(ℓ^2-a^2)+4*a^2.*(cos(alpha)).^2);
mbf=a.*sin(alpha)-2*a^2.*cos(alpha).*sin(alpha)./den;
[mbfmax,i]=max(mbf);
vt=alpha(i);
mbfMax=mbfmax
plot(alpha,mbf)
grid on
xlabel('Angle alpha (degree)')
ylabel('M/F ( meter)')
fprintf(('a) Alpha when M/F is a maximum is = %3.0f degrees and M/F =
%6.4f(m)\n',vt,mbfmax)
%%%%%%%%%%%%%%%%%%%%%%%%%%%%%
```

Output comes like this including figure ES 3.18 (*b*).

(*a*) Alpha when *M*/*F* is a maximum is = 93 degrees and *M*/*F* = 0.0501(m)>> *vt*=alpha(*i*);

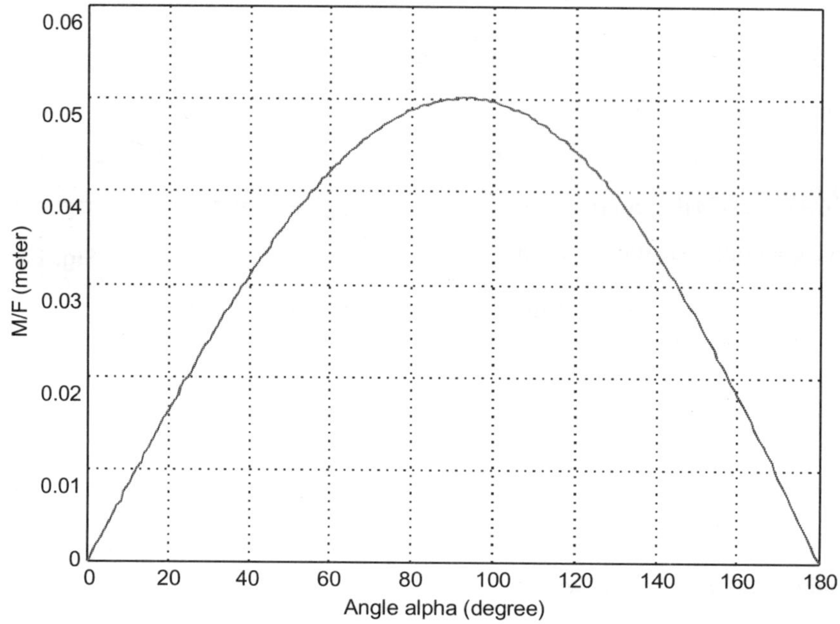

Fig. ES3.18(b)

ES 3.19: The coefficient of friction μ, can be determined by $\mu = F/mg$ where *F* is the measured force (N), *m* is the mass (kg) and *g* = acceleration due to gravity (9.81 m/s^2). The following table gives the experimental data. Determine

(*a*) the coefficient of friction in each test

(*b*) the average from all tests.

Test #	1	2	3	4	5	6	7
Mass m (kg)	2	4	5	10	20	50	100
Force F (N)	12.4	23.2	30.5	60.8	116.5	293.8	597.3

Solution: System under equilibrium is shown in Fig. ES3.19.

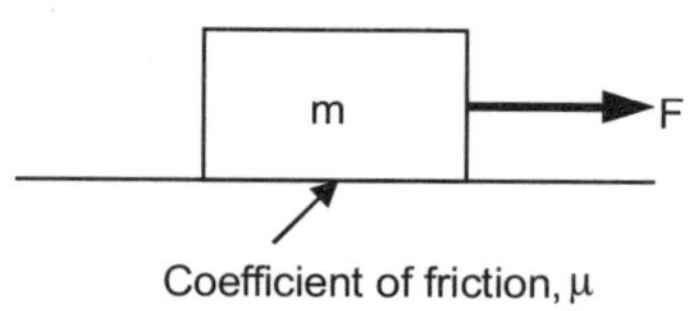

Fig. ES3.19

This is simple application of mathematics.

Program is given below:

```
% Force and mass values in vector form
F=[12.4 23.2 30.5 60.8 116.5 293.8 597.3];
m=[2 4 5 10 20 50 100];
%Coefficient of friction, mu
mu=F./(m*9.81)
```

Output is given below:

mu = 0.6320 0.5912 0.6218 0.6198 0.5938 0.5990 0.6089

Average is obtained as follows:

```
averagemu=mean(mu)
```

This gives output as

```
averagemu = 0.6095
```

ES3.20: Figure ES3.20 shows three flat blocks positioned on an inclined oriented at angle α. Write a MATLAB program to plot the maximum value of P (if no slipping occurs) versus α. Assume only positive values of P and indicate the regions over which

(*i*) the 40 kg block slides alone,

(*ii*) the 40 kg and 30 kg blocks slide together.

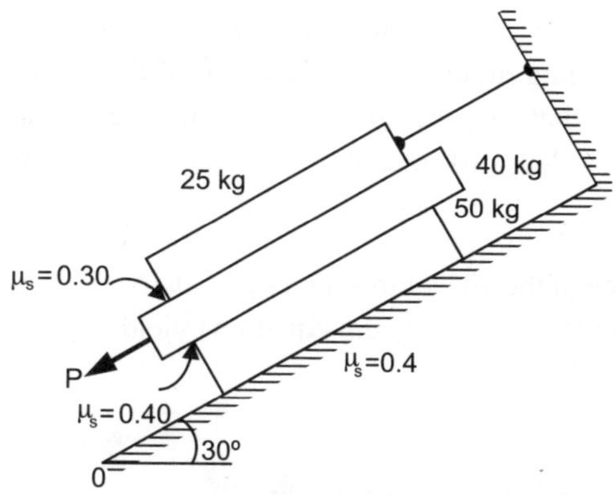

Fig. ES3.20

Solution: The free-body diagrams for the three blocks are shown in Fig. ES3.20 (*a*). To obtain the required plot it will be convenient to use a slightly different approach than that used in the sample problem in your text. We start by writing down the equilibrium equations without making any assumptions about where sliding occurs.

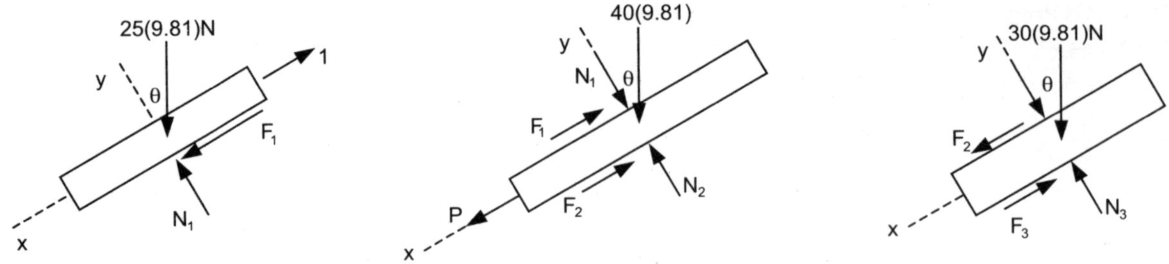

Fig. ES3.20 (a)

$[\Sigma F_y = 0]$: $\quad$ 25 kg ; $N_1 - 25(9.81)\cos\theta = 0$

40 kg ; $N_2 - N_1 - 40(9.81)\cos\theta = 0$

30 kg ; $N_3 - N_2 - 30(9.81)\cos\theta = 0$

These equations can be readily solved for the normal forces.

$$N_1 = 25(9.81)\cos\theta\,;\; N_2 = 65(9.81)\cos\theta\,;\; N_3 = 95(9.81)\cos\theta$$

$[\Sigma F_x = 0]$: $\quad$ 40 kg ; $P - F_1 - F_2 + 40(9.81)\sin\theta = 0$

30 kg ; $F_2 - F_3 + 30(9.81)\sin\theta = 0$

Now we have two equations with four unknowns, P, F_1, F_2 and F_3. Note that we have not written the equation for the summation of forces in the x-direction for the 25 kg block. The reason is that this equation introduces an additional unknown (T) that we are not interested in determining.

The next step is to make assumptions about which block(s) slide. As will be seen, either of the two possible assumptions about impending motion will reduce two of the friction forces to functions of θ only. This will result in two equations that may be solved for P and the remaining friction force. The forces calculated will be designated P_1 or P_2 to distinguish the two cases for impending slip.

Case 1: Only the 40 kg block slips.

Impending slippage at both surface of the 40 kg block gives $F_1 = 0.3N_1 = 73.575\cos\theta$ and $F_2 = 0.4N_2 = 255.06\cos\theta$. Substituting these results into the equilibrium equations yields

$$P_1 = 328.635\cos\theta - 392.4\sin\theta$$

Case 2: The 30 and 40 kg blocks slide together.

Impending slippage at the upper surface of the 30 kg block and lower surface of the 40 kg block gives $F_1 = 0.3N_1 = 73.575\cos\theta$ and $F_3 = 0.4N_3 = 372.78\cos\theta$. Substitution of these results into the equilibrium equations gives,

$$P_2 = 446.355\cos\theta - 686.7\sin\theta$$

which of these two values of P represents the maximum load that can be applied without slippage on any surface is best illustrated by plotting the two expressions as a function of θ. This plot will be generated in the script below. The basic idea is that at any specified angle θ, the critical or maximum value of P will be the smaller of two values calculated.

MATLAB Program:

```
%%%%%%%%%%%%%%%%%%%%%%%%%%%%%%%%%%%%%%
theta=0:0.01:pi/4;
P1=328.635*cos(theta)-392.4*sin(theta);
P2=446.335*cos(theta)-686.7*sin(theta);
plot(theta*180/pi, P1, theta*180/pi , P2)
Legend(‘P1’, ‘P2’)
xlabel(‘Theta  (degrees)’)
ylabel(‘P  (N)’)
%%%%%%%%%%%%%%%%%%%%%%%%%%%%%%%%%%%%%%
```

Figure 3.20(*b*) shows P_1 and P_2 plotted as a function of θ. For each θ, the critical or maximum value of *P* will be the smaller of two values calculated. By setting $P_1 = P_2$ we find that the two curves intersect at $\theta = 0.503$ radians (28.8°). Thus, for $\theta \leq 28.8°$ *P* controls and the 50 kg block slides by itself while for $\theta \leq 28.8°$, P_2 controls and the 40 and 50 kg blocks slide together.

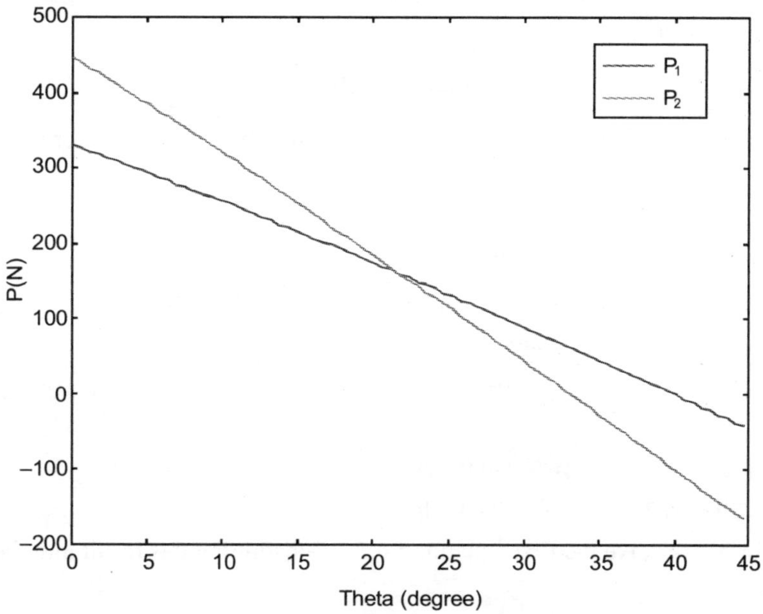

Fig. ES 3.20 (b)

ES 3.21: In Fig.ES3.21, the horizontal position of the rectangular block is adjusted by the wedge under the action of the force *P*. The wedge angle is α. μ_1 and μ_2 are the coefficient of the static friction at the two wedge surfaces and between the block and the horizontal surfaces respectively.

(*a*) Obtain a general expression for *P* (the least force required to move the block) in terms of α, μ_1 and μ_2.

(*b*) Write a MATLAB program to plot *P* as a function of μ_1 for $\alpha = 15°, 20°$, and $25°$. $\mu_2 = 0.5$.

(*c*) For $\alpha = 10°$, plot *P* as a function of μ_1 for $\mu_2 = 0.2, 0.4, 0.6$ and 0.8.

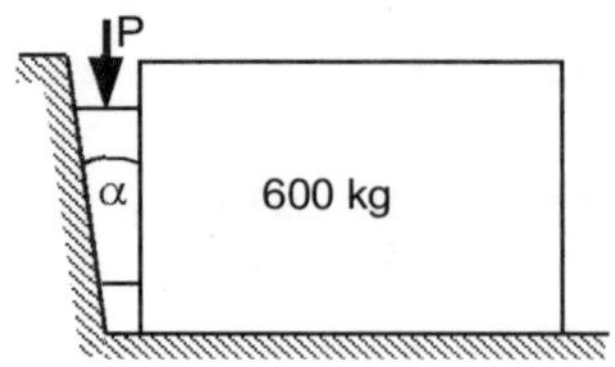

Fig. ES3.21

Solution: Free body diagram of wedge and block is shown in Fig. ES3.21 (*a*).

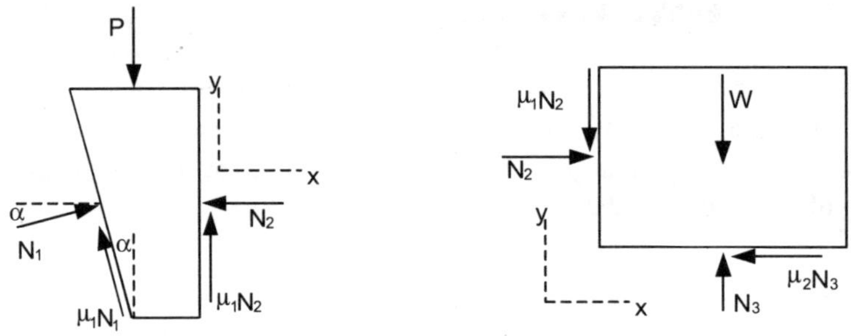

Fig. ES3.21(a)

(*a*) First we write the equilibrium equations from the free-body diagrams for the wedge and for the block.

For the Block:

$$\Sigma F_x = 0 = N_2 - \mu_2 N_3,\ \Sigma F_y = 0 = N_3 - mg - \mu_1 N_2$$

These two equations are readily solved for N_2 and N_3.

$$N_2 = \frac{\mu_2 mg}{1-\mu_1\mu_2}, \quad N_3 = \frac{mg}{1-\mu_1\mu_2}$$

For the Wedge:

$$\Sigma F_x = 0 = N_1 \cos\alpha - \mu_1 N_1 \sin\alpha - N_2$$

$$\Sigma F_y = 0 = N_1 \sin\alpha + \mu_1 N_1 \cos\alpha + \mu_1 N_2 - P$$

After substituting for N_2 we solve the first equation for N_1. Substituting this result into the second yields an expression for P.

$$P = \frac{\mu_2 mg(2\mu_1 + (1-\mu_1^2)\tan\alpha)}{(1-\mu_1\tan\alpha)(1-\mu_1\mu_2)}$$

This result is used to generate the plots for parts (*b*) and (*c*) in the worksheet below.

MATLAB program:

```
% This script produces the plot for part(b)
% Note the conversion factor on theta in the following
% function. This allows us to use degrees rather than
% radians when we use P. Also note the division by 1000,
```

```
% converting to kN.
P=inline('.5*600*9.81*(tan(th*pi/180)*(1-mu1^2)+2*mu1)/(1-mu1*tan(th*pi/
180))/(1-mu1*.6)/1000');
P=vectorize(P);
mu1=0:0.01:0.8;
plot(mu1,P(mu1,15),mu1,P(mu1,20),mu1,P(mu1,25))
legend('mu_1=15','mu_1=20','mu_1=25')
xlabel('mu_1')
ylabel('P(kN)')
%%%%%%%%%%%%%%%%%%%%%%%%%%%%%%%%%%
```

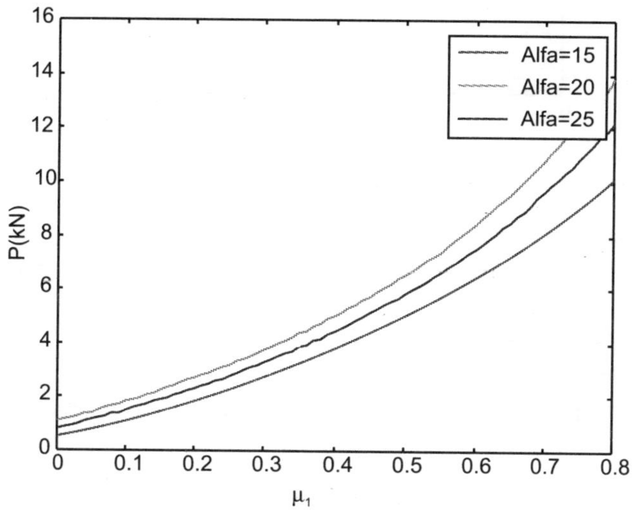

Fig. ES 3.21(b)

```
% This script produces the plot for part(c).
% Note that P is a function of mu1 and mu2 in this case.
% Also note that the division by 1000, converting to kN.

P=inline('mu2*600*9.81*(tan(10*pi/180)*(1-mu1^2)+2*mu1)/(1-mu1*tan(10*pi/
180))/
1-mu1*mu2)/1000');
P=vectorize(P);
mu1=0:0.01:0.8;
plot(mu1,P(mu1,.2),mu1,P(mu1,.4),mu1,P(mu1,.6),mu1,P(mu1,.8))
legend('mu_2=0.2','mu_2=0.4','mu_2=0.6','mu_2=0.8')
xlabel('mu_1')
ylabel('P(kN)')
```

The output is shown in Fig. ES 3.21 (*c*).

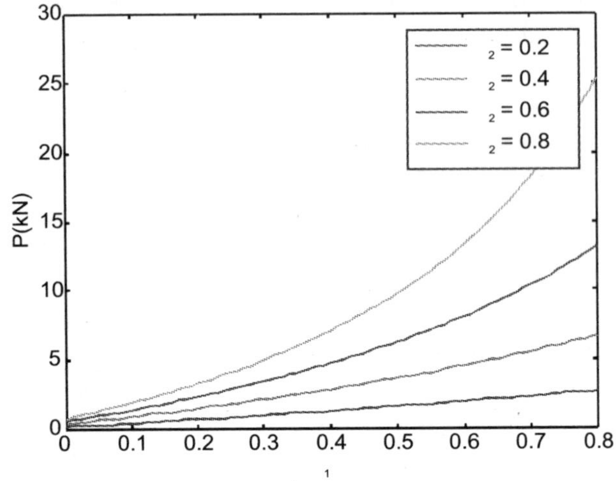

Fig. ES 3.21 (c)

ES 3.22: Figure ES3.22 shows a large turnbuckle which supports a cable tension of 12,000 N. The mean diameter of the two 1.0 mm screws is 1.15 mm and has five square threads per mm. Both screws have single start threads.

(*a*) Determine the moments M_T and M_L that must be applied to the body of the turnbuckle in order to tighten and loosen it respectively.

(*b*) Write a MATLAB program to plot the moments M_T and M_L as functions of μ for $0 \leq \mu \leq 1$, where μ is the coefficient of friction for the threads.

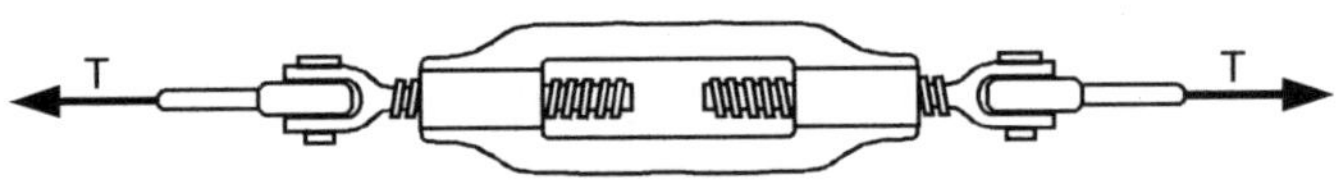

Fig. ES3.22

Solution: M_T and M_L can be obtained as

$$M_T = 2T\,r\tan(\phi + \alpha) \qquad \text{...(1)}$$

and

$$M_L = 2T\,r\tan(\phi - \alpha) \qquad \text{...(2)}$$

where $T = 12{,}000$ N and the lead $L = 1/5$ mm/rev.

The mean radius is given by $r = 1.15/2 = 0.575$ mm

We also have

$$\alpha = \tan^{-1}\left(\frac{L}{2\pi r}\right) \text{ and } \phi = \tan^{-1}(\mu) \qquad \text{...(3)}$$

From Eqs. (1), (2) and (3), we can compute M_L and M_T explicitly.

MATLAB Program:

```
% Input T,L, and r%%%%%%%%%%%%%%%%%%%
T=12000;
L=1/5;
r=1.15/2;
alpha = atan(L/2/pi/r);
mu = 0:0.01:1;
phi = atan(mu);
MT = 2*T*r*tan(alpha + phi);
ML=2*T*r*tan(phi-alpha);
% Place a horizontal line at x = 0
M=0*mu;
plot(mu,MT,mu,ML,mu,M)
xlabel('Coefficient of friction')
ylabel('Moment (lb-in)')
text(0.5,5000,'To loosen')
text(0.3,8000,'To tighten')
%%%%%%%%%%%%%%%%%%%%%%%%%%%%%%%%
```

The plot of moments M_T and M_L as functions of μ for $0 \le \mu \le 1$ is shown in Fig.ES3.22(*a*).

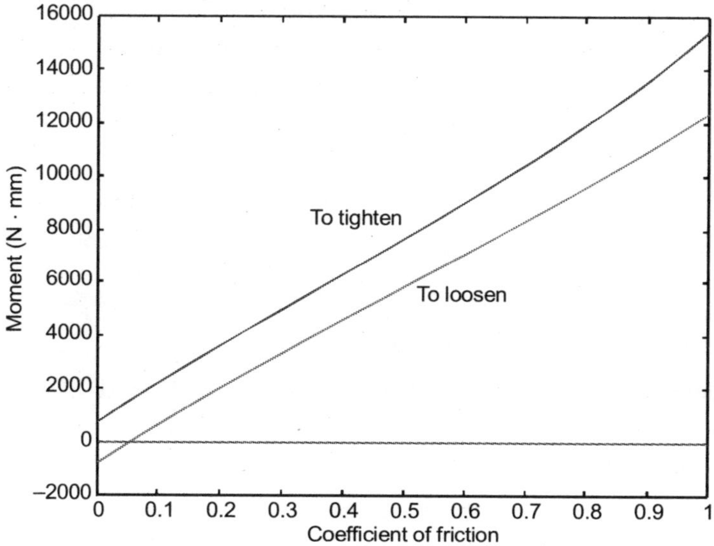

Fig. ES3.22 (a)

ES 3.23: Figure ES3.23 shows a flexible cable which supports the 100 kg load and passes over a circular drum and is subjected to a force P to maintain equilibrium.

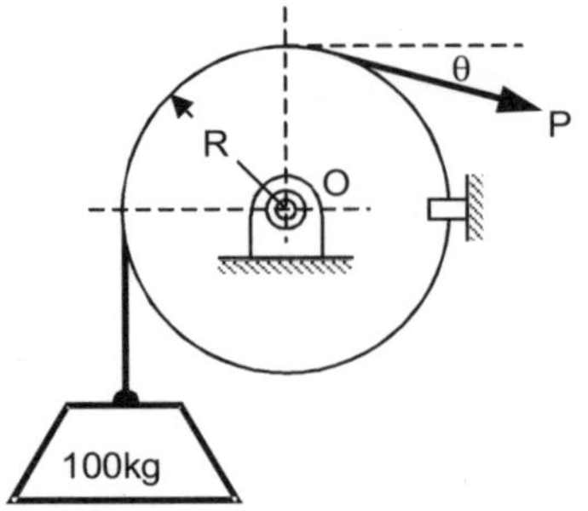

Fig. ES3.23

(*a*) For $\theta = 0$, determine the maximum and minimum values of P may have in order to raise or lower the load.

(*b*) Write a MATLAB program to plot P_{max} and P_{min} versus μ for $0 \le \mu \le 1$, where μ is the coefficient of static friction between the cable and the fixed drum.

(*c*) For $P = 550$ N, determine the minimum value for which the angle θ may have before the load begins to slip.

(*d*) Plot θ_{min} versus μ for $0 \le \mu \le 1$.

Limit the variation of θ between $-60°$ and $360°$.

Solution: Free body diagram of the circular drum is shown in Figures ES3.23 (*a*) and (*b*).

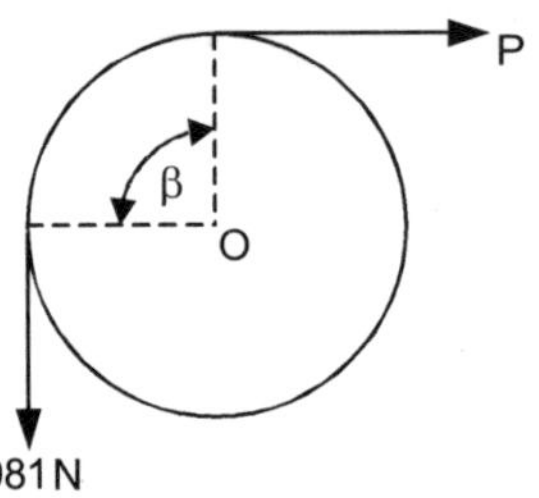

Fig. ES3.23(a) $\theta = 0$, $\beta = \pi/2$

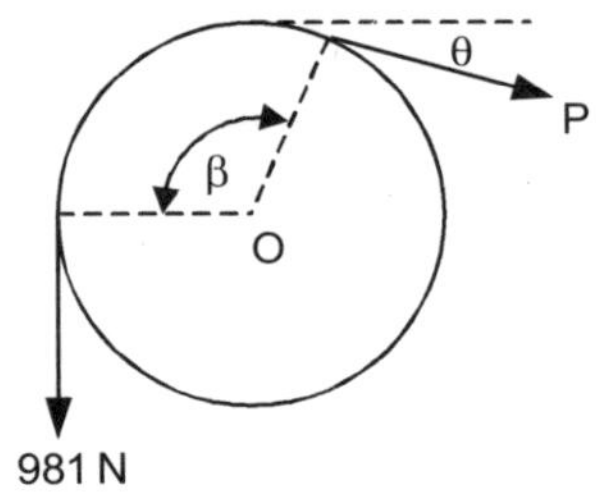

Fig. ES3.23(b) P = 550 N, $\beta = \theta + \pi/2$

Here we have $T_2 = T_1 e^{\mu\beta}$ (belt friction)

Recall that in deriving this formula it was assumed that $T_2 > T_1$.

(*a*) With $\theta = 0$ the contact angle is $\beta = \pi/2$ rad. For impending upward motion of the load we have $T_2 = T_{max}$ and $T_1 = 981$ N. Hence

$$P_{max} = 981 e^{\mu\pi/2}$$

For impending downward motion of the load we have $T_2 = 981$ N and $T_1 = P_{min}$.

$$981 = P_{min} e^{\mu\pi/2} \quad \text{or} \quad P_{min} = 981 e^{-\mu\pi/2}$$

(*b*) With $P = 550$ N we have $\beta = \pi/2 + \theta$, $T_2 = 981$ N and $T_1 = P = 550$ N. Therefore,

$$981/550 = e^{\mu(\theta+\pi/2)}$$

The plots are shown in Fig.ES3.23(*c*) as output of following MATLAB program.

MATLAB Solution:

```
% Plots of Pmax and P min versus coefficient of friction
mu=0:0.005:1;
pmax=981*exp(mu*pi/2)/1000;
pmin=981*exp(-mu*pi/2)/1000;
plot(mu,pmax,mu,pmin)
grid on
xlabel('Coefficient of friction')
ylabel('Force P (kN)')
text(0.7,2.5,'Pmax')
text(0.6,0.5,'Pmin')
```

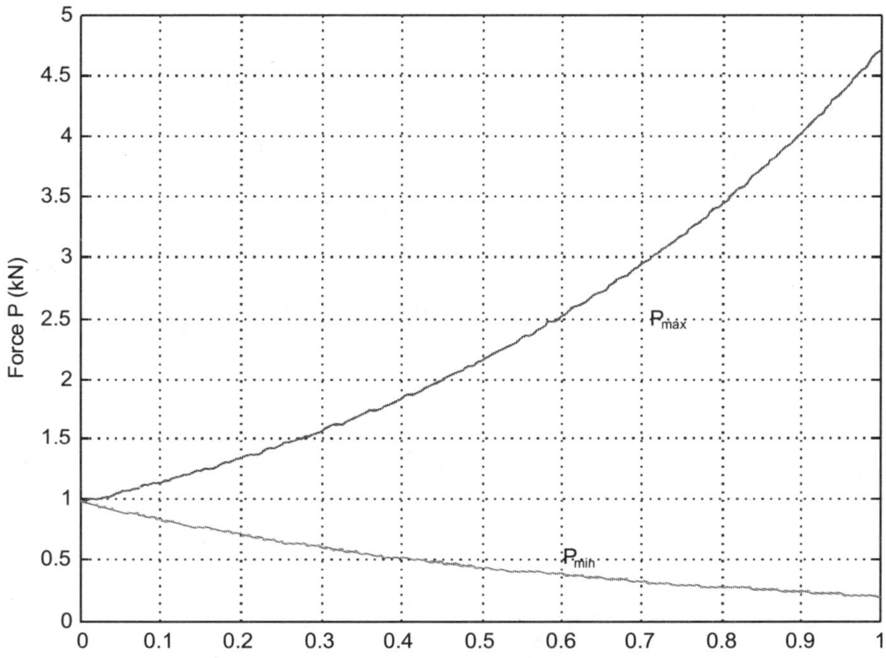

Fig. ES3.23 (c)

(*c*) Taking the natural logarithms on both sides of the above equation and solving for θ gives,

$$\theta = \frac{\ell n(981/550)}{\mu} - \frac{\pi}{2}$$

(*d*) Following program plots minimum value of θ versus μ.

```
% Plot of minimum angle theta versus mu
mu=0:0.005:1;
thet=log(981/550)./mu-pi/2
```

```
plot(mu,thet*180/pi)
axis([1 1 -60 360])
xlabel('Coefficient of friction')
ylabel('Theta (degrees)')
```

Figure ES3.23 (*d*) shows the output.

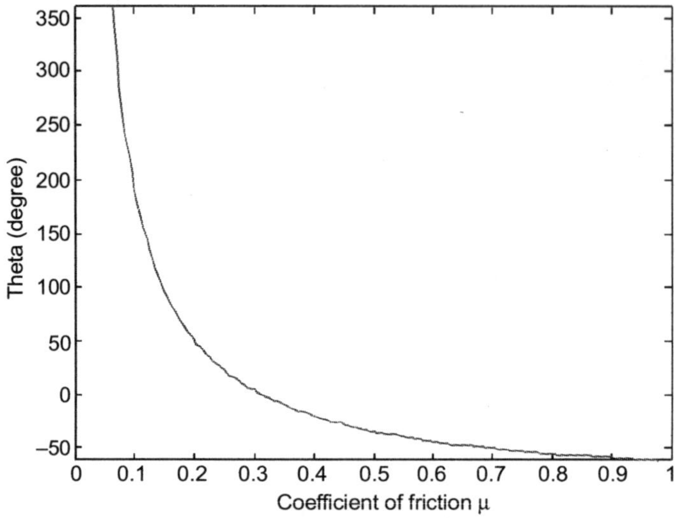

Fig. ES3.23(d)

ES 3.24: Figure ES3.24 shows axle pulley system where the coefficient of friction between cable *ABCD* and the pulley varies between 0 and 0.60. Write a MATLAB program to determine,

(*a*) the values of α for the system to remain in equilibrium

(*b*) the reactions at *A* and *D*

(*c*) Plot α as a function of the coefficient of friction.

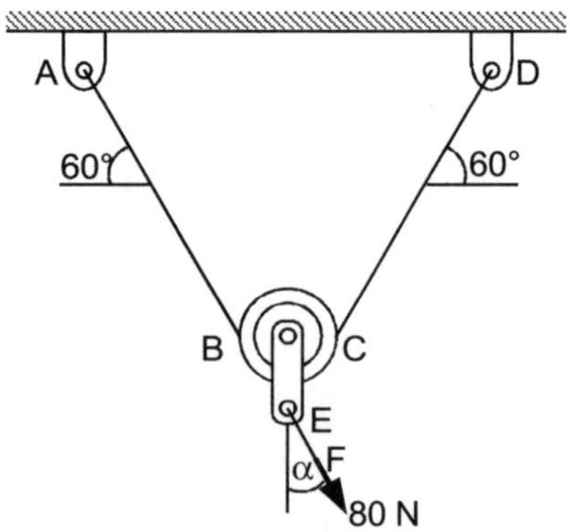

Fig. ES3.24

Solution: Free body diagram and force triangle are given in Figs.ES3.24 (*a*) and (*b*).

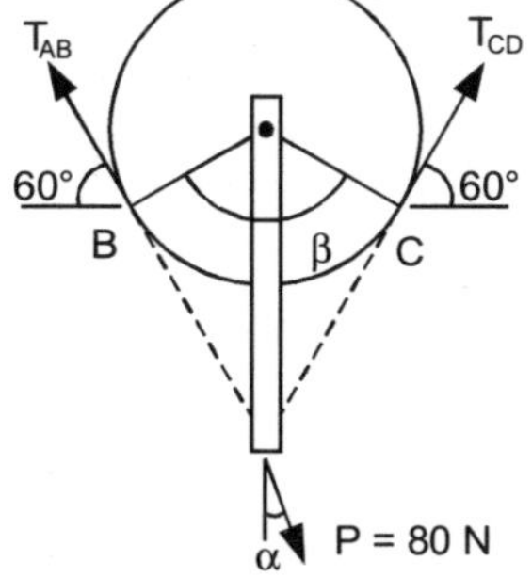

Fig. ES3.24(a)

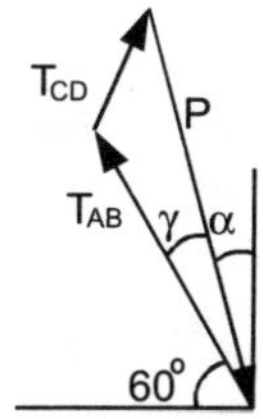

Fig. ES3.24(b)

Since the 80 N force tends to rotate the pulley counterclockwise, the cable tends to slip relative to the pulley clockwise and we have

$$T_1 = T_{CD},\ T_2 = T_{AB},\ \mu_s = \text{static friction}$$

$$\beta = 120^\circ = \frac{2\pi}{3} \text{ radian.}$$

From the ratio of belt-tension relations:

$$\frac{T_2}{T_1} = e^{\mu_s \beta} \Rightarrow \frac{T_{AB}}{T_{CD}} = e^{\frac{2\pi}{3}\mu_s}$$

or $$T_{AB} = e^{\frac{2\pi}{3}\mu_s}\, T_{CD} \qquad ...(1)$$

From the force triangle we have using the law of cosines

$$P^2 = T_{AB}^2 + T_{CD}^2 - 2T_{AB}T_{CD}\cos\beta$$

$$= \left(e^{\frac{2\pi}{3}\mu_s} T_{CD}\right) + T_{CD}^2 - 2\left(e^{\frac{2\pi}{3}\mu_s} T_{CD}\right)T_{CD}\left(-\frac{1}{2}\right)$$

$$= \left[\left(e^{\frac{2\pi}{3}\mu_s}\right)^2 + 1 + e^{\frac{2\pi}{3}\mu_s}\right]T_{CD}^2$$

$$P^2 = FT_{CD}^2$$

where $F = \left[\left(e^{\frac{2\pi}{3}\mu_s}\right)^2 + 1 + e^{\frac{2\pi}{3}\mu_s}\right]$.

Hence, we have that

$$T_{CD} = \frac{1}{\sqrt{F}} P \qquad ...(2)$$

(*a*) The corresponding values of α for the system to remain in equilibrium. Using the law of sines we have

$$\frac{\sin\gamma}{T_{CD}} = \frac{\sin\beta}{P}, \qquad \sin\gamma = \frac{T_{CD}}{P}\sin\beta$$

$$\sin\gamma = \frac{1}{\sqrt{F}}\sin\beta, \qquad \gamma = \sin^{-1}\left(\frac{1}{\sqrt{F}}\sin\beta\right)$$

$$\alpha = 90^\circ - (60^\circ + \gamma)$$

(*b*) The reactions at *A* and *D* are as follows:
Substituting $P = 80$ N in Eq.(2), then

$$D = T_{CD} = \frac{1}{\sqrt{F}}(80)\ \text{N}$$

and from Eq.(1), $A = T_{AB} = e^{\frac{2\pi}{3}\mu_s}\left(\frac{1}{\sqrt{F}}\right)(50)\ \text{N}$

MATLAB program for this problem is given below:

```
mu=0;
%Output Headings
fprintf('Friction Angle\n')
fprintf('\n')
ang=120.*pi/180.;
N=1;
while mu<0.60
  Y(N)=mu;
  a=exp(pi*ang*mu);
  E=sqrt(a^2+1.+a);
  ooe=1./E;
  sgamma=ooe*sin(ang);
  Gamma=asin(sgamma);
  alpha(N)=pi/2.-(1.0472+Gamma);
  X(N)=alpha(N)*180/pi;
  D(N)=ooe*80.;
  A(N)=a*D(N);
  fprintf('%5.3f    %5.3f\n',Y(N),X(N))
  mu=mu+.05;
  N=N+1;
end%while
fprintf('\n')
fprintf('\n')
fprintf('Friction    Reac.A    Reac.D\n')
fprintf('\n')
for I=1:N-1
  fprintf('%5.3f    %5.3f    %5.3f\n',Y(I),A(I),D(I))
end
figure(1)
plot(Y,X)
xlabel('Coefficient of friction')
```

```
ylabel('Angle alpha')
grid on
figure(2)
plot(Y,D,Y,A)
xlabel('Coefficient of friction')
ylabel('Reactions')
legend('Reaction at A','Reaction at D',2)
grid on
```

Output is as follows:

Friction	Angle
0.000	–0.000
0.050	5.377
0.100	10.391
0.150	14.780
0.200	18.425
0.250	21.332
0.300	23.582
0.350	25.286
0.400	26.558
0.450	27.497
0.500	28.186
0.550	28.687

Friction	Reac.A	Reac.D
0.000	46.188	46.188
0.050	53.481	38.488
0.100	59.860	31.001
0.150	65.068	24.251
0.200	69.105	18.535
0.250	72.125	13.922
0.300	74.336	10.326
0.350	75.934	7.591
0.400	77.083	5.546
0.450	77.907	4.033
0.500	78.498	2.925
0.550	78.921	2.116

Also following Figs. ES3.24(*c*)and (*d*) are obtained.

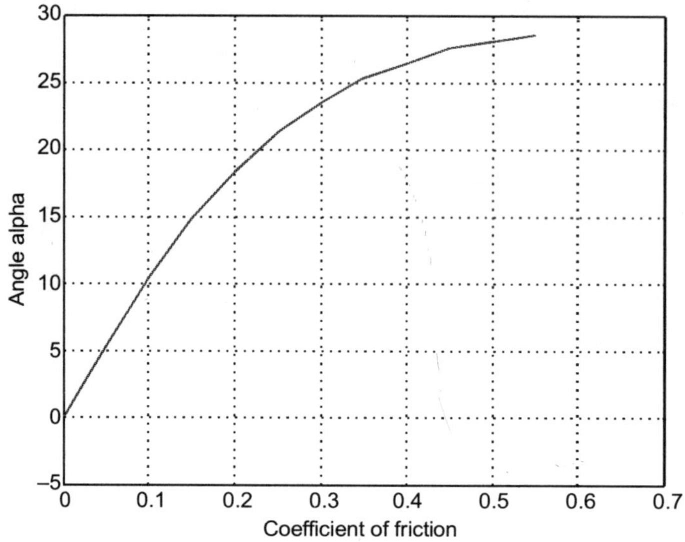

Fig. ES3.24(c)

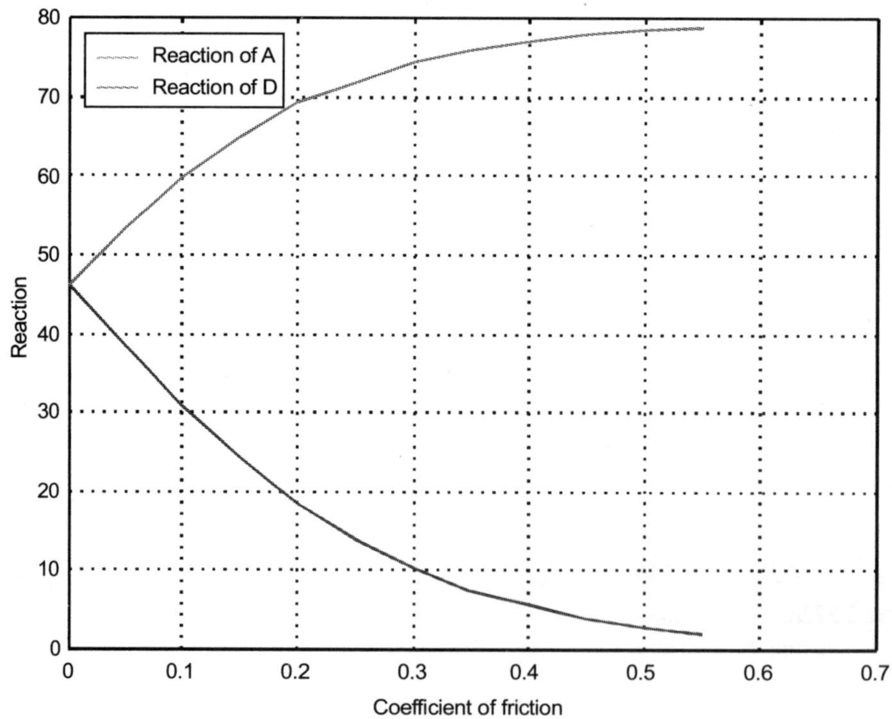

Fig. ES3.24(d)

ES 3.25: Figure ES3.25 shows a cylindrical silo where H = height of the cylindrical portion, r = radius of the cylindrical silo, R = radius of the spherical cap roof and V = volume of the silo

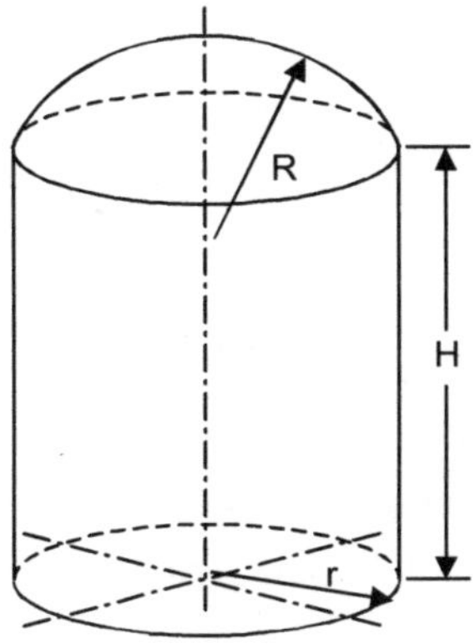

Fig. ES3.25

Write a MATLAB program to compute

(*a*) the height for given values of r, R, and V

(*b*) the surface area of the silo, S

Use the program to determine the height and surface area of a silo, given $r = 32$ cm, $R = 50$ cm and $V = 125{,}000$ cm^3.

Solution: The total volume of the silo is the sum of the volumes of the cylindrical part and the spherical cap.

$$V_{\text{total}} = V_{\text{cyl}} + V_{\text{cap}}$$

$$= \pi r^2 H + \frac{1}{3}\pi h^2 (3R - h)$$

where,

$$h = R - R\cos\theta = R(1 - \cos\theta) \text{ (see Fig. ES3.25}(a))$$

and $r = R\sin\theta$ or $\theta = \sin^{-1}(r/R)$

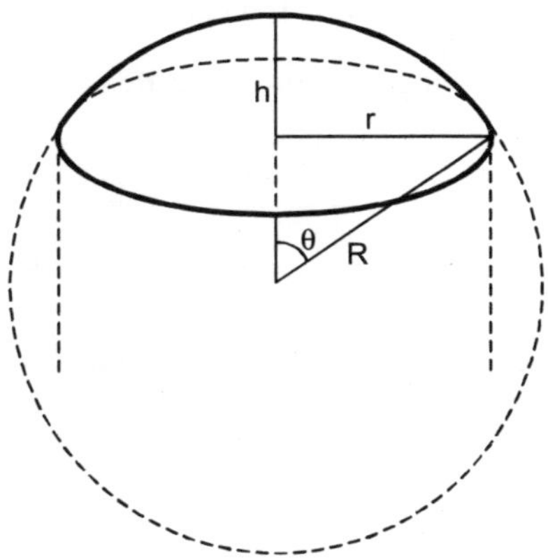

Fig. ES3.25 (a)

The height H, of the cylindrical part is given by

$$H = \frac{V - V_{\text{cap}}}{\pi r^2}$$

The surface area of the silo is the sum of the surface areas of the cylindrical part and the spherical cap.

$$S = S_{\text{cyl}} + S_{\text{cap}} = 2\pi r H + 2\pi R H$$

%MATLAB Solution:

```
r=32;
R=50;
V=125000;
```

```
Theta=asin(r/R);
h = R*(1-cos(Theta));
Vcap=pi*h^2*(3*R-h)/3;
H=(V-Vcap)/(pi*r^2);
S=2*pi*(r*H+R*h);
fprintf('The height H = %f cm',H);
fprintf('The surface area of the silo S = %f square cm.',S);
```

Output comes as follows:

The height $H = 32.812738$ cm. The surface area of the silo $S = 10235.750764$ square cm.

ES 3.26 : Figure ES3.26 shows a trapezoid.

(*a*) Obtain the x and y coordinates of the centroid when $H_1 = a/n^2$ and $H_2 = a/n$.

(*b*) Write a MATLAB program to plot the values of $\bar{x}$ and $\bar{y}$ for $1 \le n \le 5$ and $a = 9$ mm.

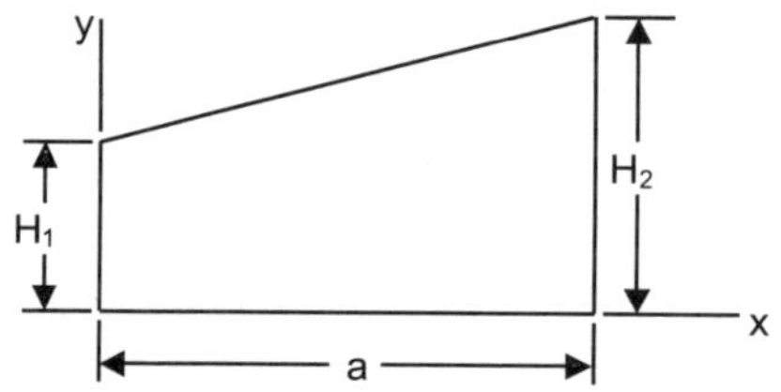

Fig. ES3.26

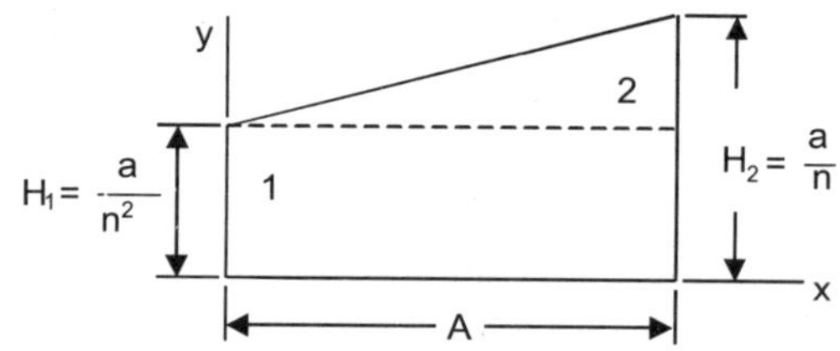

Fig. ES3.26 (a)

Solution: Entire trapezoid is divided into two triangles 1 and 2 as shown in Fig. ES3.26 (*a*).

Following table gives the area, centroid information of each area and total area.

#	Area	$\bar{x}$	$\overline{Ax}$	$\bar{y}$	$\overline{Ay}$
1.	aH_1	$\frac{a}{2}$	$\frac{1}{2}a^2H$	$\frac{1}{2}H_1$	$\frac{1}{2}aH_1^2$
2.	$\frac{1}{2}a(H_2-H_1)$	$\frac{2}{3}a$	$\frac{1}{3}a^2(H_2-H_1)$	$h_1+\frac{1}{3}(H_2-H_1)$ $=\frac{1}{3}(2H_1+H_2)$	$\frac{a}{6}\left(H_2^2+H_1H_2-2H_1^2\right)$
Σ	$\frac{a}{2}(H_1+H_2)$		$\frac{a^2}{6}(H_1+2H_2)$		$\frac{a}{6}\left(H_2^2+H_1H_2-2H_1^2\right)$

Hence, $$\bar{x} = \frac{a}{3}\left(\frac{H_1+2H_2}{H_1+H_2}\right);\quad \bar{y} = \frac{1}{3}\left(\frac{H_2^2+H_1H_2+H_1^2}{H_1+H_2}\right)$$

Substituting $H_1 = a/n^2$, $H_2 = a/n$ and $a = 9$ mm yields

$$\bar{x} = \frac{a}{3}\left(\frac{1+2n}{1+n}\right) = 3\left(\frac{1+2n}{1+n}\right)$$

$$\bar{y} = \frac{a}{3n^2}\left(\frac{n^2+n+1}{1+n}\right) = \frac{3}{n^2\left(\frac{n^2+n+1}{1+n}\right)}$$

MATLAB Solution:

```
n=[1:.5:5];
xbar=3*(1+2*n)./(1+n);
ybar=(3./n.^2).*(n.^2+n+1)./(n+1);
plot(n,xbar,'-p',n,ybar,'-*')
text(3,5,'xbar')
text(3,1.4,'ybar')
xlabel('n')
ylabel('Centroid (mm)')
grid on
```

Output is shown in Fig. ES3.26(*b*).

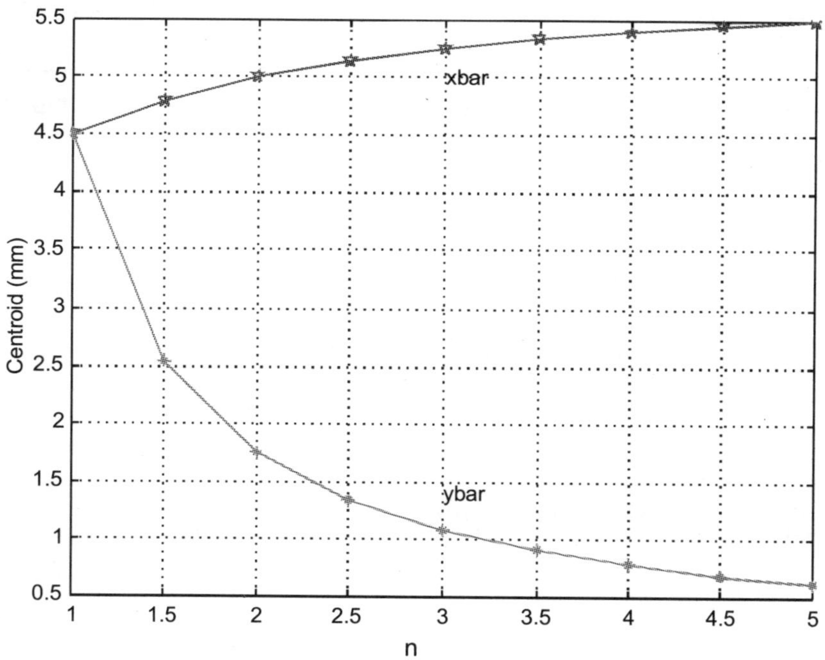

Fig. ES3.26(b)

ES 3.27: Determine the y coordinate of the centroid of the body shown in Fig.ES3.27 when $h = nb$ and $h = n^2b$.

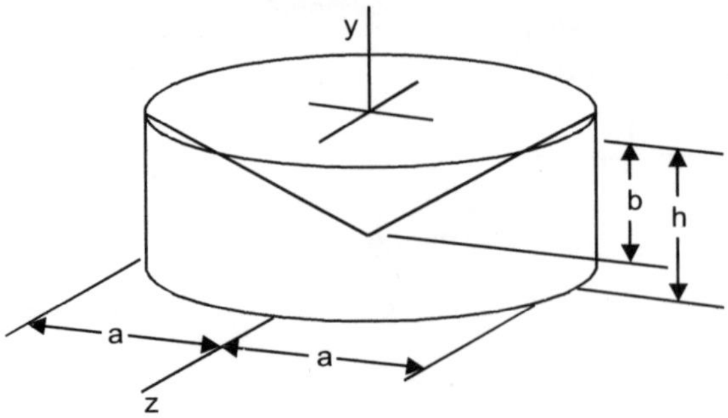

Fig. ES3.27

Write a MATLAB program to plot $\bar{y}$ as a function of n for both cases using $1 \le n \le 12$ and (*i*) b =5mm, (*ii*) b = 7 mm and (*iii*) b = 9 mm.

Solution: This contains two parts as shown in Figures ES3.27 (*a*), (*b*) and (*c*).

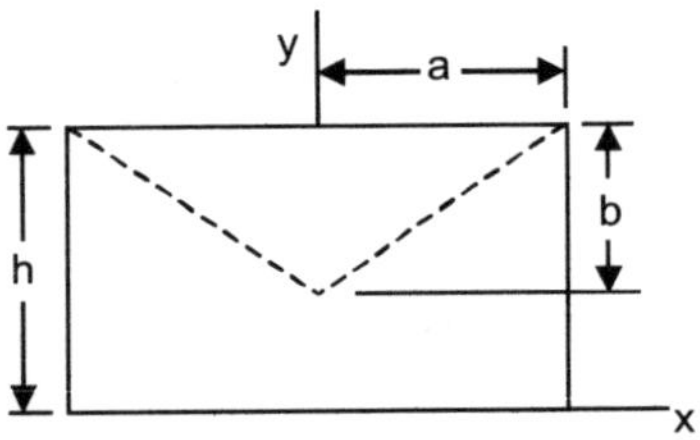

Fig. ES3.27(a)

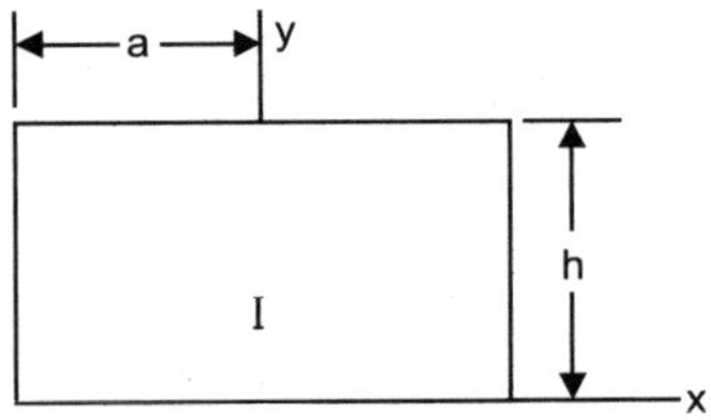

Fig. ES3.27(b)

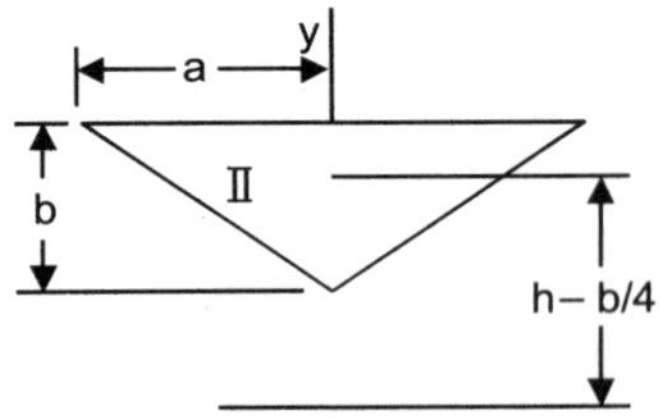

Fig. ES3.27(c)

Based on the Fig.ES3.27(*a*), the following table is constructed.

	V	$\bar{y}$	$\bar{y}V$
I	$\pi a^2 h$	$\frac{1}{2}h$	$\frac{1}{2}\pi a^2 h^2$
II	$-\frac{1}{3}\pi a^2 b$	$h-\frac{1}{4}b$	$-\frac{\pi}{3}a^2bh+\frac{\pi}{12}a^2b^2$

$$\bar{y}\Sigma V = \Sigma \bar{y} V \text{ gives}$$

$$\bar{y}\left(\pi a^2 h - \frac{\pi}{3}a^2 b\right) = \frac{\pi}{12}a^2b^2 + \frac{\pi}{2}a^2h^2 - \frac{\pi}{3}a^2bh$$

and
$$\bar{y} = \frac{6h^2 - 4hb + b^2}{4(3h-b)}$$

For $h = nb$

$$\bar{y} = \frac{b(6n^2 - 4n + 1)}{4(3n-1)}$$

For $h = n^2 b$

$$\bar{y} = \frac{b(6n^2 - 4n^2 + 1)}{4(3n^2 - 1)}$$

The plot of $\overline{y}$ as a function of n for $1 \leq n \leq 12$ for the cases is shown in Figs.ES3.27(*d*) and (*e*).

MATLAB Solution:

```
% Use symbolic notation in MATLAB%%%%%%%%%%%%%%%%%%
syms b n
 ybarn=inline('b*(6*n^2-4*n+1)/(4*(3*n-1))') ;
 ybarn2=inline('b*(6*n^4-4*n^2+1)/(4*(3*n^2-1))');
  ybarn=vectorize(ybarn)
 ybarn2=vectorize(ybarn2)
 n=[1:1:12];
 subplot(2,1,1)
 plot(n,ybarn(5,n),'-p',n,ybarn(7,n),'-*',n,ybarn(9,n),'-o')
 xlabel('n')
 ylabel('Centroid  (mm)')
 grid on
 title('\bfwhen h=nb')
 legend('b=5mm','b=7mm','b=9mm',2);
 subplot(2,1,2)
 plot(n,ybarn2(5,n),'-p',n,ybarn2(7,n),'-*',n,ybarn2(9,n),'-o')
 xlabel('n')
 ylabel('Centroid (mm)')
 grid on
 title('\bfwhen h=n^2b')
%%%%%%%%%%%%%%%%%%%%%%%%%%%%%%%%%%
```

Output of the program is shown in Fig. ES3.27(d).

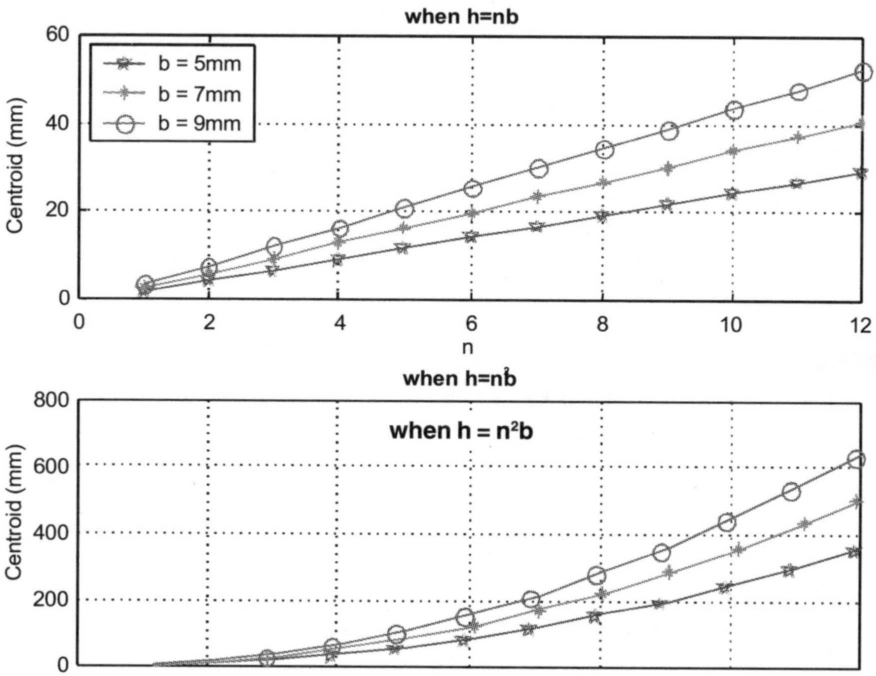

Fig. ES3.27(d) MATLAB output

ES 3.28: For the area shown in Fig.ES3.28, write a MATLAB program to plot I_{xy} as a function of N for values of N from 1 to 12 knowing that $A = 100$ mm and $B = 80$ mm.

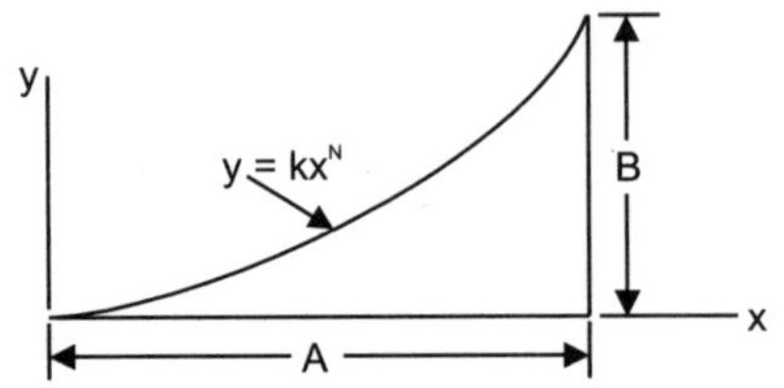

Fig. ES3.28

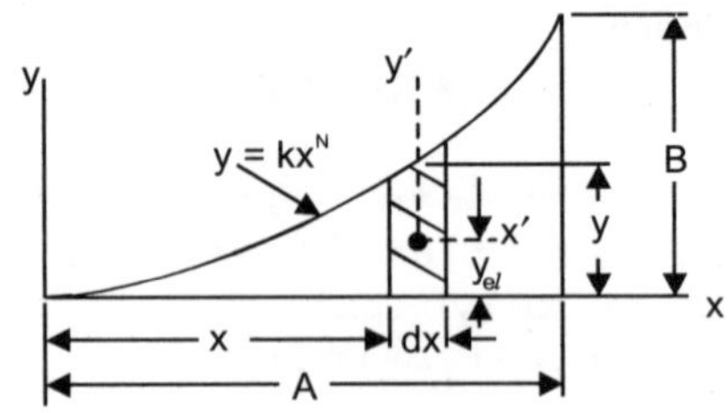

Fig. ES3.28(a)

Solution: Consider an elemental length dx as shown in Fig. ES3.28.

When $x = A$, $B = ka^N$

Therefore, $k = \dfrac{B}{A''}$

and $y = \dfrac{B}{A^N} x^N$

Also $dI_{xy} = \overline{dI}_{x'y'} + \overline{x}_{el}\overline{y}_{el} dA$

But due to symmetry

$$\overline{dI}_{x'y'} = 0$$

Since $\overline{x}_{el} = x$, $\overline{y}_{el} = 1/2y$ and $dA = ydx$, we have

$$I_{xy} = \int dI_{xy} = \int_0^a x\left(\frac{1}{2}y\right) ydx = \frac{1}{2}\int_0^A x\left(\frac{B}{A^N}x^N\right)^2 dx$$

$$I_{xy} = \frac{1}{2}\frac{B^N}{A^{2N}}\int_0^A x^{2n+1}dx = \frac{1}{2}\frac{B^N}{A^{2N}}\frac{A^{2N+2}}{2N+2}$$

$$I_{xy} = \frac{A^2B^2}{4(N+1)}$$

MATLAB Solution:

```
% Input A and B [in mm] %%%%%%%%%%%%%%%%%
A=100;
B=80;
fprintf('Enter A and B\n');
fprintf('A=%g mm, and B=%gmm\n',A,B);
fprintf('I(xy) N\n')
for N=1:15
   Ixy(N)=A^2*B^2/(4.*(N+1));
  NN(N)=N;
  fprintf('%5.2e %5.2f\n',Ixy(N),NN(N))
```

```
end
plot(NN, Ixy)
xlabel('N')
ylabel('Computed moment of inertia I(xy)(mm^4)')
grid on
%%%%%%%%%%%%%%%%%%%%%%%%%%%%%%
```

Output comes like this:

Enter A and B

$A = 100$ mm, and $B = 80$mm

$I(xy)$	N
8.00e+006	1.00
5.33e+006	2.00
4.00e+006	3.00
3.20e+006	4.00
2.67e+006	5.00
2.29e+006	6.00
2.00e+006	7.00
1.78e+006	8.00
1.60e+006	9.00
1.45e+006	10.00
1.33e+006	11.00
1.23e+006	12.00
1.14e+006	13.00
1.07e+006	14.00
1.00e+006	15.00

Plot is shown in Fig. ES3.28(*b*).

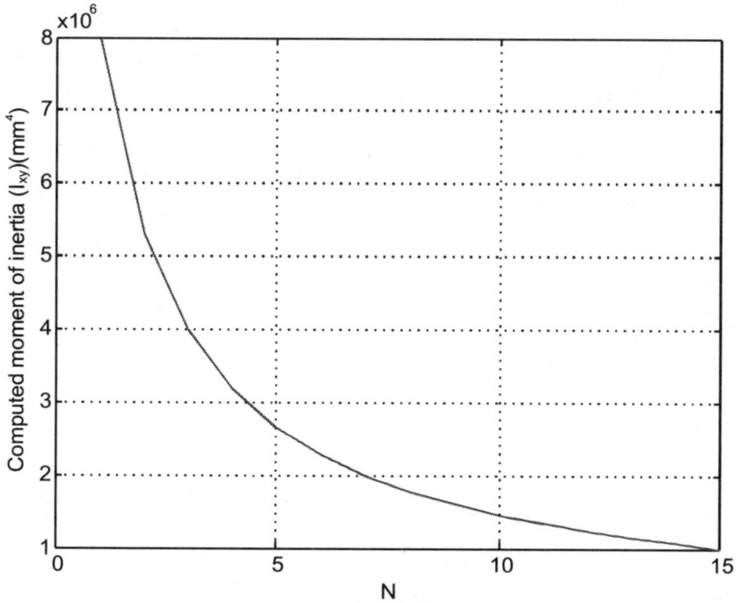

Fig. ES3.28(b)

3.2.2 Dynamics

Problems on dynamics of particles is covered before the dynamics of rigid bodies. That is particle kinematics is followed by particle kinetics. Next, examples on the use of work-energy and impulse-momentum principles are presented. The same sequence is repeated for planar rigid bodies. MATLAB Problems on dynamics of rigid bodies in three dimensions are given at the end. Use of MATLAB commands like 'int', 'diff', 'solve', 'dsolve' is presented through simple examples.

Particle Kinematics

ED 3.1: The motion of a particle is defined by the equation

$$x = 35t^2 - 110t$$

and $$y = 115t^2 - 42t^3$$

where x and y = displacement of the particle (mm)

t = time (sec.) for the time interval $0 \le t \le 25$,

Write a MATLAB program to plot:

(*a*) the path of the particle in the *x-y* plane,
(*b*) the components of the velocity v_x and v_y and the magnitude of the velocity v,
(*c*) the components of the acceleration a_x and a_y and the magnitude of the acceleration a.

Solution:

Given: $x = 35t^2 - 110t$; $y = 115t^2 - 42t^3$ as the position vector of the particle.

Hence $v_x = \dot{x} = 70t - 110$; $v_y = \dot{y} = 230t - 126t^2$ are component of velocities and

$a_x = \ddot{x} = 70$; $a_y = \ddot{y} = 230 - 252t$ are components of accelerations.

Here particle path refers to the plot of x and y positions at various instants of time.
So first x and y are found by varying t from 0 to 25 seconds.

MATLAB Program is given below:

```
t=[0:0.5:25];% Vary the time from 0 to 25 in steps of 0.5
x=35*t.^2-110*t; % compute x
y=115*t.^2-42*t.^3; % compute y
v_x=70*t-110; % compute vx
v_y=230*t-126*t.^2; % compute vy
v=sqrt(v_x.^2+v_y.^2); % compute v
a_x=70;% compute ax
a_y=230-252*t; % compute ay
a=sqrt(a_x.^2+a_y.^2); % compute a
%%%%%%%%%%%%%%% PLOTING THE CALCULATED DATA%%%%%%%%%%%%%%
figure(1)
plot(x,y);
xlabel('x(mm)')
ylabel('y(mm)')
```

```
legend('Trajectory')
grid on
axis([0 10000 -2.5e5 0])
figure(2)
plot(t,v_x,t,v_y,t,v)
xlabel('t(sec)')
ylabel('v_x,v_y,v(mmps)')
legend('v_x','v_y','v',2)
grid on
figure(3)
plot(t,a_x,t,a_y,t,a)
xlabel('t(sec)')
ylabel('a_x,a_y,a(mmps^2)')
legend('a_x','a_y','a',2)
grid on
```

The plots are shown in Figs. ED3.1(*a*), (*b*) and (*c*) respectively.

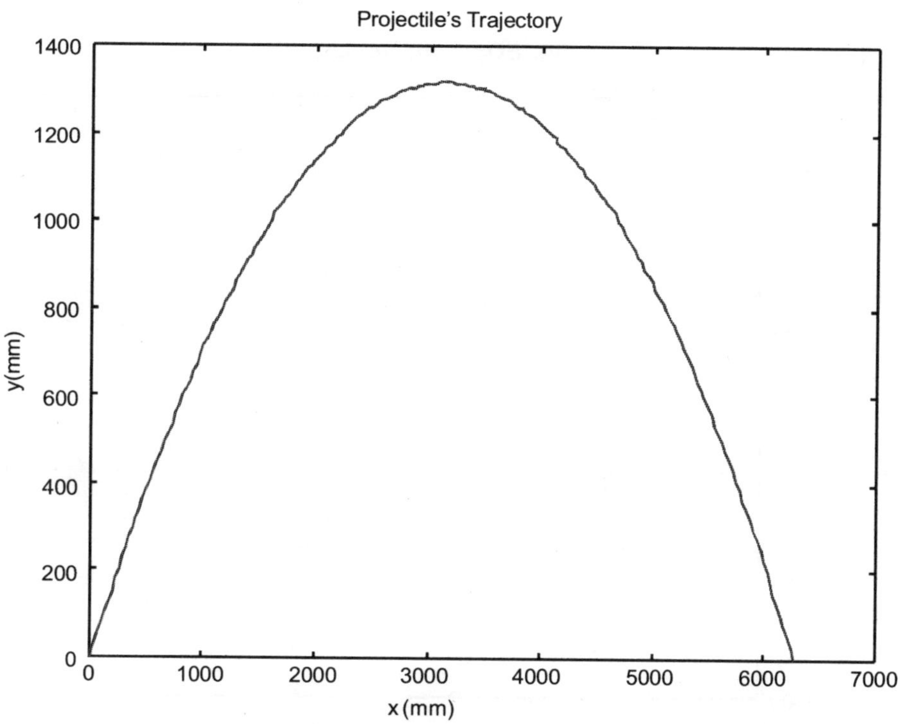

Fig. ED 3.1 (a) Position

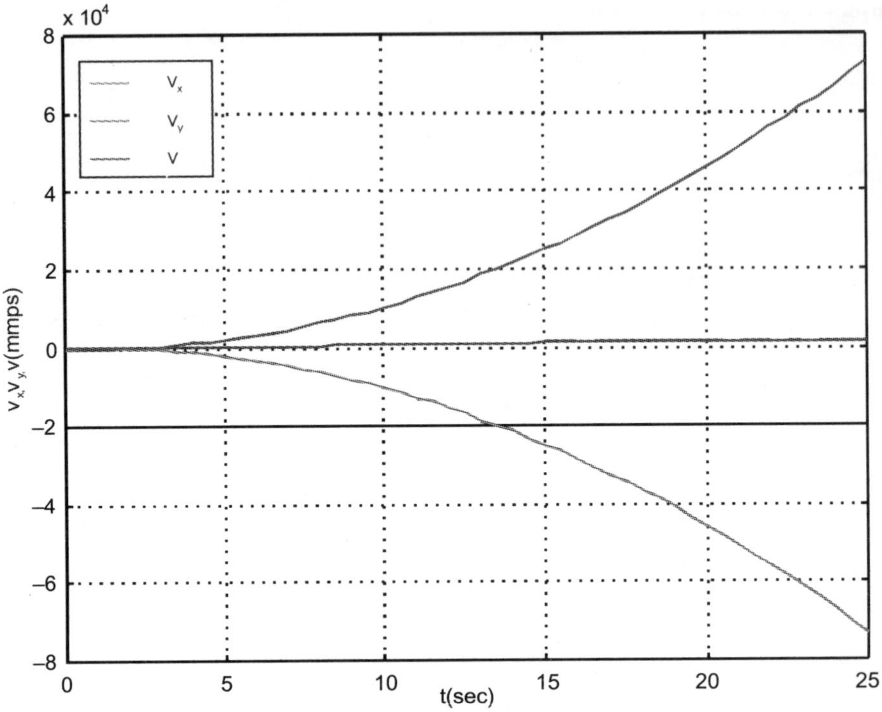

Fig. ED3.1 (b) Velocities

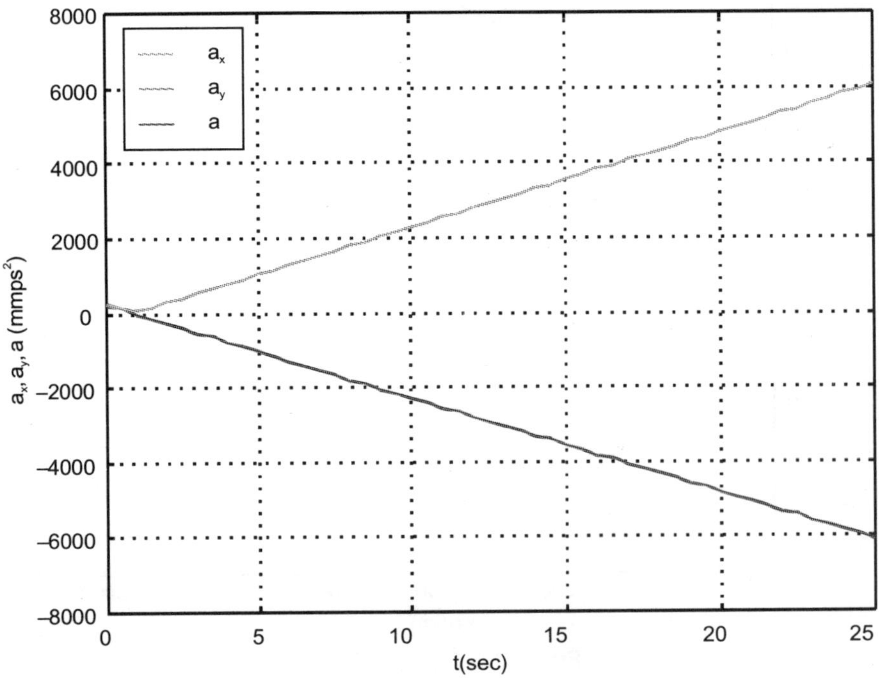

Fig. ED3.1 (c) Accelerations

ED 3.2: A particle is fired vertically downwards with a velocity 30 m/s in a fluid. Due to resistance of the fluid the particle experiences a deceleration equal to $a = -(0.7v^3)$ m/s^2, where v is velocity in m/s. Plot a graph of velocity versus time and distance versus time using MATLAB.

Solution: Given $a = f(v)$, so the velocity is determined as a function of time using $a = dv/dt$, since this equation relates v, a and t. Thus $a = \frac{dv}{dt} = -0.7v^3$.

Integrating both sides

$$-\int_{30}^{v} \frac{dv}{0.7v^3} = \int_{0}^{t} dt \quad \text{or,} \quad t = \frac{1}{1.4}\left[\frac{1}{v^2} - \frac{1}{30^2}\right] \text{or, } v = \left(1.4t + \frac{1}{30^2}\right)^{-0.5}$$

Position s is given by

$$\frac{ds}{dt} = v = \left(1.4t + \frac{1}{30^2}\right)^{-0.5} \quad \text{or, integrating} \int_{0}^{s} ds = \int_{0}^{t}\left(1.4t + \frac{1}{30^2}\right)^{-0.5} dt$$

or

$$s = \frac{2}{1.4}\left[\left(1.4t + \frac{1}{30^2}\right)^{0.5} - \frac{1}{30}\right]$$

Using MATLAB, the graphical representation of velocity and displacement as a function of time t is given with following simple program.

MATLAB Program:

```
t=0:0.05:1.5;
v0=30; %INITIAL VELOCITY
v=(1.4*t+1/v0^2).^-0.5;
s=2/1.4*(sqrt(1.4*t+1/v0^2)-1/30);
figure(1)
plot(t,v);
xlabel('t(sec)')
ylabel('v(mps)')
grid on
figure(2)
plot(t,s)
xlabel('t(sec)')
ylabel('s(m)')
grid on
```

The output of the program is shown in Figs. ED 3.2 (*a*) and (*b*).

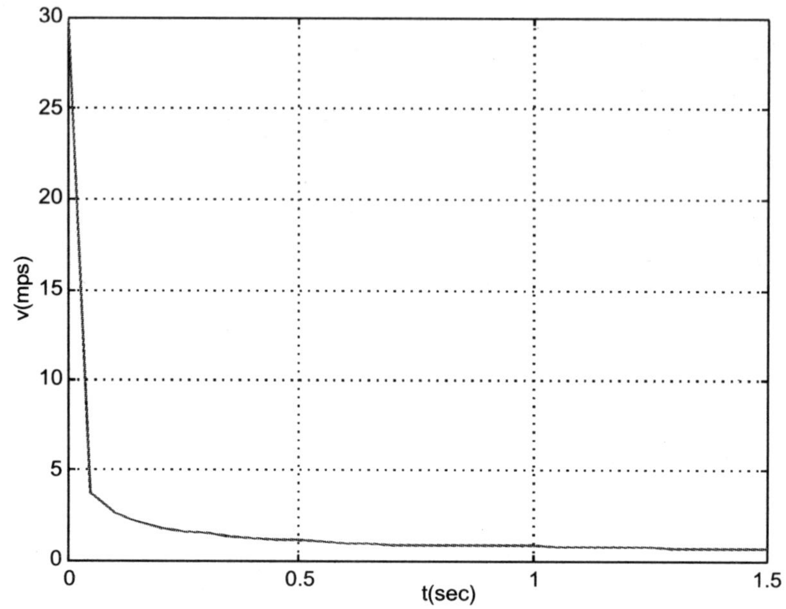

Fig. ED3.2 (a) Velocity variation

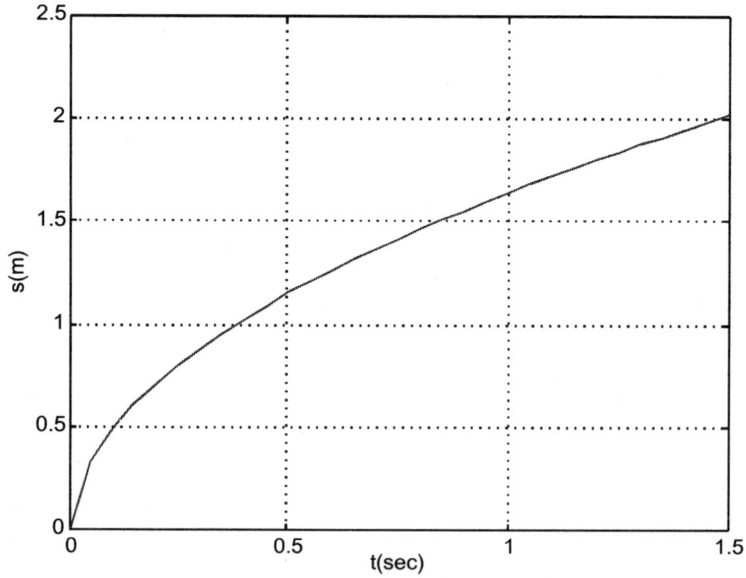

Fig. ED3.2 (b) Displacement variation

ED3.3: If position of a car is defined as $s = (5t - 3t^2)$, construct *s-t*, *v-t* and *a-t* graphs over the interval $0<t<10$ sec.

Solution: As the given function is a simple one, derivatives can be obtained easily. However, initially the derivatives are obtained with following commands in MATLAB.

```
s=sym('5*t-3*t^2');
v= diff(s, 't')
a=diff(v,'t')
```

The 'inline' function is used to generalize the variables in terms of time coordinate *t*.

The entire script is given below:

```
% The script plots s-t, v-t and a-t curves for specified values of t
s=inline('5*t-3*t.^2');
v=inline('5-6*t');
a=inline('-6');
t=0:0.5:10;
figure(1);
subplot(3,1,1);            %SUB-PLOT-1
plot(t,s(t));
title('\bfDisplacement Plot');
ylabel('\bfDisplacement (m)');
grid on;
subplot(3,1,2);            % SUB-PLOT-2
plot(t,v(t));
title('\bfVelocity Plot');
ylabel('\bfVelocity (mps)');
grid on;
subplot(3,1,3);            % SUB-PLOT-3
plot(t,a(t),'*');
title('\bfacceleration Plot');
ylabel('\bfAcceleration (mps^2)');
grid on;
```

The output of the program is shown in Fig. ED3.3.

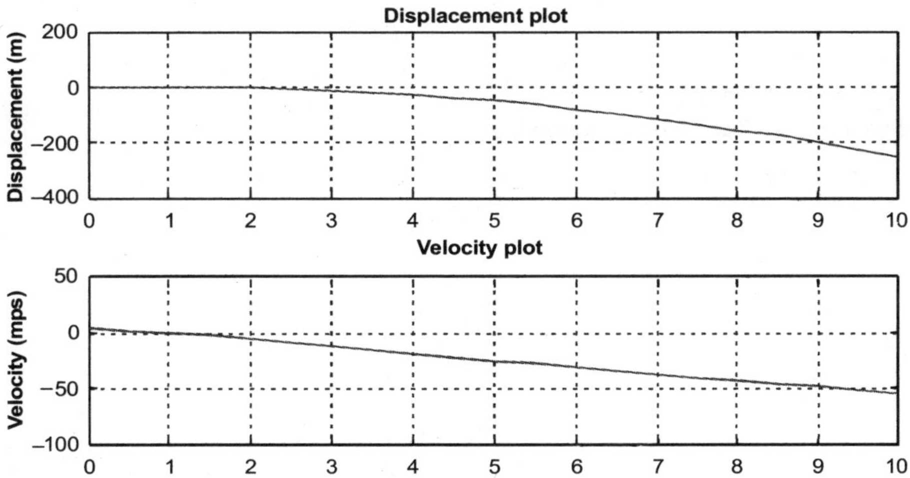

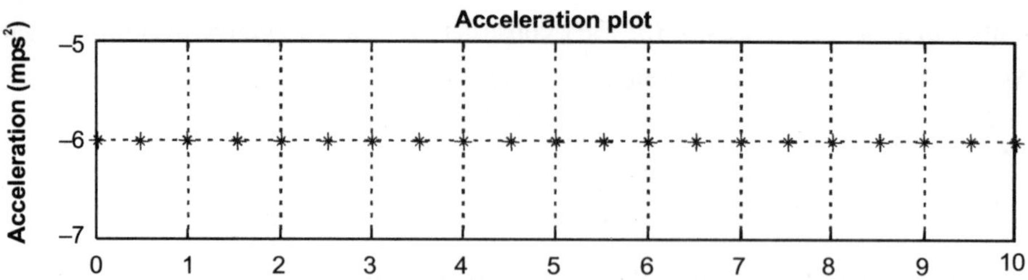

Fig. ED3.3 MATLAB output

ED 3.4: Figure ED3.4 shows the motion of a particle. The initial velocity and the angle at which the projectile is fired are known. Write a MATLAB program to calculate and plot the maximum height and distance. Use the program to calculate and plot the trajectory of a projectile that is fired at a velocity of 250 m/s at an angle of 40°.

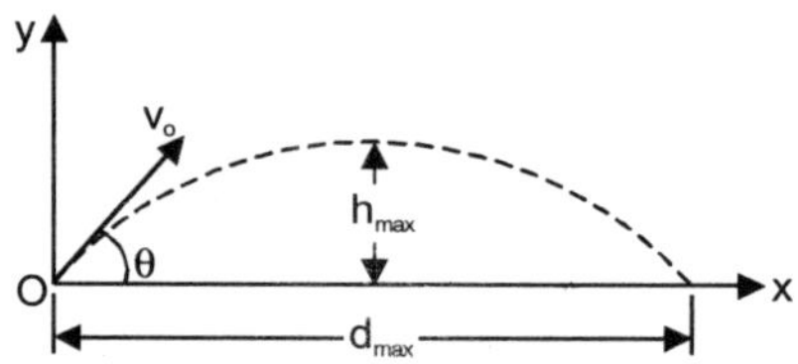

Fig. ED3.4

Solution: Components of velocity $v_{ox} = v_o \cos\theta$ and $v_{oy} = v_o \sin\theta$.

Height $$h_{\max} = \frac{v_{oy}^2}{2g}$$

and corresponding time

$$t_{\max} = \frac{v_{oy}^2}{g}$$

But to draw trajectory as a function of distance(x), use

$$y = v_{oy}t - \frac{1}{2}gt^2 \text{ and } x = v_{ox}t$$

MATLAB script is shown below:

Open the file "trajectory.m" and type the following

```
function[hmax,dmax]=trajectory(v0,theta)
% Trajectory calculates the max height and distance of a projectile,
% makes a plot of the trajectory.
% Input arguments are:
%v0=initial velocity in(m/s).
%theta=angle in degrees.
%Output arguments are:
%hmax=maximum height in (m).
%dmax=maximum distance in(m).
```

```
%The function creates also a plot of the trajectory.
g=9.81;
v0x=v0*cos(theta*pi/180);
v0y=v0*sin(theta*pi/180);
thmax=v0y/g;
hmax=v0y^2/(2*g);
ttot=2*thmax;
dmax=v0x*ttot;

%Creating a trajectory plot
tplot=linspace(0,ttot,200);
x=v0x*tplot;
y=v0y*tplot-0.5*g*tplot.^2;
plot(x,y)
xlabel('Distance(m)')
ylabel('Height(m)')
title('Projectile's Trajectory')
```

To execute it type the following at the MATLAB command prompt

```
>> [h d] = trajectory(250,40)
```

Then the output comes as:

```
h=
    1.3162e+003
d=
    6.2743e+003
```

Followed by the Fig. ED 3.4 (*a*).

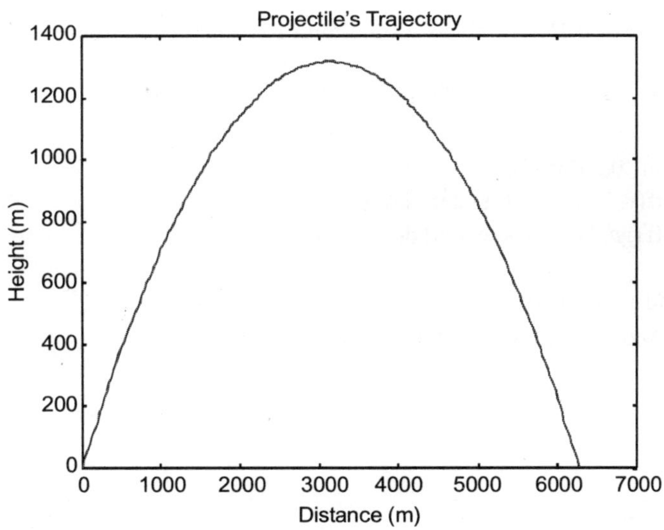

Fig. ED 3.4(a) MATLAB output

ED 3.5: A jet plane is going in a parabolic path described by $y=0.05x^2$. At a point in the path, it has a velocity of 200 m/s, which is increasing at the rate of 0.8 m/s^2. Find the resultant acceleration and plot the variation of acceleration as a function of its horizontal position x.

Solution: The motion of the plane is described in Fig. ED3.5.

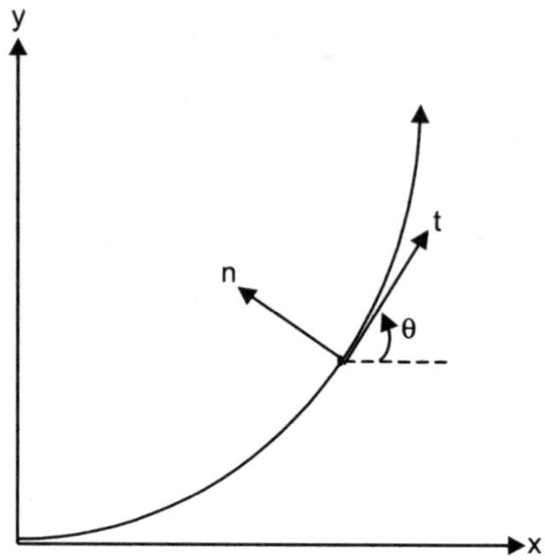

Fig. ED3.5

The components of acceleration along n and t directions are as follows:

$$a_t = dv/dt = 0.8 \text{ m/s}^2,\quad a_n = v^2/(\rho g) = 200^2/(\rho g), \text{ with } \rho = \frac{\left[1+\left(\frac{dy}{dx}\right)^2\right]^{3/2}}{\left|\frac{d^2y}{dx^2}\right|}$$

So, it requires computation of differentials as a function of x. Then using these differentials, the resultant acceleration $a=\sqrt{a_r^2+a_n^2}$ is plotted as a function of x in MATLAB.

As usual first define

y = sym('0.05*x^2');
yd = diff(y, 'x'); %% first derivative
ydd = diff(yd, 'x');%% second derivative

They give yd = 0.10*x and ydd = 0.10

Now use the following program to plot variation of acceleration with x.

MATLAB Program:

```
ac=0.8;
v=200;vs=v*v;
y=inline('0.05*x.^2');
```

```
yd=inline('0.1*x');
ydd=inline('0.1');
x=0.1:0.1:20;
rho=((1+yd(x).*yd(x)).^1.5)/abs(ydd(x));
an=vs./rho;
at=ac./x.*x;
a=sqrt(at.^2+an.^2);
plot(x,an,x,at,'*')
xlabel('x-position');
ylabel('Acceleration in m per s^2');
legend('Normal acceleration','tangential acceleration')
```

The output is shown in Fig. ED3.5 (*a*).

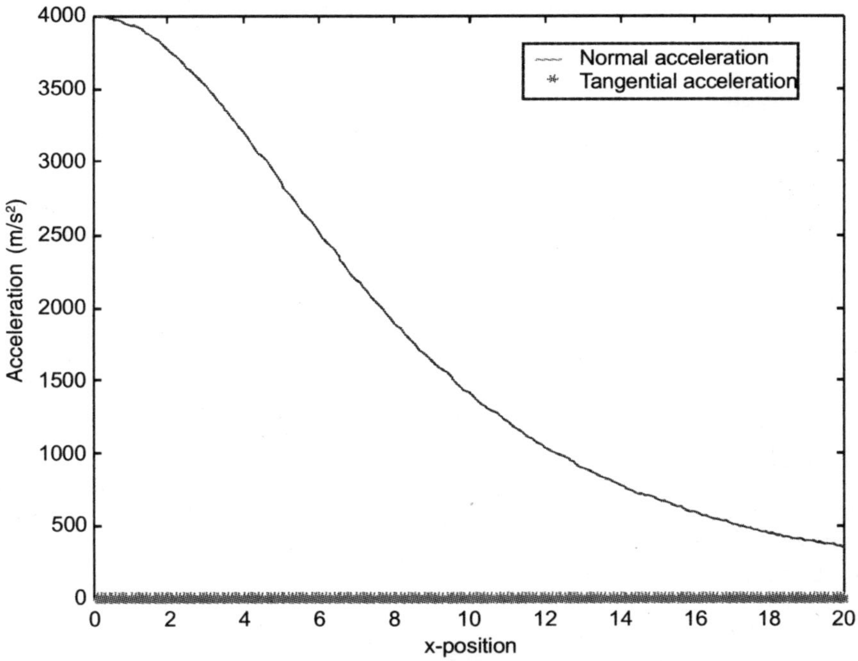

Fig. ED3.5 (a) MATLAB Output

ED3.6: Path of a particle *P* is ellipse defines with $r = 1.75/(1 - 0.75\cos\pi t)$ and $\theta = \pi t$. Derive expressions for velocity and acceleration. Plot the components of velocities and accelerations as a function of time.

Solution: This problem should be solved in cylindrical coordinate system. Thus

Radial velocity $v_r = \dot{r}$ and transverse velocity $v_\theta = r\dot{\theta}$,

where $\dot{\theta} = \pi$ and $\dot{r} = 1.75 \times 0.75 \times \pi \sin\pi t/(1 - 0.75\cos\pi t)^2$

Radial acceleration $a_r = \ddot{r} - r\dot{\theta}^2$ and transverse acceleration $a_\theta = r\ddot{\theta} + 2\dot{r}\dot{\theta}$. Here $\ddot{\theta} = 0$, but $\ddot{r}$ should be found from the MATLAB's diff command. This is done as follows:

```
%%%%%%%%%%%%
y=sym('1.75/(1-0.75*cos(3.14*t))')
yd= diff(y, 't');
ydd=diff(yd,'t');
%%%%%%%%%%%%%%%%%
```

Output:

```
y =1.75/(1-0.75*cos(3.14*t))
 yd=-4.121250/(1-.75*cos(3.14*t))^2*sin(3.14*t)
 ydd=
19.4110875000/(1-.75*cos(3.14*t))^3*sin(3.14*t)^2-12.94072500/
(1-.75*cos(3.14*t))^2*cos(3.14*t)
```

This defines the differentials $\dot{r}$ and as follows:

$$\dot{r} = -4.121 \sin\pi t/(1-0.75\cos\pi t)^2 \text{ and}$$

$$\ddot{r} = \frac{19.4\sin^2 \pi t}{(1-0.75\cos \pi t)^3} - \frac{12.94\cos \pi t}{(1-0.75\cos \pi t)^2}$$

The complete MATLAB program for plotting the velocities and acceleration is given below:

```
pi=3.14;
r =inline('1.75./(1-0.75*cos(3.14*t))');
rd=inline('-4.121*sin(3.14*t)./(1-0.75*cos(3.14*t)).^2');
rdd=inline('19.4*sin(3.14*t).^2/(1-0.75*cos(3.14*t)).^3-12.94*cos(3.14*t)./
(1-.75*cos(3.14*t)).^2');
theta=inline('3.14*t');
thetad=inline('3.14*t./t');
t=0.5:0.1:2;
vr=rd(t);vt=r(t).*theta(t);
ar=(rdd(t)-r(t).*thetad(t).^2);
at=2*rd(t).*thetad(t);
figure(1);
subplot(2,1,1); %SUB-PLOT-1
plot(t,vr,t,vt,'-p');
title('\bfVelocity Plot');
ylabel('Velocity (m per s)');
legend('radial velocity','transverse velocity');
grid on;
subplot(2,1,2); % SUB-PLOT-2
plot(t,ar,t,at,'-p');
title('\bfAcceleration Plot');
ylabel('Acceleration (mps^2)');
xlabel('Time in s');
legend('radial acceleration','transverse acceleration');
grid on;
```

The output plot is shown in Fig. ED3.6.

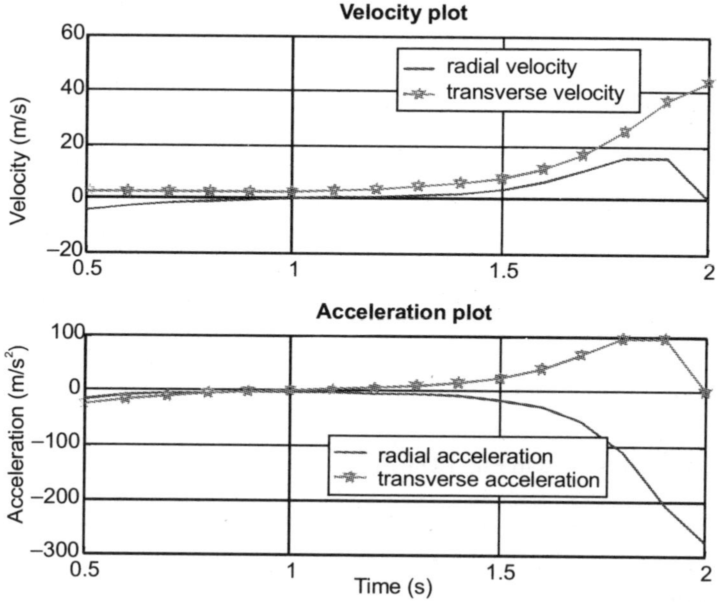

Fig. ED3.6 Velocity and Acceleration of particle in ellipse

ED3.7: Two planes *A* and *B* are flying at a speed of 600 km/h with *A* in a straight line path while *B* is along a circular path of radius ρ. If *A* is accelerating at 50 km/h^2, while *B* decelerates at 100 km/h^2, plot the resultant acceleration of *B* with respect to *A* as a function of radius of curvature ρ.

Solution: The motion of two planes *A* and *B* are given in Fig. ED3.7.

Acceleration of *B* is

$$a_B = a_{Bt}\hat{u}_t + a_{Bn}\hat{u}_n,$$

where $a_{Bt} = -100$ km/h^2, $a_{Bn} = 600^2/\rho$

Likewise for *A* $a_{At} = 50$ km/h^2 and $a_{An} = 0$

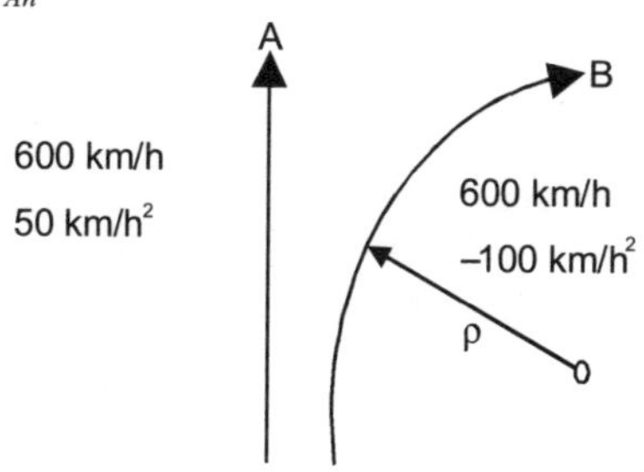

Fig. ED3.7

The relative velocity is then given by $a_{BA} = a_B - a_A = (a_{Bt} - a_{At})\hat{u}_t + (a_{Bn} - a_{An})\hat{u}_n$

$$= (a_{Bt} - a_{At})\hat{u}_t + (a_{Bn})\hat{u}_n$$

It's magnitude is given by $\sqrt{(a_{Bt} - a_{At})^2 + a_{Bn}^2}$.

Following MATLAB code can be readily employed for this problem.

```
%%%%%%%%%%%%%%%%
v0=600;
aA=50;
aBt=-100;
r=100:20:600;
aB=sqrt((aBt-aA).^2+(v0*v0./r).^2);
plot(r,aB);
xlabel('Radius of curvature');
ylabel('Relative acceleration of plane B');
grid on;
%%%%%%%%%%%%%%%%%%
```

The output of the program is presented in Fig. ED3.7 (*a*).

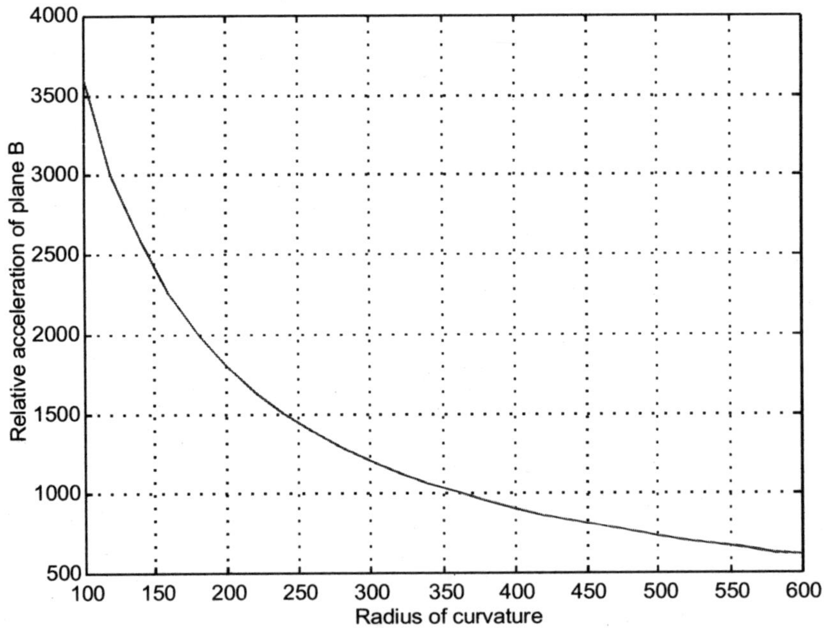

Fig. ED3.7 (a)

Particle Kinetics

ED 3.8: A particle of mass 10 kg is projected vertically from the ground with a velocity of 50 m/s. Find the maximum height reached by the particle when air resistance is $0.01v^2$ newton, where v is speed at any instant. Plot a graph of height as a function of velocity using MATLAB.

Solution: Here the particle is under the influence of forces shown in Fig. ED3.8(*a*)

Applying Newton's second law along y-direction: $\sum F_y = ma_y$,
where $\sum F_y = -F - mg = (-0.01v^2 - mg)$

Thus $ma_y = (-0.01v^2 - mg)$ or $a_y = (-0.01v^2 - mg)/m$

As the distance *vs* velocity plot is required use the relation $vdv = ads$

Fig. ED3.8

i.e., $$ds = \frac{-mvdv}{0.01v^2 + mg} \text{ or, } \frac{ds}{dv} = -\frac{mv}{0.01v^2 + mg}.$$

This differential equation is solved with initial conditions at $s=0$, $v=50$ m/s with the following program.

```
%%%%%%%%%%%%%%%%%%%%%%%%%%%%%
s0=0;
v0=50;
vspan=[v0:-1:0];% a vector that specifies the interval of the solution
v s]=ode45('resist',vspan,s0);% Solving the ODE
plot(s,v)
xlabel('x(m)');ylabel('Velocity(m/s)')
grid on;
%%%%%%%%%%%%%%%%%%%%%%%%%%%%%
```

The function file with the differential equation named 'resist.m' is listed below:

```
%%%%%%%%%%%%%%%%%%%%%%%%%
function dsdv=resist(v,s)
m=10; g=9.81;
dsdv=-m*v/(0.01*v^2+m*g);
%%%%%%%%%%%%%%%%%%%%%%%%%
```

The output of the program is shown in Fig. ED3.8(b).

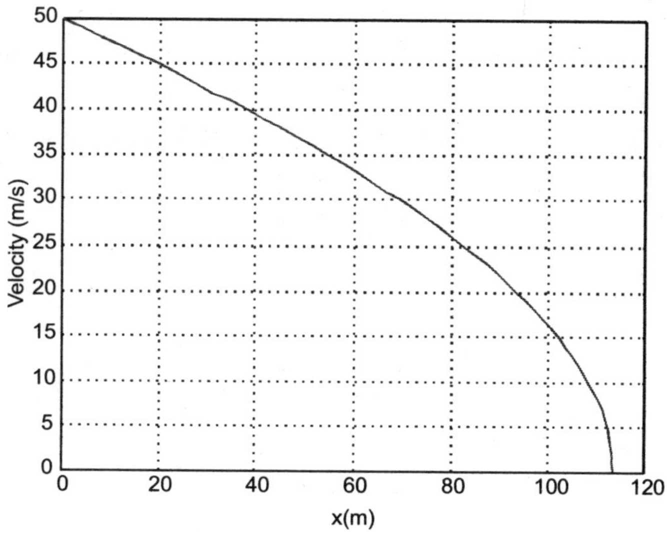

Fig. ED3.8(b)

ED 3.9 Figure ED3.9 shows a safety bumper placed at the end of a racetrack to stop out-of-control vehicles. The force that the bumper applies to the vehicle is given by

$$F = Kv^3(x + 1)^3$$

where $K = 32$ kg-s/m^5 (a constant)

x = displacement of the front edge of the bumper

v = velocity of the front edge of the bumper.

A vehicle of mass 2000 kg hits the bumper at a speed of 100 km/h. Write a MATLAB program to determine and plot the velocity of the vehicle as a function of x for $0 \le x \le 5$m.

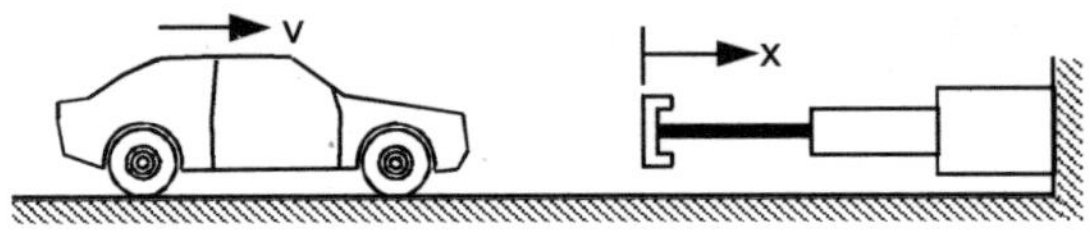

Fig. ED3.9

Solution: The deceleration of the car once it hits the bumper can be calculated from Newton's second law of motion.

$$ma = -Kv^3(x + 1)^3$$

which can be solved for the acceleration a as a function of v and x:

$$a = \frac{-Kv^3(x+1)^3}{m}$$

The velocity as a function of x can be calculated by substituting the acceleration in the equation:

$$vdv = adx$$

which gives:

$$\frac{dv}{dx} = \frac{-Kv^2(x+1)^3}{m}$$

The last equation is a first order ODE that needs to be solved for the interval $0 \le x \le 3$ with the initial condition:

$$v = 90 \text{ km/h at } x = 0.$$

Numerical solution of differential equation is shown in the following program written in a script file (*m*-file), which should be executed by typing the file name in command prompt.

MATLAB Program:

```
%%%%%%%%%%%%%%%%%
global k m
k=32;m=2000;v0=100;
xspan=[0:0.2:5];% a vector that specifies the interval of the solution
v0mps=v0*1000/3600; % changing velocity in m/s
[x v]=ode45('bumper',xspan,v0mps);% Solving the ODE
table=[x,v];
disp('disp velocity')
disp(table);
plot(x,v)
```

```
xlabel('x(m)');
ylabel('velocity(m/s)')
grid on;
%%%%%%%%%%%%
```

The function file with the differential equation named bumper.m is listed below:

```
function dvdx=bumper(x,v)
global k m
dvdx=-(k*v^2*(x+1)^3)/m;
```

The output of the above program is the vectors x and v as follows along with plot.

disp	velocity
0	27.7778
0.2000	24.8173
0.4000	21.1176
0.6000	17.1814
0.8000	13.5109
1.0000	10.4127
1.2000	7.9665
1.4000	6.0685
1.6000	4.6826
1.8000	3.6212
2.0000	2.8257
2.2000	2.2268
2.4000	1.7734
2.6000	1.4268
2.8000	1.1588
3.0000	0.9505
3.2000	0.7855
3.4000	0.6552
3.6000	0.5501
3.8000	0.4651
4.0000	0.3959
4.2000	0.3389
4.4000	0.2919
4.6000	0.2526
4.8000	0.2197
5.0000	0.1920

Output is given in Fig.ED3.9(*a*).

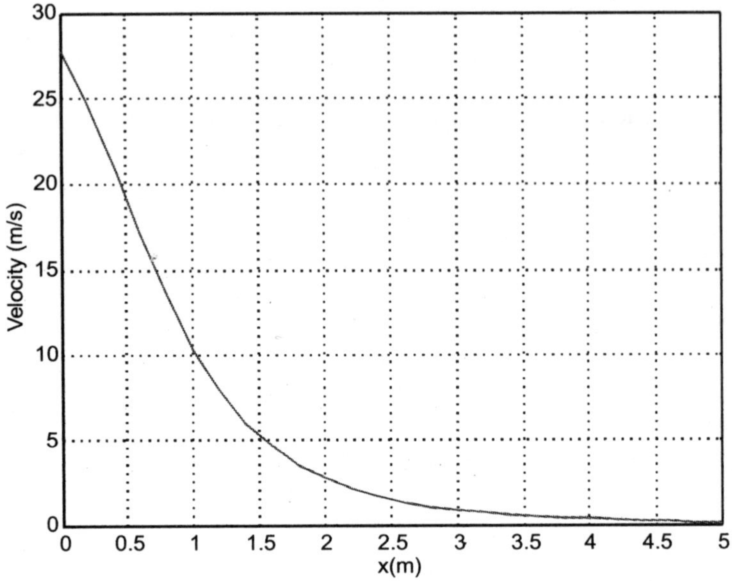

Fig. ED3.9(a)

ED 3.10: Figure ED3.10 shows a block of mass m which is at rest on the top of a cylindrical surface. When an initial velocity v_0 is given to the block towards right, it starts to slide on the cylindrical surface. Write a MATLAB program to plot the variation of velocity of block and α as a function of time at a friction coefficient $\mu_k = 0.2$. Also find the angle at which the block leaves the surface. Assume mass of the block = 0.60 kg and initial velocity of the block, $v_0 = 4$ m/s.

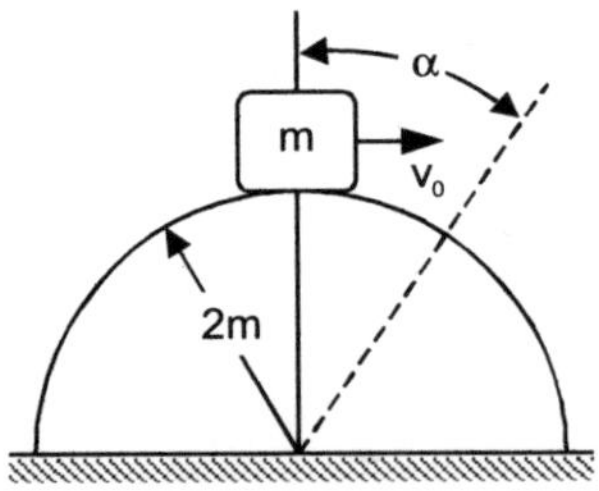

Fig. ED3.10

Solution: The freebody diagram of block when moved at an angle α is given in Fig. ED3.10(*a*).

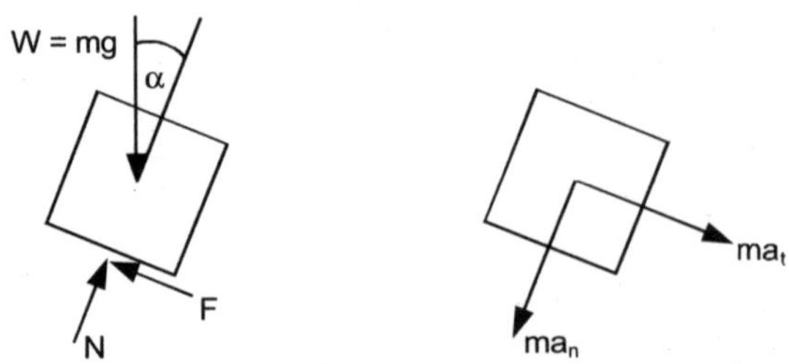

Fig. ED3.10(a) Freebody diagram

Summing forces in the normal direction gives $\Sigma F_n = ma_n$:

i.e.,
$$m\frac{v^2}{\rho} = W\cos\alpha - N$$

or
$$N = m\left(g\cos\alpha - \frac{v^2}{\rho}\right)$$

Hence frictional force $(F) = \mu_k N = \mu_k m\left(g\cos\alpha - \frac{v^2}{\rho}\right)$

Summing forces in the tangential direction gives

$$\Sigma F_t = ma_t:$$

or
$$ma_t = W\sin\alpha - F$$

or
$$a_t = g\sin\alpha - \frac{1}{m}F$$

Substituting for F in the above expression yields

$$a_t = g\sin\alpha - \frac{1}{m}\mu_k m\left(g\cos\alpha - \frac{v^2}{\rho}\right)$$

$$= g\sin\alpha - \mu_k g\cos\alpha + \frac{\mu_k v^2}{\rho}$$

Now
$$a_t = \frac{dv}{dt},$$

Thus
$$\frac{dv}{dt} = g(\sin\alpha - \mu_k\cos\alpha) + \mu_k\frac{v^2}{\rho} \qquad ...(1)$$

Also, we have that

$$v = r\dot{\alpha}$$

where $r = \rho$, (in polar coordinates)

or
$$\frac{d\alpha}{dt} = \frac{1}{\rho}v \qquad ...(2)$$

Thus, differential equations (1) and (2) define the motion of the block.

As the block leaves the surface, $N = 0$. Hence $\left(g\cos\alpha - \frac{v^2}{\rho}\right) = 0$, which defines the value of α at which the block leaves the surface.

The MATLAB program for this problem is given as follows:

```
%%%%%%%%%%%%%%%%%%%%%%%%%%%%%%%%%%%%%%%%%%%%%%%%%%
global g rho mk
vo=4; % initial velocity
g=9.81;  % acceleration due to gravity
```

```
rho=2.0;% radius of curvature
mk=0.2;% coefficient of kinetic friction
tspan=[0:0.01:0.35]; % initial and final times
x0=[vo,0]'; % Initial velocity and angle
[t,x]=ode45('f3',tspan,x0); % solving two differential eqs.
figure(1);
subplot(2,1,1)
plot(t,x(:,1),'-o');
xlabel('time (sec');ylabel('velocity in m/s)');
title('Coefficient friction between surface and block is 0.2')
grid on;
subplot(2,1,2)
plot(t,x(:,2)*180/pi,'-p');
xlabel('time in sec)');ylabel('angle in degrees)';
grid on;
eq1=g*cos(x(:,2))-x(:,1).^2/rho; % Expression for Normal reaction
[y,k]=min(abs(eq1)); % Minimum value and corresponding index of this function
theta=x(k,2)*180/pi; % Finding theta
fprintf('Minimum angle in degrees it leaves the surface at friction mu=%3.2f
is %4.2f\n',mk, theta);
%%%%%%%%%%%%%%%%%%%%%%%%%%%%%%%%%%%%%%%%%%%%
```

Here the subfunction $f3(\,)$ is defined in separate *m* file as follows:

```
%%%%%%%%%%%%%%%%%%%%%%%%%%%%%%%%
function vdot = f3(t,x)
global g rho mk;
vdot=[g*(sin(x(2))-mk*cos(x(2)))+mk*x(1)^2/rho;x(1)/rho];
%%%%%%%%%%%%%%%%%%%%%%%%%%%%%%%%
```

Ouput is shown in Fig. ED3.10 (*b*).

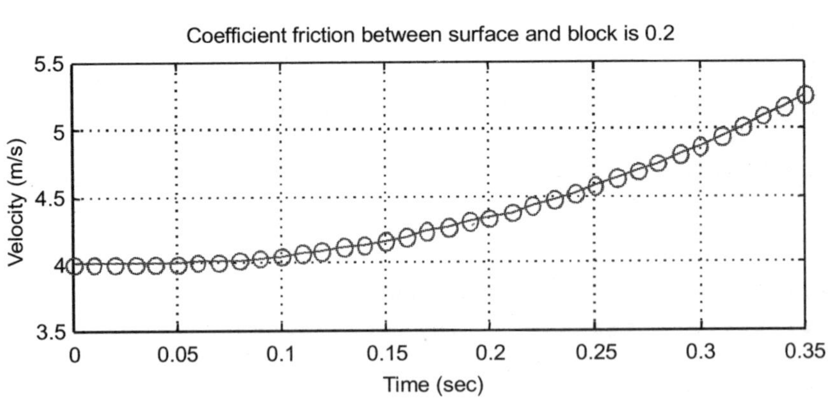

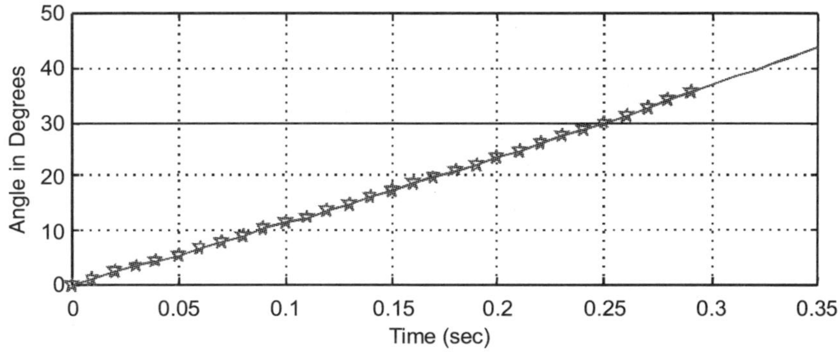

Fig. ED3.10 (b)

>> Minimum angle in degrees it leaves the surface at friction $\mu = 0.2$ is 21.03.

ED 3.11: The acceleration due to gravity due to earth at any height h above the surface of the earth is given by

$$g = \frac{-GM}{(R+h)^2} \text{ m/s}^2$$

where G = the gravitational constant (6.67×10^{-11} Nm^2/kg^2)

M = mass of the earth (5.99×10^{24} kg)

R = mean radius of the earth (6372 km)

h = height above the surface of the earth (m).

Write a MATLAB program to calculate and plot the acceleration due to earth's gravity for $0 \le h \le 45{,}000$ km in 5000 km increments.

Solution: This problem is an application of central force motion and space mechanics. MATLAB code presented with explanations is given as follows:

```
%%%%%%%%%%%%%%%%%%%%%%%%%%%%%%%%
G=6.67E-11; % Gravitational constant
m_earth=5.99e24; % Mass of earth (kg)
r_earth=6372.e3; % Radius of the earth(m)

disp('This program displays the acceleration due to gravity');
disp('as a function of height above the Earth's surface:');
fprintf('\n\n Height Acceleration\n');
fprintf(   ' (km)   (m/sec**2)   \n');
fprintf(   ' =======        ============\n');

% Now calculate values
jj=0;
for ii=0:5000000:45000000
 % Increment counter
  jj=jj+1;
 %Get height
  height(jj)=ii;
 % Calculate acceleration
  grav(jj)=-G*m_earth/(r_earth+height(jj))^2;
```

```
% Write out results
  fprintf('%7d    %8.3f\n',height(jj)/1000,grav(jj));
   end
% Now plot the acceleration due to gravity vs height
figure(1);
plot(height/1000,grav,'LineWidth',2);
title('\bfAcceleration due to gravity vs height');
xlabel('\bfHeight above Earth's surface (km)');
ylabel('\bfAcceleration (m/s^{2})');
%%%%%%%%%%%%%%%%%%%%%%%%%%%%%%%%
```

Program output:

This program displays the acceleration due to gravity as a function of height above the Earth's surface:

Height (km)	Acceleration (m/sec**2)
0	–9.840
5000	–3.089
10000	–1.491
15000	–0.875
20000	–0.574
25000	–0.406
30000	–0.302
35000	–0.233
40000	–0.186
45000	–0.151

Figure ED3.11 (*a*) shows the plot of acceleration due to gravity as a function of height.

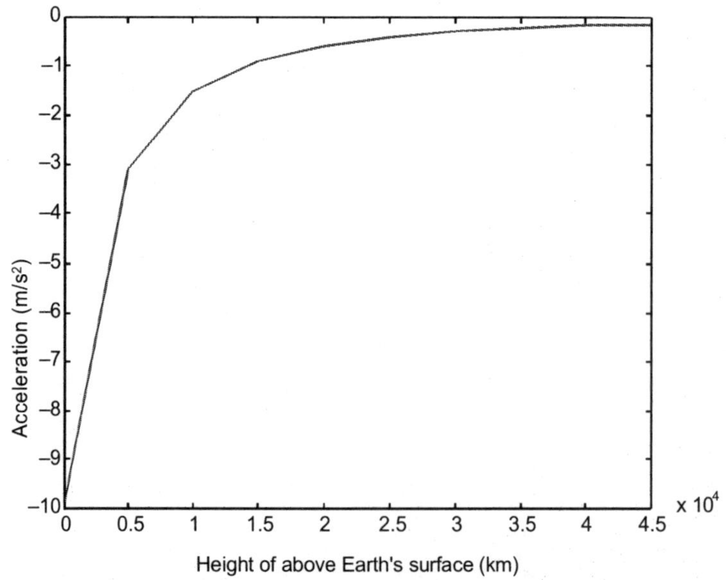

Fig. ED3.11(a)

ED3.12: A 3 kg block is subjected to two forces as shown in Fig.ED3.12. If block starts from rest, determine the distance it has moved when it attains a velocity of 10 m/s. Plot its distance as a function of coefficient of kinetic-friction between the block and floor.

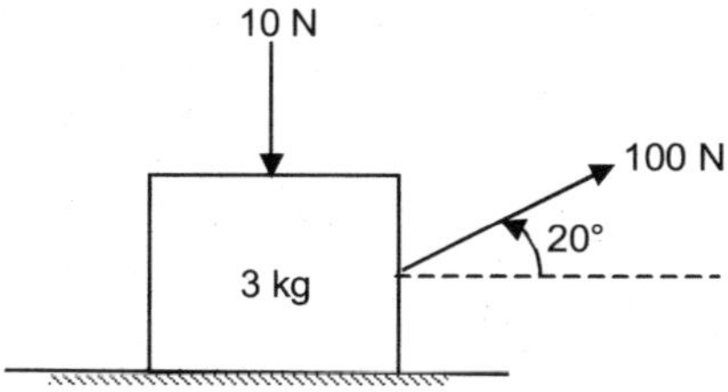

Fig. ED3.12

Solution: This is an application of work-energy principle on the motion of a particle.

The work-energy principle is

$$T_1 + \Sigma U = T_2,$$

where T_1 = Initial kinetic energy, and

T_2 = Final kinetic energy of the particle.

Here, $T_1 = 0$ and $T_2 = \frac{1}{2}mv^2$.

The forces doing the work on the particle are horizontal component of external force 100 N and friction acting on the floor. Work-done by the forces are

$$\Sigma U = (100 \cos 20° - \mu N) \times s,$$

where $N = (10 + W - 100 \sin 10°)$,

W = 3g, the weight of block.

Thus = ΣU, relates velocity and kinetic friction μ.

The program for this problem is written as follows:

```
%%%%%%%%%%%%%%%%%%%%%%%
% DEFINE ALL VARIABLES
m=3;
v=10;
g=9.81;
F1=100;
F2=10;
T=0.5*m*v^2;
mu=0:0.05:0.5;
s=150./(F1*cos(20*pi/180)-mu*(F2+m*g-F1*sin(20*pi/180)));
plot(mu,s,'-p')
xlabel('Kinetic friction');
ylabel('Distance in m)');
grid on;
%%%%%%%%%%%%%%%%%%%%%%
```

Output is given in Fig.ED3.12 (*a*).

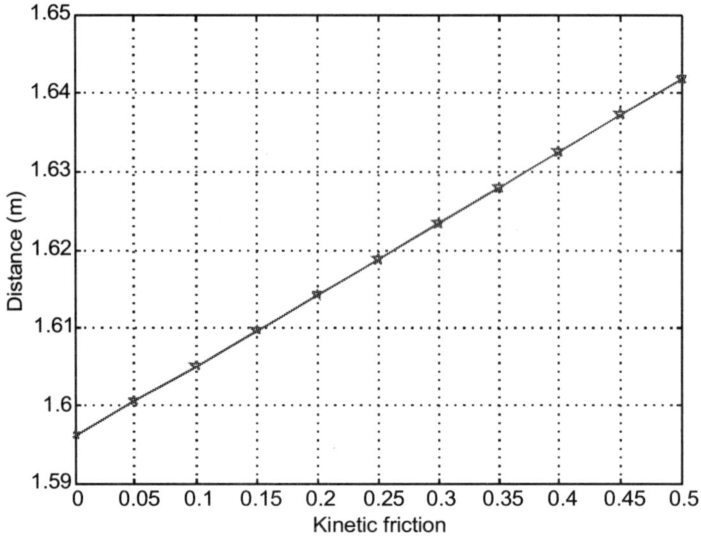

Fig. ED3.12 (a)

ED 3.13: A 5 kg block is attached to a cable and to a spring as shown in Fig.ED3.13.

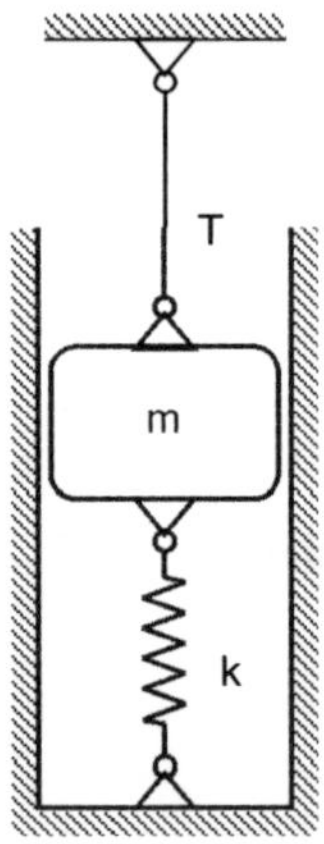

Fig. ED3.13

The constant of the spring is $k = 3$ kN/m and the tension in the cable is 30 N. When the cable is cut, (*a*) derive an expression for the velocity of the block as a function of its displacement x, (*b*) determine the maximum displacement x_m and the maximum speed v_m, (*c*) plot the speed of the block as a function of x for $0 \le x \le x_m$.

Solution: Free body diagram of the block before and after the cable is cut is shown in Fig. ED3.13 (*a*). For the static case we have entire forces are in equilibrium.

$\therefore \quad T + R - W = 0$ with $T = 30$ N, $W = mg = 50$ N from which $R = 20$ N

But,

$$R = k\delta_{st} \text{ or } \delta_{st} = \frac{R}{k} \text{ which is initial tension in spring.}$$

$$\delta_{st} = \frac{20}{3000} = 0.006667 \text{ m}$$

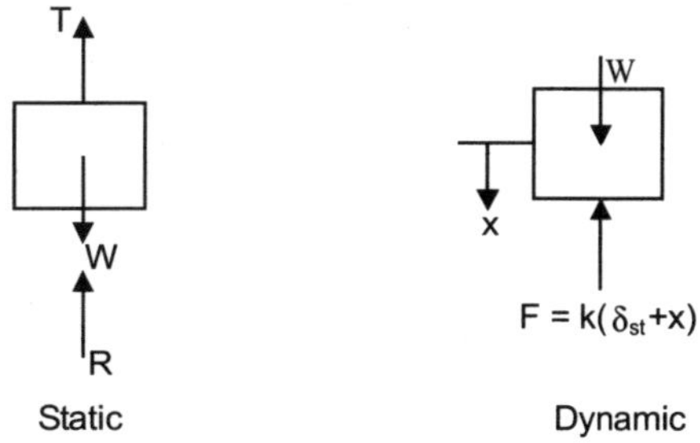

Fig. ED3.13 (a)

Using the principle of work and energy we have

$$T_1 + U_{1\to2} = T_2$$

or

$$\int_0^x [W - k(\delta_{st} + x)]dx = \frac{1}{2}mv^2$$

Substituting yields

$$\int_0^x [50 - 3000(0.006667 + x]dx = 2.5v^2$$

$$\Rightarrow \quad 50x - 20x - 1500x^2 = 2.5v^2$$

$$v^2 = 12x - 600x^2$$

At maximum displacement, the velocity is zero.

$$\therefore \quad 12x - 600x^2 = 0 \text{ or } x = 0.02 \text{ m.}$$

The MATLAB program for plotting can be written as follows:

```
%%%%%%%%%%%%%%%%%%%%%%%%%%%
xmax=0.02;
x=[0:.001:xmax];
v=sqrt(12.*x-600.*x.^2);% Expression for velocity for given values
[vmax,i]=max(v); % finding minimum value of velocity and corresponding index
fprintf('The maximum velocity is %5.4fm/s and the maximum displacement is
%5.4fm\n',vmax,xmax);
figure(1)
plot(x,v,'-o')
xlabel('x,(m)')
ylabel('Velocity(m/s)')
grid on
%%%%%%%%%%%%%%%%%%%%%%%%%%%
```

Output is shown in Fig. ED 3.13(*b*).

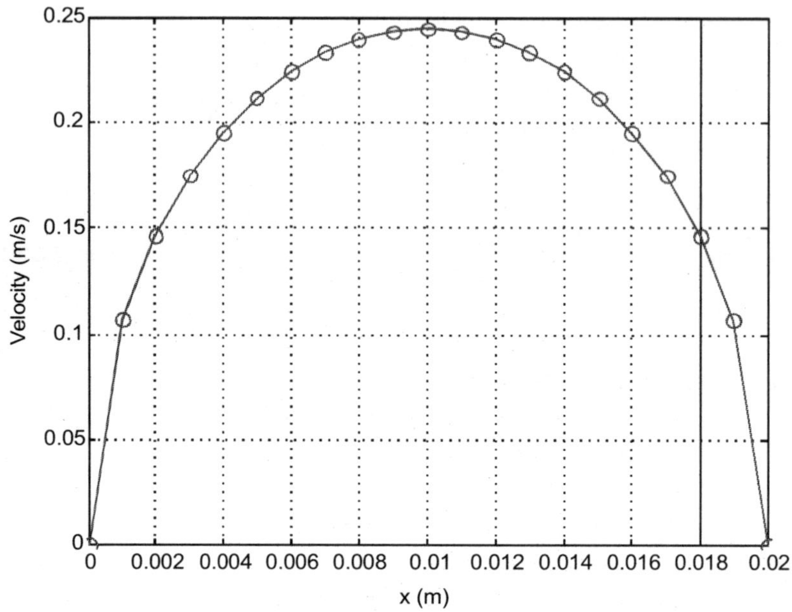

Fig. ED3.13(b)

>> The maximum velocity is 0.2449m/s and the maximum displacement is 0.0200m.

ED3.14: A block of 0.8 kg mass moves within the smooth vertical slot as shown in Fig.ED3.14. If it starts from rest when the attached spring is in unstretched position at *A*, plot a graph of velocity of block as a function of distanced moved by the block. Assume $F = 100$ N, $k = 100$ N/m, $0 \le s \le 0.4$.

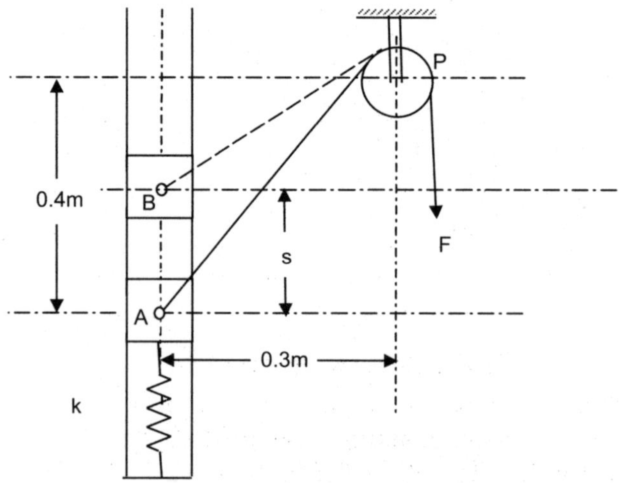

Fig. ED3.14

Solution: This is an application of work-energy principle. As surfaces are smooth, no frictional force is possible. Thus there are no non-conservative forces in the system. So work-energy principle reduces to principle of conservation of energy, mathematically

$$T_1+V_1=T_2+V_2$$

where initial state is rest. Thus, $T_1 = V_1 = 0$. Here, initially if spring is compressed then potential energy

$$V_1=\frac{1}{2}k\partial_{ini}^2$$

where ∂_{ini} is initial compression of spring.

Final energies are written as

$$T_2=\frac{1}{2}mv^2,$$

where v is velocity of block

and V_2 = elastic energy + gravitational energy + work due to external force F.

Thus $$V_2=\int_{x=0}^{\partial} kxdx+\text{mg}\times s-F\times\Delta f$$

where $\Delta f = AP - BP$ = stretch in string length, which can be expressed in terms of s.

From geometry,

$$AP=\sqrt{0.4^2+0.3^2}=0.5\text{ m}$$

and $$BP=\sqrt{(0.4-s)^2+0.3^2}.$$

Now the principle of conservation energy can be applied to relate velocity v and distance s.

i.e., $V_2+T_2=0$

MATLAB program for this problem is generalized as follows:

```
%%%%%%%%%%%%%%%%%%%%
% Set of default values
g=9.81;
m=0.8;
k=100;
F=100;
W=m*g;% weight of block
s=0:0.05:0.4;
df= 0.5-sqrt((0.4-s).^2+0.3^2); % distance moved by string
V2= (W.*s+0.5*k*s.^2-F.*df); % potential energy of block
v=sqrt((2/m).*(-V2)); % Velocity of block
plot(s,v);
xlabel('Distance moved along slot in m)');
ylabel('Velocity of block in m/s)')
grid on;
%%%%%%%%%%%%%%%%%%%%%%%%%%%%%%
```

The output is shown in Fig. ED3.14(a).

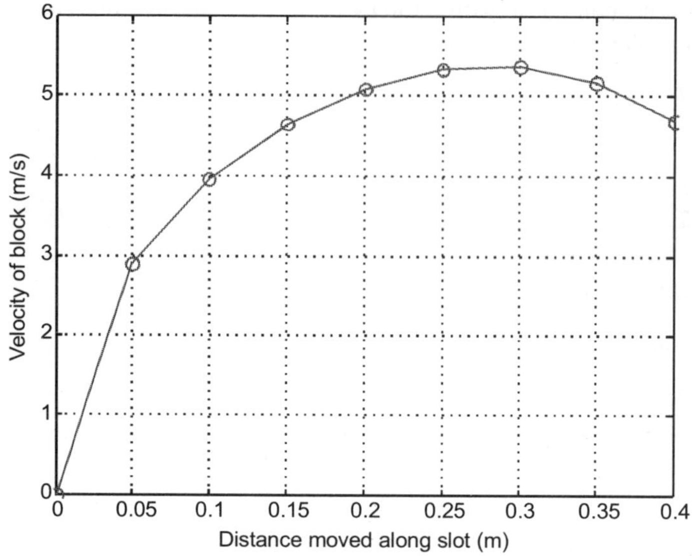

Fig. ED3.14(a) Output

ED 3.15: Figure ED3.15 shows a block B of mass m_B starts from rest and slides down an inclined plane of a wedge of mass m_A which is supported by a horizontal surface. (a) Obtain an expression for the speed of block B relative to wedge, A. (b) The speed of wedge A, (c) Write a MATLAB program to plot the speed of B relative to A and the speed of A as function of s, where s is the distance travelled by the block B down the surface of the wedge for $0 \le s \le 1.0$ m. Neglect friction between all the surfaces. Given: $m_B = 10$ kg and $m_A = 16$ kg.

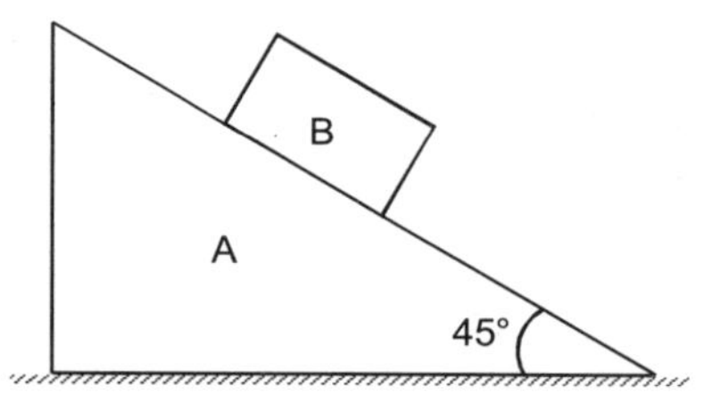

Fig. ED3.15

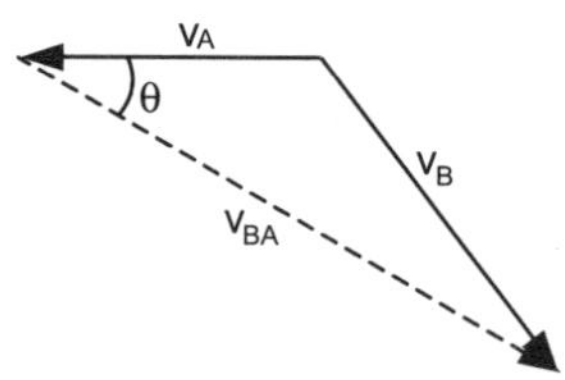

Fig. ED3.15(a)

Solution : Drawing velocity triangle as shown in Fig. ED3.15 (a).

Applying principle of conservation of momentum to the particles A and B

$$0 = m_B\, v_{BA} \cos\theta - (m_A + m_B)\, v_A$$

Therefore, speed of block B relative to wedge A:

$$v_{BA} = \frac{(m_A + m_B)\, v_A}{m_B \cos\theta}$$

Applying conservation of energy rule(as there is no friction)

$$m_B gh = \frac{1}{2}m_A v_A^2 + \frac{1}{2}m_B v_B^2$$

where $v_B^2 = v_A^2 + v_{BA}^2 - 2v_A\, v_{BA} \cos\theta$ (cosine rule from triangle Fig.ED3.15(a))

$$\Rightarrow \qquad 10 \times 9.81 \times s \cos\theta = \frac{1}{2}m_A v_A^2 + \frac{1}{2}m_B v_B^2$$

Here v_B and v_{BA} can be replaced in terms of v_A and plot of v_A and s can be drawn.

i.e., $$m_B gs \cos\theta = \frac{1}{2}m_A v_A^2 + \frac{1}{2}m_B v_A^2\left(1 + \frac{(m_A+m_B)^2}{m_B^2\cos^2\theta} - 2\frac{(m_A+m_B)}{m_B}\right)$$

$$= 0.5 \times \left[m_A + m_B\left(1 + \left\{\frac{m_A+m_B}{m_B\cos\theta}\right\}^2 - 2\left\{\frac{m_A+m_B}{m_B\cos\theta}\right\}\cos\theta\right)\right]v_A^2$$

Now speed of wedge v_A and relative velocity v_{BA} are plotted as a function of s. Complete MATLAB program is given below:

MATLAB Program:

```
%%%%%%%%%%%%%%%%%%%%%%%%%%%%%%%%%%%%%%%%
%DEFAULT DATA GIVEN
g=9.81;
ma=16;
mb=10;
theta=45*pi/180; % ANGLE OF WEDGE
mul=(ma+mb)/(mb*cos(theta)); % MULTIPLICATION FACTOR
den=0.5*ma+0.5*mb*(1+mul^2-2*mul*cos(theta)));
s=[0:0.02:1];
pe=mb*s*cos(theta)*g; % POTENTITAL ENERGY
va=sqrt(pe./den); % EXPRESSION FOR VELOCITY OF A
vba=mul*va; % RELATIVE VELOCITY
plot(s,va,'-*',s,vba,'-p')
xlabel('s(m)')
ylabel('Velocity (m/s)')
legend('Velocity of A','Velocity of B relative to A',2);
grid on
%%%%%%%%%%%%%%%%%%%%%%%%%%%%%%%%%%%%%%%%
```

Output is shown in Fig. ED3.15(*b*).

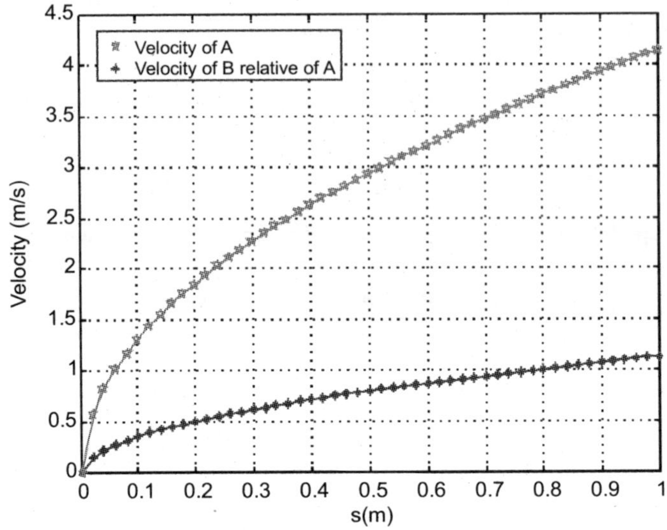

Fig. ED3.15 (b)

ED 3.16: A 10 kg block is held at rest on the smooth inclined plane by the stop block at *A* (see Fig.ED3.16). If a 10 gm bullet travelling at velocity *v* strikes and embedded in the block, plot the distance moved-up by the block along the plane as a function of velocity before it comes to momentary stop.

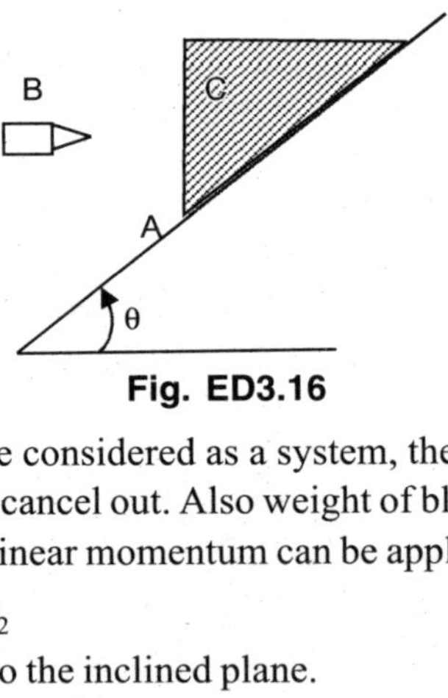

Fig. ED3.16

Solution: If block (*C*) and bullet (*B*) are considered as a system, the impulsive force *F*, caused by impact is internal to the system. Therefore, it will cancel out. Also weight of block and bullet are non-impulsive forces. Hence the principle of conservation of linear momentum can be applicable.

$$m_B \times v_{Bx} = (m_B + m_C) \times v_2$$

where *x* is in upward direction parallel to the inclined plane.

Here $v_{Bx} = v \cos \theta$, the component of bullet velocity along the positive *x*-direction. Here the datum is set as block's initial position as shown in Fig. ED3.16 (*a*).

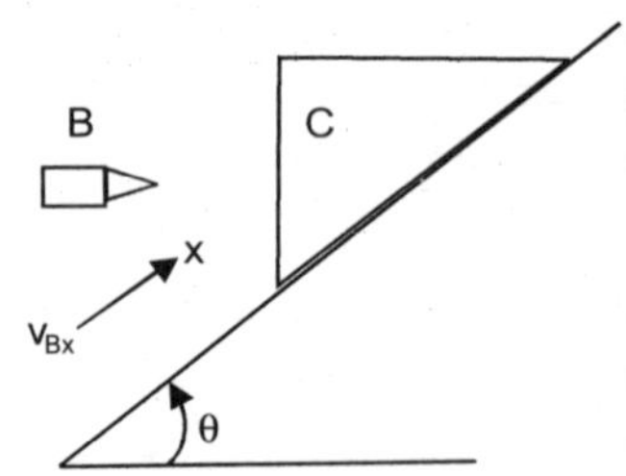

Fig. ED3.16(a)

To find the distance moved by the block, apply the work-energy principle.

i.e., $$\Sigma U = -s \times (m_B + m_C) g \sin \alpha = T_2 - T_1 = -\frac{1}{2}(m_B + m_C) v_2^2 \ .$$

Thus *s* can be plotted as a function of velocity.

MATLAB program for this problem is given below:

```
%%%%%%%%%%%%%%%%%%%%%%%%%%
mB=10e-3; % Mass of bullet
mC=10; % Mass of block
angle=[30*pi/180;45*pi/180;60*pi/180];
g=9.81;
v=50:50:500; % Bullet velocity variation
for k=1:length(angle)
  v2=mB.*v*cos(angle(k))/(mB+mC);
  dT=0.5*(mB+mC)*v2.^2;
  s(k,:)=dT./((mB+mC)*g*sin(angle(k)));
end
plot(v,s(1,:),'-o',v,s(2,:),'-p',v,s(3,:),'-*');
xlabel('Velocity of bullet (m/s)');
ylabel('Distance moved over the plane in m)');
grid on;
legend('angle=30deg','angle=45deg','angle=60deg')
%%%%%%%%%%%%%%%%%%%%%%%%%%
```

The output of this program is shown in Fig. ED3.16 (*b*).

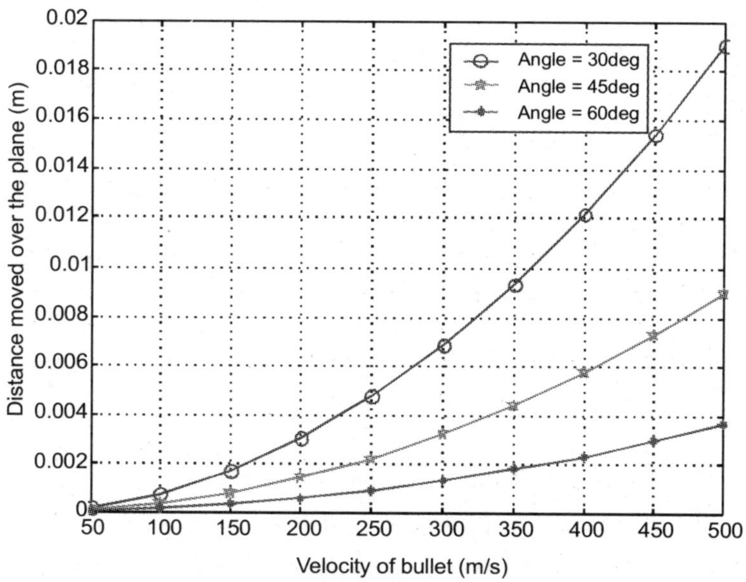

Fig. ED3.16 (b) Output of MATLAB

ED 3.17: A system consists of *n* particles A_i of mass m_i and coordinates x_i, y_i and z_i having velocity components $(v_x)_i$, $(v_y)_i$ and $(v_z)_i$. Derive expression for components of angular momentum about origin *O* of coordinates. Use MATLAB to solve the following system:

Particle	*Mass (kg)*	*Position (m)*	*Velocity(m/s)*
A_1	3	$3\hat{j}$	$4\hat{i}+2\hat{j}+2\hat{k}$
A_2	2	$1.2\hat{i}+2.4\hat{j}+3\hat{k}$	$4\hat{i}+3\hat{j}$
A_3	4	$3.6\hat{i}$	$-24\hat{i}+4\hat{j}+2\hat{k}$

Solution: Angular momentum (H_O) about O of system of n-particles is defined as:

$$(H_O) = \sum_{i=1}^{n} r_i \times m_i v_i \text{. This is a vector (cross) product.}$$

$$= r_1 \times m_1 v_1 + r_2 \times m_2 v_2 + r_3 \times m_3 v_3$$

$$= \left[m_1 \begin{vmatrix} r_{1y} & r_{1z} \\ v_{1y} & v_{1z} \end{vmatrix} + m_2 \begin{vmatrix} r_{2y} & r_{2z} \\ v_{2y} & v_{2z} \end{vmatrix} + \cdots\cdots\cdots \right]\hat{i} - \left[m_1 \begin{vmatrix} r_{1x} & r_{1z} \\ v_{1x} & v_{1z} \end{vmatrix} + m_2 \begin{vmatrix} r_{2x} & r_{2z} \\ v_{2x} & v_{2z} \end{vmatrix} + \cdots\cdots\cdots \right]\hat{j}$$

$$+ \left[m_1 \begin{vmatrix} r_{1x} & r_{1y} \\ v_{1x} & v_{1y} \end{vmatrix} + m_2 \begin{vmatrix} r_{2x} & r_{2y} \\ v_{2x} & v_{2y} \end{vmatrix} + \cdots\cdots\cdots \right]\hat{k}$$

MATLAB program for this problem is written as follows:

```
%%%%%%%%%%%%%%%%%%%%%%%%%%%%%
n=3; % NUMBER OF PARTICLES
m=[3 2 4]; % MASSES OF THREE PARTICLES
Hx=0;Hy=0;Hz=0; % INITIAL ANGULAR MOMENTA
r=[0 3 0;1.2 2.4 3;3.6 0 0]; % POSITION VECTORS OF PARTICLES
% HERE COLUMNS SHOW THE X, Y AND Z coordinate & rows the points
v=[4 2 2;4 3 0;-2 4 2]; % VELOCITY VECTORS
or k=1:n
 P=[r(k,2) r(k,3);v(k,2) v(k,3)];
 Q=[r(k,1) r(k,3);v(k,1) v(k,3)];
 R=[r(k,1) r(k,2);v(k,1) v(k,2)];
 Hx=Hx+m(k)*det(P);
 Hy=Hy-m(k)*det(Q);
 Hz=Hz+m(k)*det(R);
end
disp('Total angular momentum is');
fprintf('%5.2f i+%5.2f j+%5.2f k\n' ,Hx,Hy,Hz);
%%%%%%%%%%%%%%%%%%%%%%%%%%%%%%
```

Output of the program:
Total angular momentum is

$0.00\, i$ +$-4.80\, j$+$9.60\, k$

ED 3.18: A rocket has a mass 960 kg including 800 kg of fuel, which is consumed at the rate of 10 kg/s and ejected with a relative velocity of 3600 m/s. Plot the magnitude of velocity of rocket as a function of time, neglect air resistance.

Solution: This an example of systems of particles with variable mass flow. Rocket is an example of a system loosing the mass. Only external force acting on the system is its weight $W = mg$, but here m reduces as a function of time.
Figure ED3.18 shows the initial and final states of the rocket in a short time dt.

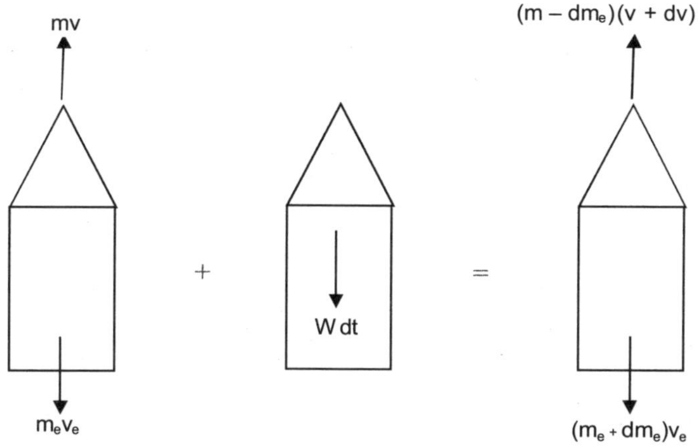

Fig. ED3.18 Initial momentum+ Impulse = Final momentum

Applying principle of impulse and momentum to this system gives:

$$mv - m_e v_e - mg\,dt = (m - dm_e)(v + dv) - (m_e + dm_e)v_e$$

or

$$-mg = m\,(dv/dt) - u\,(dm_e/dt) \qquad ...(1)$$

where $u = v + v_e$ is relative velocity of exhaust.

Substituting the given values:

$$-mg = m\,(dv/dt) - 3600\,(10)$$

Here m = mass of rocket at an instant of time t during flight

$$= m_o - (dm/dt)t = (960 - 10t) \qquad ...(2)$$

Substituting (2) in (1), we obtain

$$-(960 - 10t)g = (960 - 10t)(dv/dt) - 36000$$

At $t = 0$, $v = 0$. This gives initial conditions for the problem.

So it requires to solve the differential equation:

$$(dv/dt) = \frac{36000}{960 - 10t} - g$$

Time required to consume all the fuel is given by $(960 - 10t) = 0$ or $t = 96$ seconds.

MATLAB program for ploting the variation of velocity of rocket as a function of time is given below:

```
v0=0;
    tspan=[0:5:95];% a vector that specifies the interval of the solution
    t v]=ode45('rock',tspan,v0);% Solving the ODE
    plot(t,v,'-p')
    xlabel('t(sec)');
    ylabel('velocity(m/s)')
    grid on;
    %%%%%%%%%%%%
```

The function file with the differential equation named rock.m is listed below:

```
function dvdt=rock(t,v)
m0=960;g=9.81;mr=10; u=3600;
dvdt=-g+mr*u/(m0-mr*t);
```

The output is shown in Fig. ED3.18 (*a*), below:

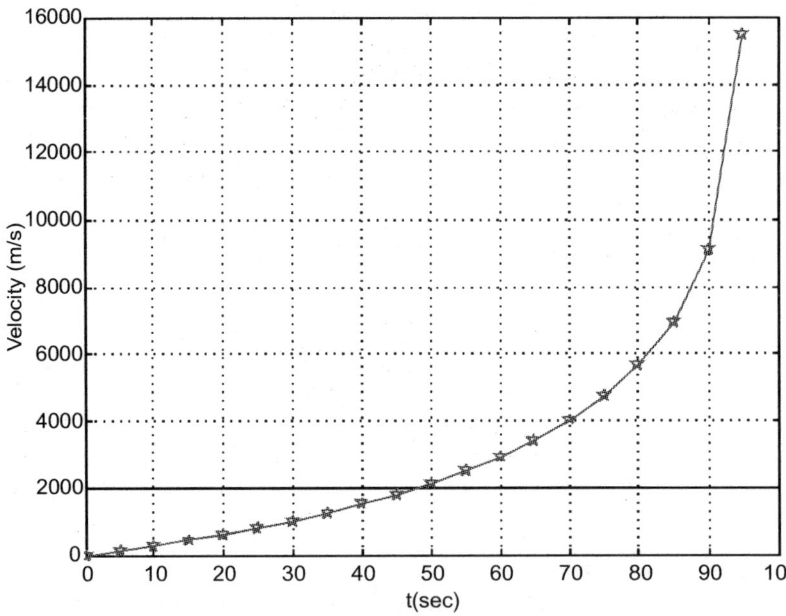

Fig. ED3.18(a) MATLAB output

Plane Kinematics of Rigid bodies

ED 3.19: Motion of an oscillating flywheel is defined by the relation $\theta = \theta_0 \, e^{-3\pi t} \cos 4\pi t$ with $\theta_0 = 0.5$ radian. Plot the angular velocity and acceleration of flywheel as a function of time. Take $0 \le t \le 0.5$s.

Solution: As θ is defined explicitly as a function of time it is easy to find ω and α from their definitions. The derivatives are evaluated in the worksheet below.

```
%%%%%%%%%%%%%%%%
syms t
theta=0.5*exp(-3*pi*t)*cos(4*pi*t);
omega=diff(theta,t);
alpha=diff(omega,t);
th=vectorize(theta)
om=vectorize(omega)
al=vectorize(alpha)
%%%%%%%%%%%%%%%%
```

Output of this program will be as follows:

```
th =
1./2.*exp(-3.*pi.*t).*cos(4.*pi.*t)

om =
-3./2.*pi.*exp(-3.*pi.*t).*cos(4.*pi.*t)-2.*exp(-3.*pi.*t).*sin(4.*pi.*t).*pi
al =
-7./2.*pi.^2.*exp(-3.*pi.*t).*cos(4.*pi.*t)+12.*pi.^2.*exp(-3.*pi.*t).*sin(4.*pi.*t)
```

These results are used further for plotting. They can be pasted in the final program.

MATLAB code of plotting is written as follows:

```
%%%%%%%%%%%%%%%%%%%%
t=0:0.05:0.5;
th =1./2.*exp(-3.*pi.*t).*cos(4.*pi.*t);
om=-3./2.*pi.*exp(-3.*pi.*t).*cos(4.*pi.*t)-2.*exp(-3.*pi.*t).*sin(4.*pi.
*t).*pi;
al =-7./2.*pi.^2.*exp(-3.*pi.*t).*cos(4.*pi.*t)+12.*pi.^2.*exp(-3.*pi.*t).
*sin(4.*pi.*t);
subplot(3,1,1);
plot(t,th,'-p');
ylabel('Theta (rad)');
grid on;
subplot(3,1,2);
plot(t,om,'-*');
ylabel('Omega (rad/s)');
grid on;
subplot(3,1,3);
plot(t,al,'-o');
ylabel('Alpha (rad/s^2)');
grid on;
xlabel('Time (s)');
%%%%%%%%%%%%%%%%%%%%%
```

Output is shown in Fig. ED3.19.

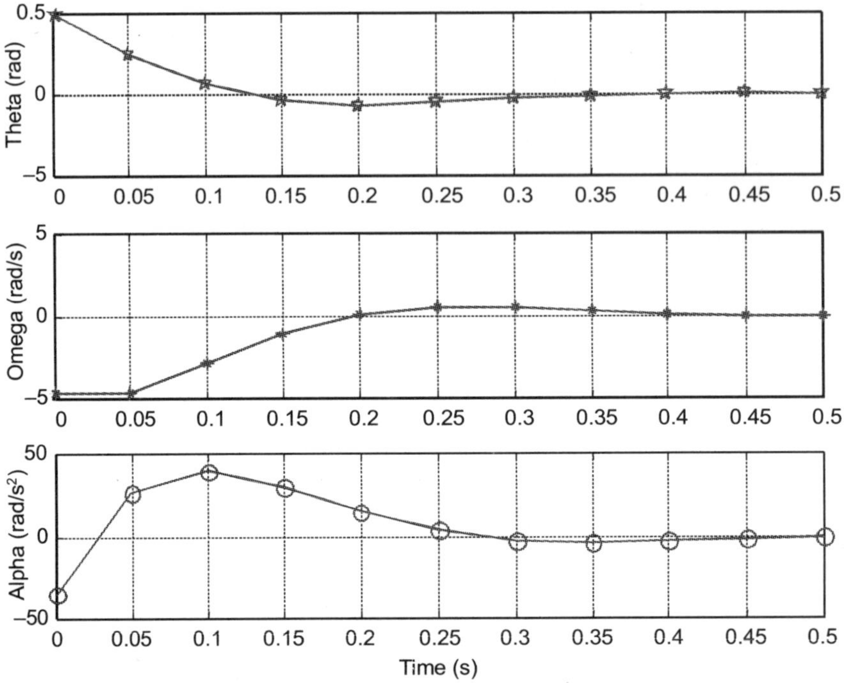

Fig. ED3.19 MATLAB output

ED 3.20: Figure ED3.20 shows the slider crank mechanism. Write a MATLAB program that calculates and plots the position, velocity and acceleration of the piston for one full revolution of the crank. Assume that the crank is rotating at a constant speed of 550 rpm. Given radius of crank = 125 mm and radius of crank shaft = 250 mm.

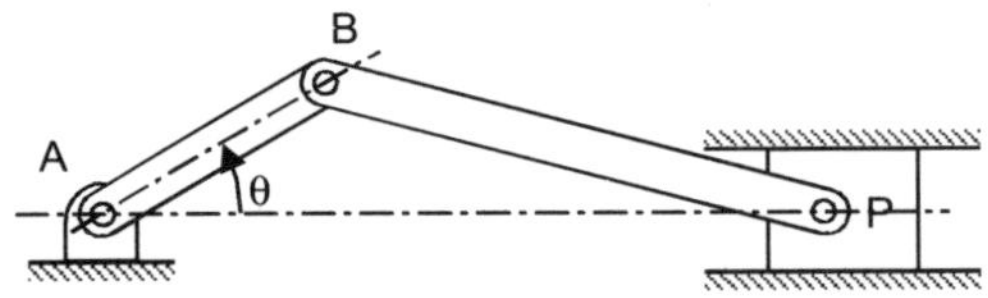

Fig. ED3.20

Solution: This problem can be done with either absolute motion analysis or relative motion analysis. Let us do it with absolute motion analysis, where the coordinates of points *B* and *P* are defined first with common origin *A* and then differentiated with respect to time to obtain velocities. Figure ED3.20 (*a*) shows the line diagram of the mechanism.

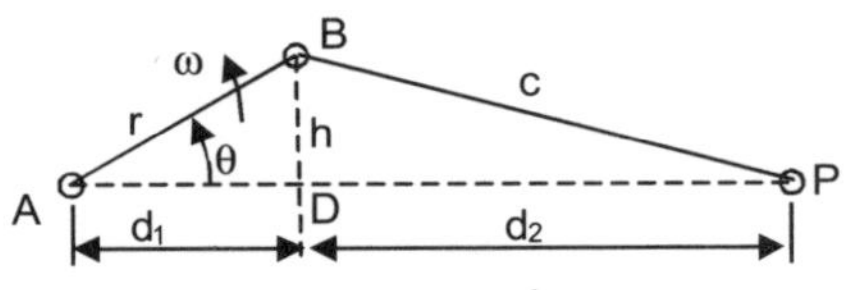

Fig. ED3.20(a)

The crank is rotating with a constant angular velocity $\omega = \dot{\theta}$.

When $t = 0$, $\theta = 0^{\circ}$.

At time t, the angle θ is given by

$$\theta = \omega t = \dot{\theta} t \text{, and}$$

that $\ddot{\theta} = 0$ at all times.

The distances d_1 and h are given by

$$d_1 = r\cos\theta \text{ and } h = r\sin\theta$$

Knowing h, the distance d_2 is obtained as:

$$d_2 = \left(c^2 - h^2\right)^{1/2} = \left(c^2 - r^2\sin^2\theta\right)^{1/2}$$

The position x of the piston P with respect to A (common origin) is given by

$$x = d_1 + d_2 = r\cos\theta + \left(c^2 - r^2\sin^2\theta\right)^{1/2}$$

The velocity of the piston is given by

$$\dot{x} = -r\dot{\theta}\sin\theta - \frac{r^2}{2\left(c^2 - r^2\sin^2\theta\right)^{1/2}}$$

The acceleration of the piston is given by

$$\ddot{x} = -r\dot{\theta}^2\cos\theta - \frac{4r^2\dot{\theta}^2\cos 2\theta\left(c^2 - r^2\sin^2\theta\right) + \left(r^2\dot{\theta}\sin 2\theta\right)^2}{4\left(c^2 - r^2\sin\theta\right)^{3/2}}$$

Complete MATLAB program for this problem is given below:

% MATLAB Solution:

```
% Define TD, r, and c%%%%%%%%%%%%%%%%%%%%%%%%%%%%%%%%%%%
N=550; % Speed in rpm
TD=N*2*pi/60; % Speed in radians/sec
tf=2*pi/TD;
r=0.125; % radius of crank in meters
c=0.250; % length of connecting rod in meters
t=linspace(0,tf,200); % Create a vector with 200 elements
TH=TD*t; % Compute Theta for each t
d2s=c^2-r^2*sin(TH).^2;    % d2 squared
x=r*cos(TH)+sqrt(d2s);     % Calculate x for each Theta
xd=-r*TD*sin(TH)-(r^2*TD*sin(2*TH))./(2*sqrt(d2s)); % Velocity
% Acceleration
xdd=-r*TD^2*cos(TH)-(4*r^2*TD^2*cos(2*TH).*d2s+(r^2*sin(2*TH)*TD).^2)./
(4*d2s.^(3/2));
subplot(3,1,1)
plot(t, x)% Plot x versus t
grid
```

```
xlabel('Time (s)')
ylabel(' Position (m)')
subplot(3,1,2)
plot(t, xd) % Plot Velocity vs. t
grid
xlabel(' Time (s)')
ylabel('Velocity (m/s)')
subplot(3,1,3)
plot(t, xdd) % Plot Acceleration Vs. t
grid
xlabel('Time (s)')
ylabel('Acceleration (m/s^2)')
%%%%%%%%%%%%%%%%%%%%%%%%%%%%%%%%%%%%%%%%%%%%%%%%%%%%%%%%%%%
```

Output of the program is shown in Fig. ED3.20(*b*).

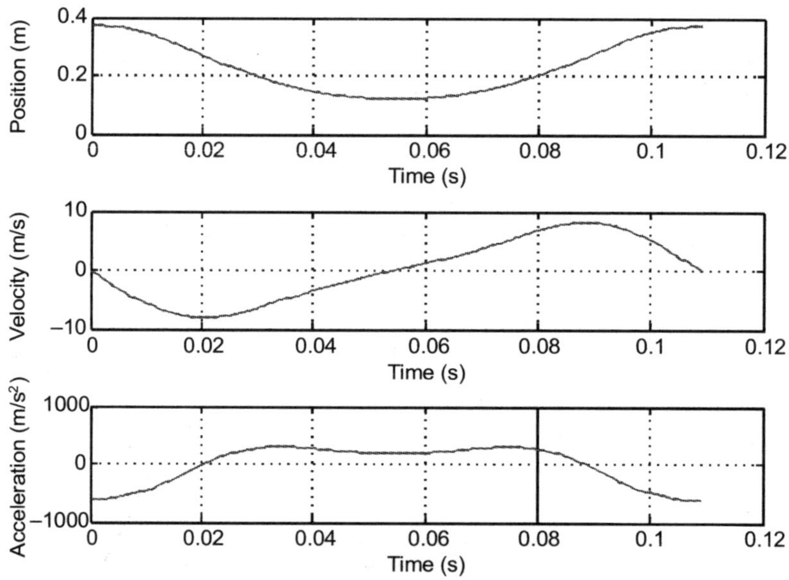

Fig. ED 3.20(b) MATLAB output

ED 3.21: Figure ED3.21 shows an engine system where the crank *AB* rotates with a constant angular velocity ω_{AB} clockwise. Write a MATLAB program to determine the plot for values of θ from 0 to 180° (*a*) the angular velocity and angular acceleration of connecting rod *BD*, (*b*) the velocity and acceleration of piston *P*. The length of the connecting rod, $\ell = 10$ cm, crank length, $AB = 3.5$ cm and $\omega_{AB} = 1000$rpm.

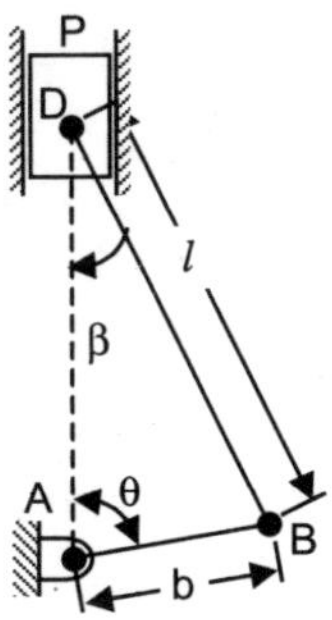

Fig. ED3.21

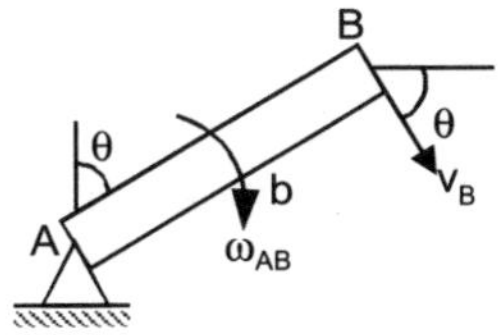

Fig. ED3.21(a)

Solution: This problem is attempted with relative motion analysis method.

Motion of Crank:

Kinematic diagram showing the motion of crank is illustrated in Fig. ED3.21(*a*).

The velocity for the rod *AB*: $v_B = b\omega_{AB}$

Motion of connecting rod BD:

From the law of sines, we have

$$\frac{\sin\beta}{b} = \frac{\sin\theta}{\ell}$$

$$\sin\beta = \frac{b}{\ell}\sin\theta = \frac{\sin\theta}{n} \quad \text{where, } n = \ell/b.$$

The velocity v_D of the point *D* where the rod is attached to the piston must be vertical, while the velocity of point *B* is equal to the velocity v_B obtained above. Resolving the motion of *BD* into a translation with *B* and a rotation about *B*, we obtain the relation:

$$v_D = v_B + v_{DB}$$

Figure ED3.21(*b*), shows this motion.

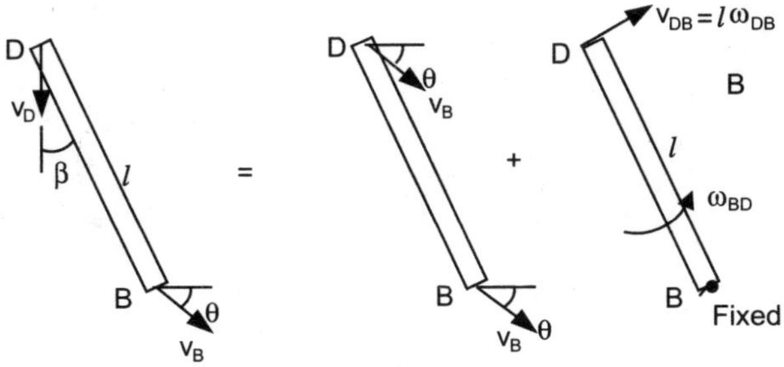

Fig. ED3.21(b)

The vector diagram corresponding to this equation is presented in Fig. ED3.21(*c*).

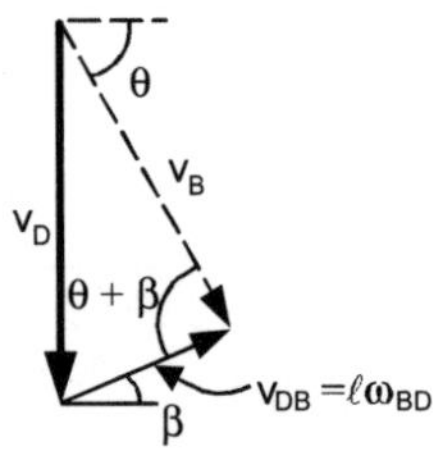

Fig. ED3.21 (c)

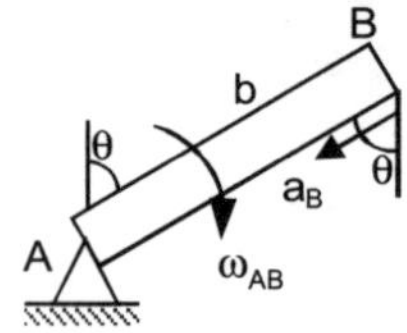

Fig. ED3.21 (d)

Using the law of sines in this triangle:

$$\frac{v_D}{\sin(\theta+\beta)} = \frac{v_B}{\sin(90^\circ-\beta)} = \frac{\ell\omega_{BD}}{\sin(90^\circ-\theta)}$$

$$v_D = v_B \frac{\sin(\theta+\beta)}{\sin(90^\circ-\beta)} = v_B \frac{\sin(\theta+\beta)}{\cos\beta}$$

$$\omega_{BD} = \frac{v_B}{\ell}\frac{\sin(90^\circ-\theta)}{\sin(90^\circ-\beta)} = \frac{v_B}{\ell}\frac{\cos\theta}{\cos\beta}$$

For the acceleration we have for rod AB

$$a_B = b\omega^2_{AB} \text{ (parallel to } AB \text{ towards } A\text{) (see Fig. ED3.21 (}d\text{))}$$

For rod BD:

$$(a_{D/B})_t = \ell\alpha BD$$

$$(a_{D/B})_n = \ell\omega^2_{BD}$$

For the acceleration we again look at plane motion with the translation of B plus a rotation about B. For B we have

$$a_D = a_B + a_{DB}$$

$$a_D = a_B + (a_{DB})_t + (a_{DB})_n$$

Figure ED3.21(e) shows the plane motion comprising of translation and rotation.

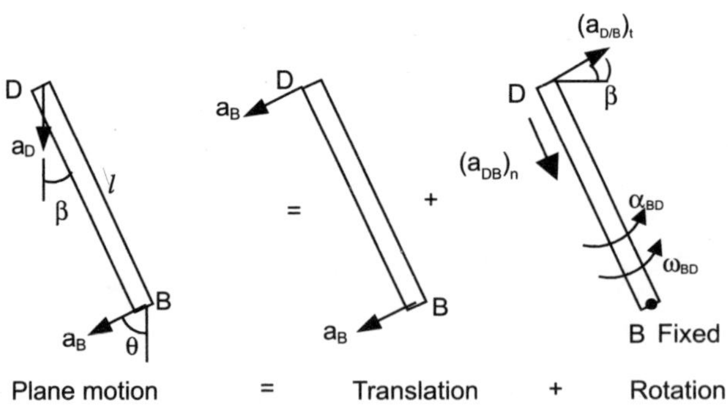

Fig. ED3.21(e)

Using the acceleration vector diagram (Fig. ED3.21(f)) :

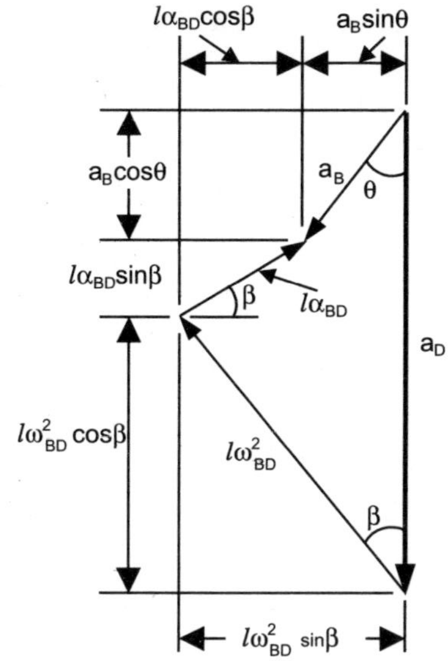

Fig. ED3.21(f)

We obtain $\quad a_B \sin\theta = \ell\omega_{BD}^2 \sin\beta - \ell\alpha_{BD}\cos\beta$

$$\alpha_{BD} = \frac{\ell\omega_{BD}^2 \sin\beta - a_B \sin\theta}{\ell\cos\beta}$$

$$a_B \cos\theta = a_D - \ell\omega_{BD}^2 \cos\beta + \ell\alpha_{BD}\sin\beta$$

$$a_D = a_B \cos\theta + \ell\omega_{BD}^2 \cos\beta + \ell\alpha_{BD}\sin\beta$$

Based on this formulation, following MATLAB script is used to obtain variation of velocities and accelerations as a function of angle θ.

MATLAB Program :

```
L=10; % Length of connecting rod in cm
b=3.5; % crank radius in cm
N=1000; % speed in rpm
%Velocity
omega_AB=N*2*pi/60; % angular velocity in rad/s
v_B=b*omega_AB; % linear velocity of crank pin in cm/s
Theta=[0:20:180];
t=Theta*pi/180;
beta=asin((b/L)*sin(t));% inclination of connecting rod with axis of piston
v_D=v_B*sin(t+beta)./cos(beta);% velocity of piston in cm/sec
omega_BD=v_B*cos(t)./(L*cos(beta));% angular velocity of connecting rod
% acceleration
a_B=b*omega_AB^2; % acceleration of point B
alpha_BD=(L.*omega_BD.^2.*sin(beta)-a_B.*cos(t))./(L.*cos(beta));
```

```
a_D=a_B.*cos(t)+L.*omega_BD.^2.*cos(beta)+L.*alpha_BD.*sin(beta);
% Determine and plot values
z=[Theta;omega_BD;alpha_BD;v_D;a_D];
fprintf('Theta         Angular        Angular          Piston          Piston\n')
fprintf('              velocity     acceleration       velocity      acceleration\n')
fprintf('(deg)         rad/s)         (rad/s^2)        (cm/s)          (cm/s^2)\n')
fprintf('\n');
fprintf('%5.3f         %6.3f          %6.3f            %6.3f           %6.3f\n',z);
fprintf('\n');
%fprintf('The two values of theta for zero collar speed are %5.3f and %5.3f
degrees\n',a1,a2)
figure(1)
plot(Theta,omega_BD,'-p')
xlabel('Theta(degree)')
ylabel('Angular Velocity Rod BD (rad/s)')
grid on
figure(2)
plot(Theta,alpha_BD, '-*')
xlabel('Theta (degree)')
ylabel('Angular acceleration rod BD(rad/s^2)')
grid on
figure(3)
plot(Theta,v_D, '-p')
xlabel('Theta(degrees)')
ylabel('Piston Velocity (cm/s)')
grid on
figure(4)
plot(Theta,a_D,'-*')
xlabel('Theta(degrees)')
ylabel('Piston Acceleration D (cm/s^2)')
grid on
```

Output is as follows:

Theta (deg)	Angular velocity (rad/s)	Angular Acceleration (rad/s^2)	Piston velocity (cm/s)	Piston acceleration (cm/s^2)
0.000	36.652	–3838.179	0.000	51815.423
20.000	34.691	–3487.725	166.884	43840.145
40.000	28.816	–2825.851	300.422	31135.267
60.000	19.231	–1896.200	375.705	16967.547
80.000	6.780	– 693.123	384.320	4707.369
100.000	–6.780	726.881	337.581	–3727.977
120.000	–19.231	2131.457	259.125	–9206.061
140.000	–28.816	3209.296	170.766	–14091.443
160.000	–34.691	3777.938	83.829	–19596.523
180.000	–36.652	3838.179	0.000	–24948.167

Figure ED3.21(g)–(j) shows the variations of angular and linear velocities.

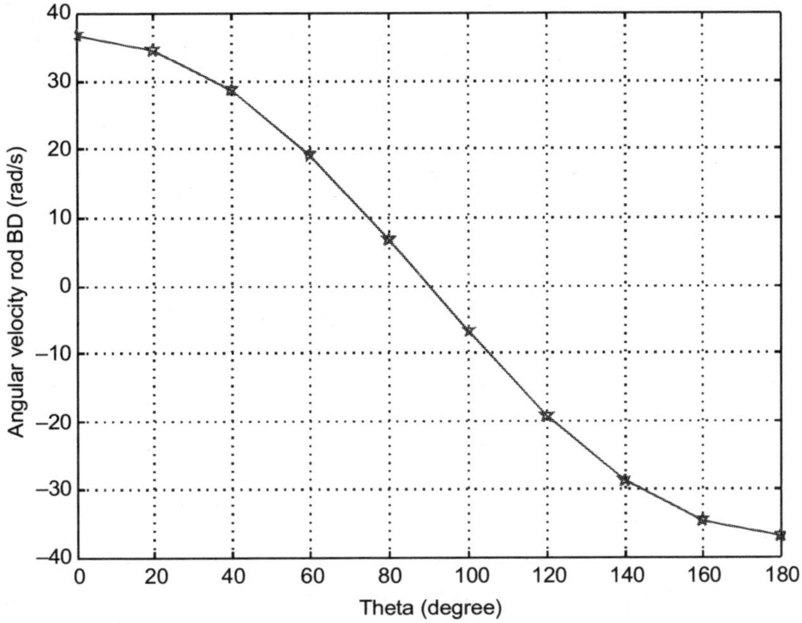

Fig. ED3.21(g) Angular velocity of connecting rod

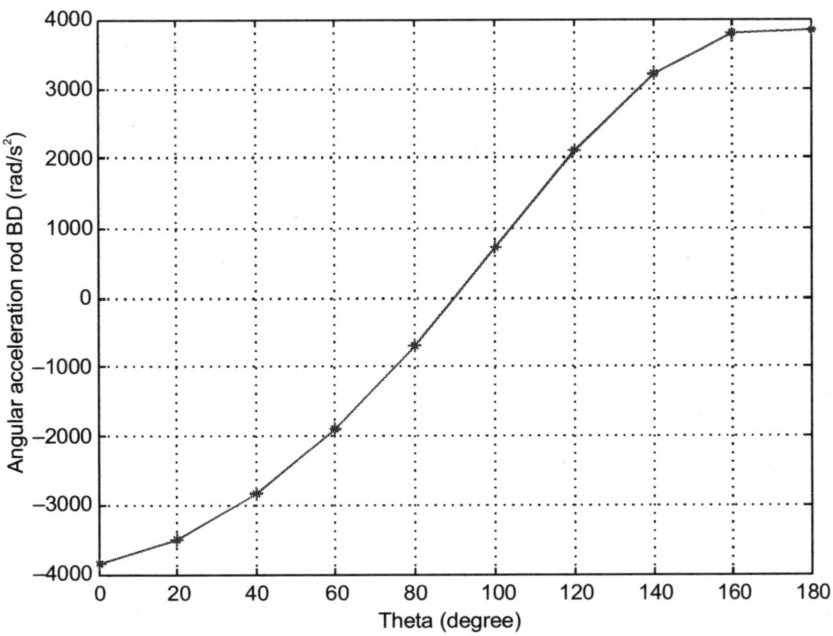

Fig. ED3.21 (h) Angular acceleration of connecting rod

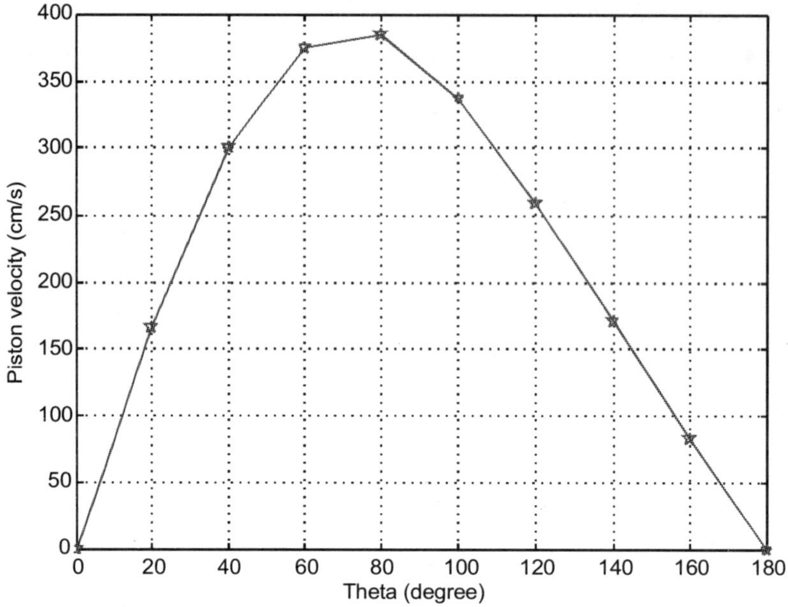

Fig. ED3.21(i) Piston velocity

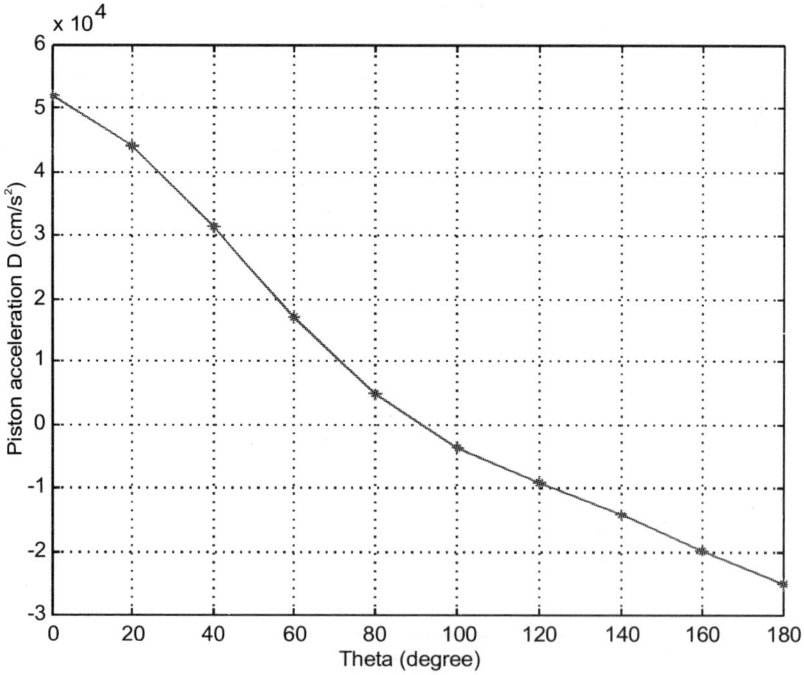

Fig. ED3.21(j) Piston acceleration

ED3.22: A disk shown in Fig. ED3.22 has a constant angular velocity 400 rpm counterclockwise. Knowing that rod *BD* is 300 mm long, use MATLAB to determine and plot the velocity of collar *D* and angular velocity of rod *BD* as a function of angle θ. Consider $0 \le \theta \le 360°$.

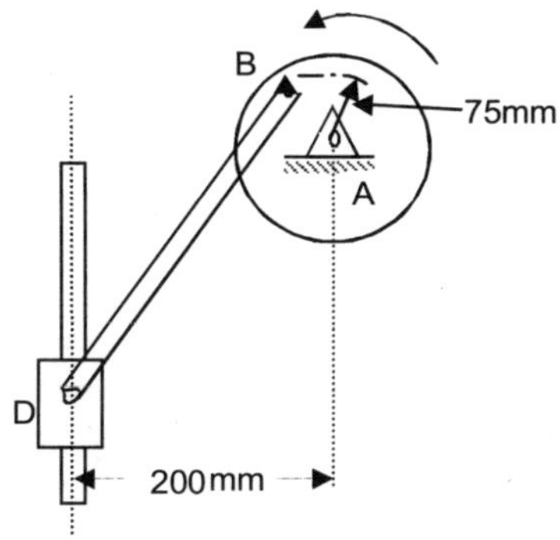

Fig. ED3.22

Solution: This problem is attempted with vector notation. Fix the origin of coordinate system at *A* and define the position vectors accordingly.
Motion of *BA*:

$$v_B = (+\omega\hat{k}) \times AB = \omega\hat{k} \times (r\cos\theta\,\hat{j} - r\sin\theta\,\hat{i}) = (-\omega r\cos\theta\,\hat{i} - \omega r\sin\theta\,\hat{j}) \quad ...(1)$$

Motion of *BD*: Consider motion of *D* with respect to *B*.

$$v_D = v_B + v_{DB}$$

i.e.,

$$-v_D\hat{j} = v_B + (+\omega_{DB}\hat{k}) \times BD \quad ...(2)$$

Here $BD = -BD_x\hat{i} - BD_y\hat{j}$, where BD_x is horizontal distance from *B* to *D* and BD_y is vertical distance.
From geometry,

$$BD_x = (200 - r\sin\theta) \text{ and } BD_y = \sqrt{\ell^2 - (BD_x)^2}\text{, where } \ell = BD = 300 \text{ mm and } r = 75 \text{ mm}$$

Substituting v_B, BD_x and BD_y in equation (2)

$$-v_D\,\hat{j} = (-\omega r\cos\theta\,\hat{i} - \omega r\sin\theta\,\hat{j}) + (+\omega_{DB}\hat{k}) \times (-BD_x\hat{i} - BD_y\hat{j})$$

$$= (-\omega r\cos\theta\,\hat{i} - \omega r\sin\theta\,\hat{j}) + (\omega_{DB}BD_y\hat{i} - \omega_{DB}BD_x\hat{j})$$

Equating $\hat{i}$ and $\hat{j}$ terms independently on both sides:

$$-\omega r\cos\theta + \omega_{DB}BD_y = 0$$

$$\Rightarrow \quad \omega_{DB} = \omega r\cos\theta/(BD_y) \quad ...(3)$$

and

$$v_D = \omega r\sin\theta + \omega_{DB}BD_x \quad ...(4)$$

Based on the eqs.(3) and (4), complete MATLAB program is written as follows:

```
% Initialization of constants%%%%%%%%%%%
L=300; %Length of BD in mm
r=75; % radius AB
N=400; % Speed in rpm
```

```
w=2*pi*N/60; % angular velocity of BA
% variation of theta in steps of 20 degrees
th=0:20:360;
thr=th.*pi/180; % theta in radians
BDx=200-r.*sin(thr);
BDy=sqrt(L^2-BDx.^2);
wDB= w*r*cos(thr)./BDy;
vD=w*r.*sin(thr)+wDB.*BDx
subplot(2,1,1);
plot(th,wDB,'-p');
ylabel('Angular velocity of rod BD in rad/s)');
grid on;
subplot(2,1,2);
plot(th,vD,'-*');
ylabel('Linear velocity of D in (mm/s)');
xlabel('Angle theta (in degrees)');
grid on;
```

The output of the program is shown below in Fig. ED3.22 (*a*).

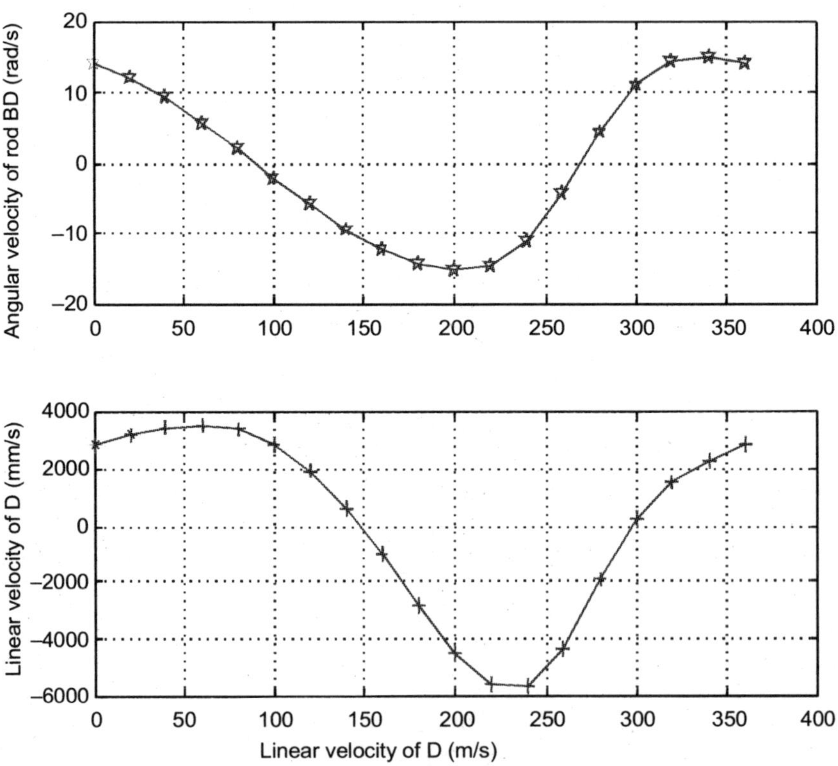

Fig. ED3.22 (a) MATLAB outputs

ED 3.23: Figure ED3.23 shows an engine system where the connecting rod BD is attached to the piston P. The crank AB rotates with a constant angular velocity of ω_{AB} rpm clockwise with no force applied to the face of the piston. Write a MATLAB program to plot the horizontal and vertical components of the dynamic reactions exerted on the connecting rod at B and D for $0 \le \theta \le 180°$. Weight of the connecting rod $BD = 5.5$ N. Weight of the piston, $W_P = 6.3$ N. Length of connecting rod = 10 cm and crank radius = 3.5 cm. Speed of rotation= 1000 *rpm*.

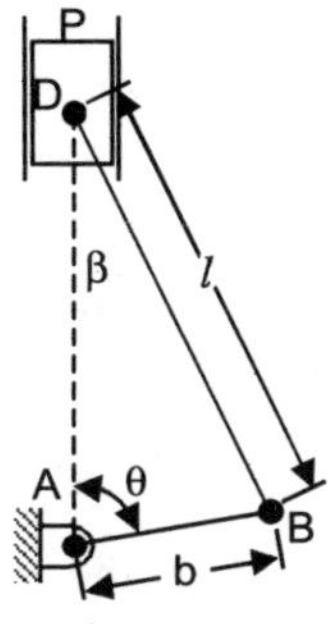

Fig. ED3.23

Solution: We have $\sin\beta = \frac{b}{\ell}\sin\theta$

or $$\beta = \sin^{-1}\left(\frac{b\sin\theta}{\ell}\right) \quad ...(1)$$

$$v_B = b\omega_{AB} \quad ...(2)$$

$$\omega_{BD} = \frac{v_B\cos\theta}{\ell\cos\beta} \quad ...(3)$$

$$a_B = b\omega_{AB}^2 \quad ...(4)$$

$$\alpha_{BD} = \frac{\ell\omega_{BD}^2\sin\beta - a_B\sin\theta}{\ell\cos\beta} \quad ...(5)$$

$$a_D = a_B\cos\theta + \ell\omega_{BD}^2\cos\beta + \ell\alpha_{BD}\sin\beta \quad ...(6)$$

Directions of velocity and accelerations is shown in Fig.ED3.23 (*a*).

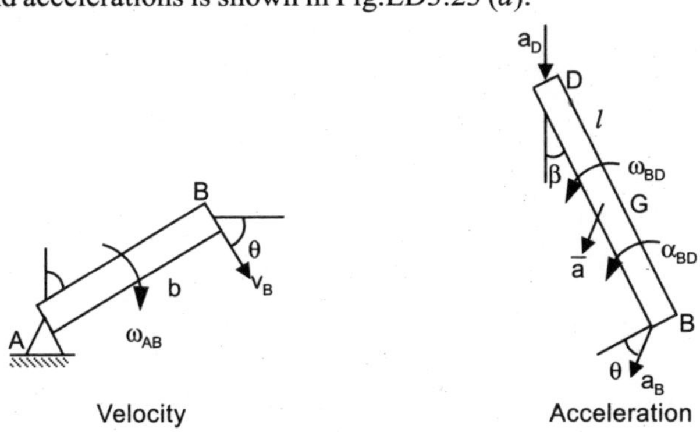

Fig. ED3.23(a)

From the figure, we have

$$\underset{\rightarrow}{+}(a_B)_x = -a_B \sin\theta$$

$$+\downarrow(a_B)_y = a_B \cos\theta$$

Since the position of center of gravity G is at the middle of BD we have

$$\underset{\rightarrow}{+}\bar{a}_x = \frac{1}{2}(a_B)_x \qquad ...(7)$$

$$+\downarrow \bar{a}_y = \frac{1}{2}[(a_B)_x + a_D] \qquad ...(8)$$

For the piston (see free-body diagram Fig. ED3.23 (*b*) we find

$$+\downarrow \sum F_y = \sum (F_y)_{\text{eff}} :$$

$$D_y = -m_P a_D \qquad ...(9)$$

Since we are after the dynamic reactions, we shall omit the weight of the piston and connecting rod. For the connecting rod we have

$$\bar{I} = \frac{1}{12} m_{BD} \ell^2 \qquad ...(10)$$

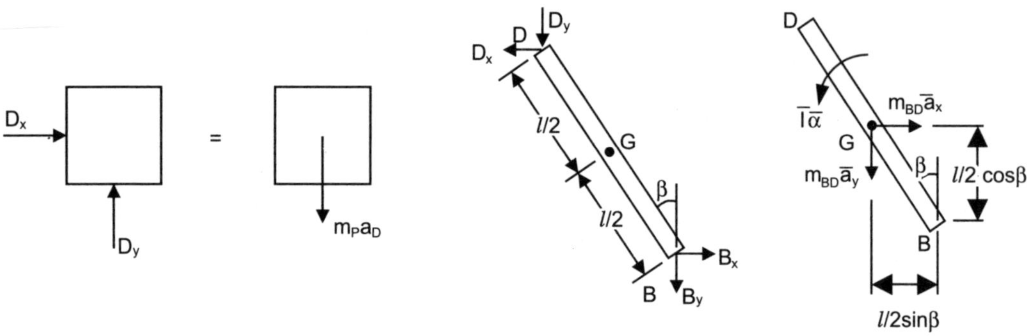

Piston Connecting rod

Fig. ED3.23 (b) Free body diagrams

Summing moments about B yields

$$\sum M_B = \sum (M_B)_{eff} :$$

$$-D_x \ell \cos\beta - D_y \ell \sin\beta = -\bar{I}\alpha + m_{BD} a_x \left(\frac{1}{2}\cos\beta\right) - m_{BD} a_y \left(\frac{1}{2}\sin\beta\right)$$

Dividing by ℓ and solving for D_x gives

$$D_x = -D_y \frac{\sin\beta}{\cos\beta} + \frac{\bar{I}\alpha}{\ell\cos\beta} - \frac{m_{BD}}{2}\bar{a}_x + \frac{m_{BD}}{2}\bar{a}_y \frac{\sin\beta}{\cos\beta}$$

$$D_x = -D_y \tan\beta + \frac{\bar{I}\alpha}{\ell\cos\beta} - \frac{m_{BD}}{2}\bar{a}_x + \frac{m_{BD}}{2}\bar{a}_y \tan\beta \qquad ...(11)$$

$$\overset{+}{\rightarrow}\sum F_x = \sum (F_x)_{\text{eff}}:$$

$$B_x - D_x = m_{BD}\bar{a}_x$$

$$B_x = m_{BD}\bar{a}_x + D_x \qquad ...(12)$$

$$+\downarrow \sum F_x = \sum (F_x)_{\text{eff}}:$$

$$B_y + D_y = m_{BD}\bar{a}_y$$

$$B_y = m_{BD}\bar{a}_y - D_y \qquad ...(13)$$

Complete MATLAB program to find these reactions as a function of angle θ is given below:

MATLAB Program:

```
g=9.81; % Acceleration due to gravity
Wbd=5.5; % Weight of the connecting rod in N
Wp=6.3; % Weight of the piston in N
mp=Wp/g; % Mass of the piston
mbd=Wbd/g;
Lcm=10; % Length of connecting rod in cm
L=0.1; % Length of connecting rod in m
b=3.5/100; % crank radius in m
I_bar=(1/12)*mbd*L^2; % mass moment of inertia in kg.m^2
omega_AB=1000*(2*pi)/60;
v_B=b*omega_AB;
Theta=[0:10:180];
t=Theta*pi/180;
beta=asin(b*sin(t)/L);
omega_BD=v_B*cos(t)./(L*cos(beta));
%acceleration
a_B=b*omega_AB^2;
alpha_BD=(L.*omega_BD.^2.*sin(beta)-a_B.*sin(t))./(L.*cos(beta));
a_D=a_B.*cos(t)+L.*omega_BD.^2.*cos(beta)+L.*alpha_BD.*sin(beta);
%
ax_bar=-0.5*a_B*sin(t);
ay_bar=0.5*a_B*cos(t)+0.5*a_D;
Dy=-mp*a_D;
Dx=-Dy.*tan(beta)+(I_bar*alpha_BD)./(L*cos(beta))-mbd*ax_bar./
2+mbd*ay_bar.*tan(beta)./2;
Bx=mbd*ax_bar+Dx;
By=mbd*ay_bar-Dy;
%Determine and plot values
z=[Theta;Bx;By;Dx;Dy];
fprintf('Theta     Bx            By Dx  Dy\n')
fprintf('(deg)     (N)    (N)    (N)   N)\n')
```

```
fprintf('\n');
fprintf('%5.0f  %8.3f   %8.3f %7.3f   %8.3f\n',z);
fprintf('\n');
figure(1)
plot(Theta,Bx,'-+',Theta,By,'-p')
xlabel('Theta(degrees)')
ylabel('Dynamic Reactions(N)')
legend('Bx','By')
grid on
figure(2)
plot(Theta,Dx,'-+',Theta,Dy,'-p')
xlabel('Theta(degrees)')
ylabel('Dynamic Reactions(N)')
legend('Dx','Dy',2)
grid on
```

The output of the program is given below:

Theta (deg)	B_x (N)	B_y (N)	D_x (N)	D_y (N)
0	0.000	−79.913	0.000	−332.760
10	15.186	−76.766	33.869	−324.273
20	26.121	−67.551	62.920	−299.305
30	29.169	−52.964	82.966	−259.353
40	21.931	−34.202	91.091	−206.966
50	3.837	−12.972	86.259	−145.755
60	−23.455	8.566	69.725	−80.269
70	−56.109	27.998	44.997	−15.620
80	−88.824	43.018	17.136	43.185
90	−116.153	51.895	−8.559	92.096
100	−133.977	53.888	−28.018	128.789
110	−140.462	49.408	−39.356	152.989
120	−136.079	39.866	−42.899	166.220
130	−122.848	27.267	−40.426	171.124
140	−103.327	13.753	−34.166	170.676
150	−79.839	1.249	−26.042	167.578
160	−54.116	−8.726	−17.317	163.942
170	−27.279	−15.116	−8.596	161.215
180	−0.000	−17.313	−0.000	160.218

Figure ED3.23 (*c*) shows the output plots.

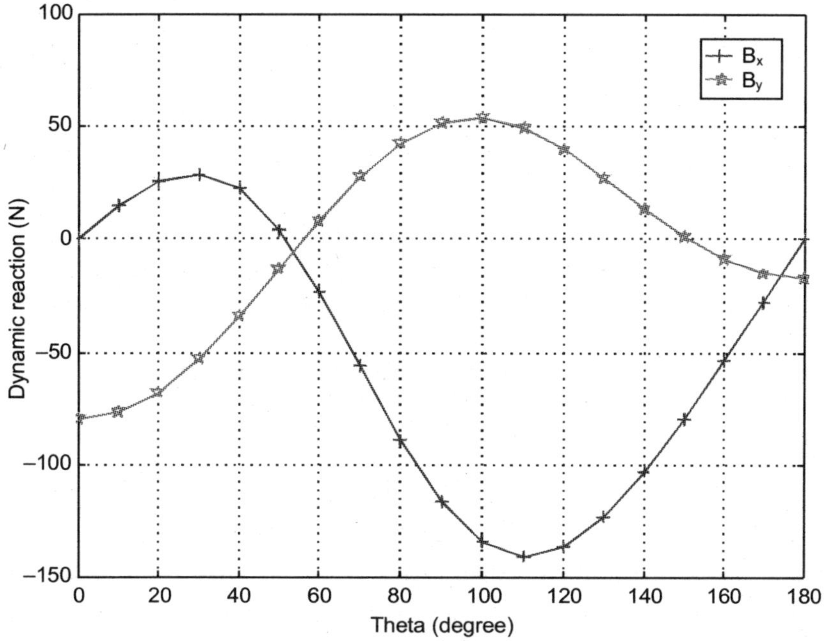

(i)

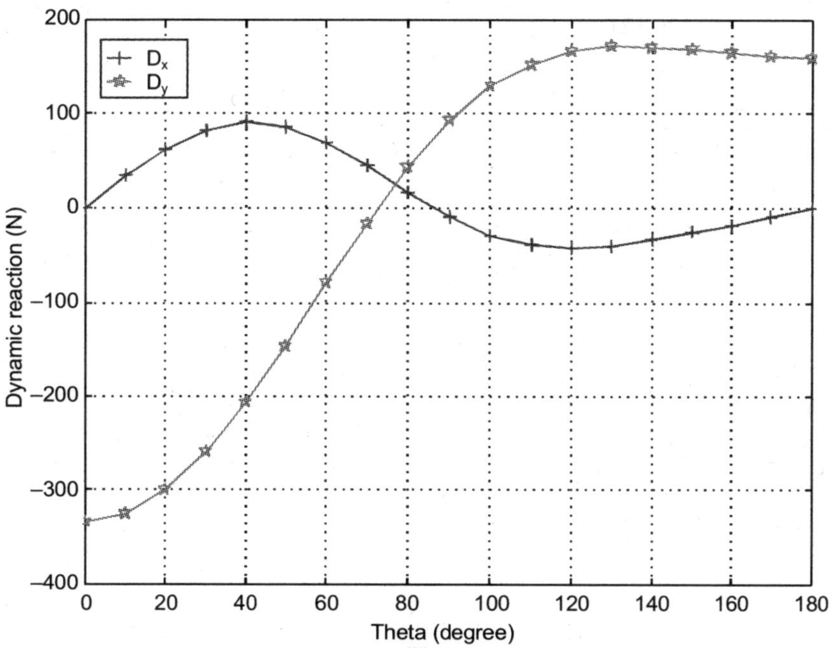

(ii)

Fig. ED3.23 MATLAB outputs

ED3.24: A 30 kg disk is pin-supported at its center. It is acted upon by a constant force $F = 10$ N which is applied to a cord wrapped around its periphery and a constant couple 5 Nm. Plot the variation of angular speed with the number of revolutions it makes. Assume the system started from rest.

Solution: Figure ED3.24 shows the configuration of the system.

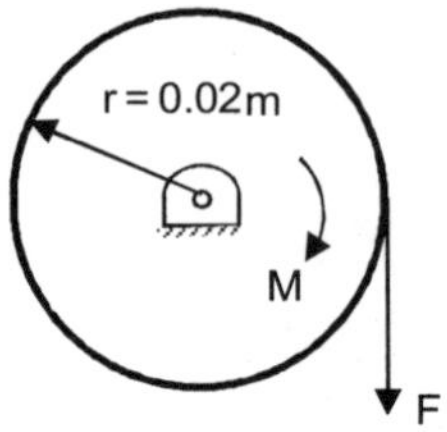

Fig. ED3.24 Configuration of the system

Here as angular speed and displacement are involved, one can apply work-energy principle. But remember that it is rigid-body motion.

i.e., $$T_1 + \Sigma U_{1\text{-}2} = T_2$$

Here T_1 = initial kinetic energy of the system = 0

T_2 = final kinetic energy $= \frac{1}{2} I\omega^2$, where $I = \frac{1}{2} mr^2$

$\Sigma U_{1\text{-}2}$ = work done by force and moment $= M\theta + Fs = (M + Fr)\theta$

Hence $$(M + Fr)\,\theta = \frac{1}{2} I\omega^2 = \frac{1}{2}\left(\frac{1}{2} mr^2\right)\omega^2 = \frac{1}{4} mr^2\omega^2$$

A simple program that relates ω and θ is given below:

```
% Initialize values
F=10;
M=5;
m=30;
r=0.02;
theta=0:10;
omega=2*sqrt((M+F*r).*theta/(m*r^2));
plot (theta,omega,'-p');
xlabel('Number of revolutions');
ylabel('Angular speed (rad/s)');
grid on;
```

Output is shown in Fig. ED3.24 (*a*)

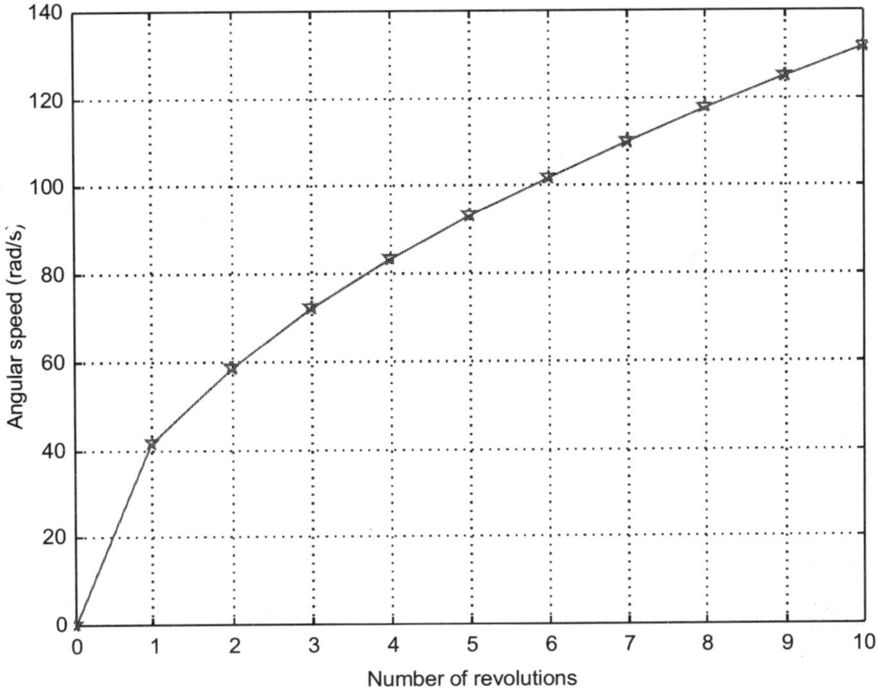

Fig. ED3.24(a) MATLAB output.

ED 3.25: Two identical slender rods are pin connected as shown in Fig. ED3.25 have lengths $L = 600$ mm. If the system is released from rest when $\beta = 60°$. Use MATLAB to plot angular velocity of rod *AB* and velocity of point *D* for various values of β ranging from 0° to 60°.

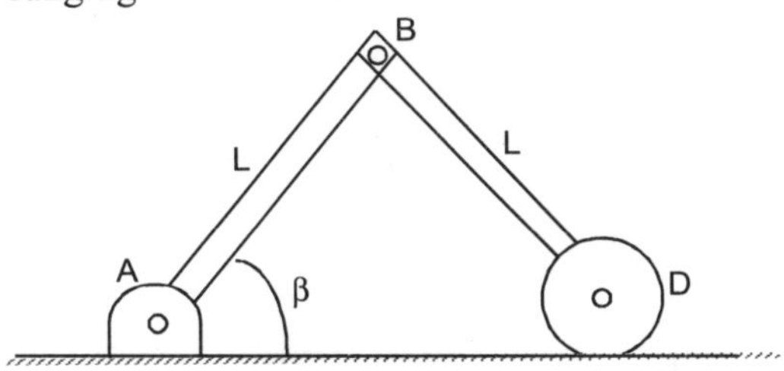

Fig. ED3.25

Solution: Here as there is no friction given, principle of conservation of energy can be applied.

Thus $T_1 + V_1 = T_2 + V_2$

where T_1 = initial kinetic energy = 0

T_2 = final kinetic energy = $\frac{1}{2} m v_{AB}^2 + \frac{1}{2} I_{AB} \omega_{AB}^2 + \frac{1}{2} m v_{BD}^2 + \frac{1}{2} I_{BD} \omega_{BD}^2$

V_1 = initial potential energy = $2mg\, y_1 = 2mg\,(L \sin 60°) = \sqrt{3}\, mg\, L$

V_2 = final potential energy = $2mg\, y_2 = 2mg\,(L \sin \beta)$

The angular velocity ω_{BD} is obtained in terms of $\omega_{AB}=\omega$ from the kinematics of the linkage at any angle β. Taking A as the fixed frame of reference, the velocity of B is

$$v_B=(-\omega\hat{k})\times AB=-\omega\hat{k}\times(L\sin\beta\,\hat{j}+L\cos\beta\,\hat{i})=(\omega L\sin\beta\,\hat{i}-\omega L\cos\beta\,\hat{j})$$

Velocity of center of mass of AB is

$$v_{AB}=(\omega(L/2)\sin\beta\,\hat{i}-\omega(L/2)\cos\beta\,\hat{j}). \quad ...(1)$$

It has magnitude $=\omega(L/2)$

Consider motion of D with respect to B.

$$v_D=v_B+v_{DB}$$

i.e., $$v_D\hat{i}=v_B+(\omega_{DB}\hat{k})\times BD$$

Here $BD=L\cos\beta\,\hat{i}-L\sin\beta\,\hat{j}$

$\therefore$ $$v_D\hat{i}=v_B+(\omega_{DB}\hat{k})\times(L\cos\beta\,\hat{i}-L\sin\beta\,\hat{j})$$

$$=v_B+\omega_{DB}(L\cos\beta\,\hat{j}+L\sin\beta\,\hat{i})$$

So equating $\hat{i}$ and $\hat{j}$ terms independently

$$v_D=\omega L\sin\beta+L\,\omega_{DB}\sin\beta,$$

and $$\omega_{DB}=\omega L\cos\beta/L\cos\beta=\omega$$

Velocity of center of gravity of BD is

$$v_{BD}=v_B+\omega_{DB}((L/2)\cos\beta\,\hat{j}+(L/2)\sin\beta\,\hat{i})$$

$$=(\omega L\sin\beta\,\hat{i}-\omega L\cos\beta\,\hat{j})+\omega_{DB}(L/2)\cos\beta\,\hat{j}+\omega_{DB}(L/2)\sin\beta\,\hat{i}$$

$$=(\omega L+\omega_{DB}(L/2))\sin\beta\,\hat{i}+(-\omega L+\omega_{DB}(L/2))\cos\beta\,\hat{j} \quad ...(2)$$

$$=(3\omega L/2)\sin\beta\,\hat{i}+(-\omega L/2)\cos\beta\,\hat{j}$$

It has magnitude $$=\sqrt{\left(3\omega L\sin\beta/2\right)^2+\left(-\omega L\cos\beta/2\right)^2}=\omega L/2\sqrt{9\sin^2\beta+\cos^2\beta}$$

Substituting all the terms in T_2,

$$T_2=\frac{1}{2}m\left(\frac{\omega L}{2}\right)^2+\frac{1}{2}\left(\frac{1}{12}mL^2\right)\omega^2+\frac{1}{2}m\frac{\omega^2L^2}{4}(9\sin^2\beta+\cos^2\beta)+\frac{1}{2}\left(\frac{1}{12}mL^2\right)\omega^2$$

$$=\frac{5}{24}mL^2\omega^2+\frac{1}{8}mL^2\omega^2\left(9\sin^2\beta+\cos^2\beta\right)$$

Thus $$T_2=V_1-V_2\Rightarrow\frac{5}{24}mL^2\omega^2+\frac{1}{8}mL^2\omega^2\left(9\sin^2\beta+\cos^2\beta\right)=\sqrt{3}mgL-2mg\,(L\sin\beta)$$

or $$\omega=\sqrt{\frac{g\sqrt{3}-2g\sin\beta}{\left(\frac{5}{24}L+\frac{1}{8}L\left(9\sin^2\beta+\cos^2\beta\right)\right)}}\text{ and }v_D=2L\omega\sin\beta$$

Complete MATLAB program for computing ω and v_D is presented below:

```
% Initial values %%%%%%%%%%%%%%%%
 g=9.81; % Acceleration due to gravity
 L=0.600 % Length of rod in m
 beta=60:-5:0; % angle beta decrement
 betar=beta*pi/180;
 num=g*(3^0.5)-2*g*sin(betar);
 den=(5/24)*L+((1/8)*L*(9*sin(betar).^2+cos(betar).^2))
 omega=sqrt(num./den);
 vD=2*L*omega.*sin(betar);
 subplot(2,1,1);
 plot(beta,omega,'-p');
 ylabel('Angular velocity (rad/s)');
grid on;
subplot(2,1,2);
plot(beta,vD, '-*');
ylabel('Linear velocity of D (m/s)');
xlabel('Angle beta degree');
grid on;
%%%%%%%%%%%%%%%%%%%%%%%%%
```

Output of the program is shown in Fig. ED3.25 (*a*)

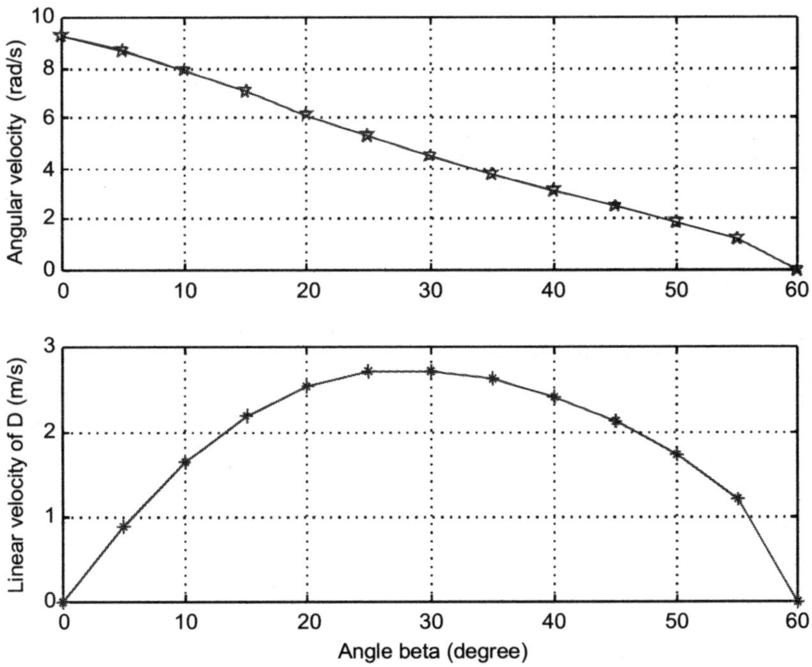

Fig. ED3.25 (a) MATLAB output

ED 3.26: Figure ED3.26 shows the 3-*D* projectile trajectory where a projectile is fired with an initial velocity of v_0 at an angle of θ relative to the ground.

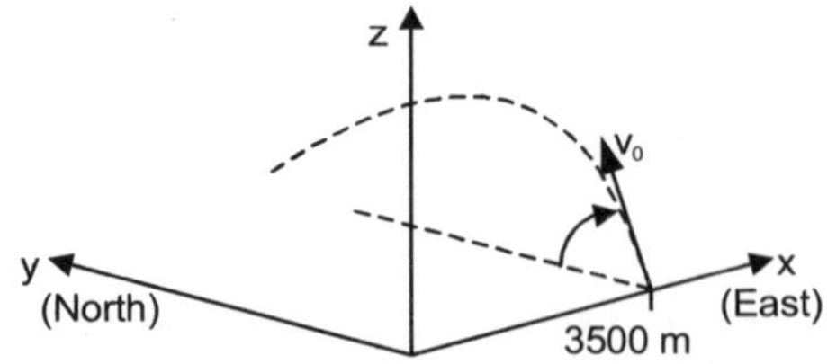

Fig. ED3.26

The projectile is aimed directly north. The projectile also moves in the western direction at a constant speed of 25 m/s. Write a MATLAB program to (*a*) determine and plot the trajectory of the projectile until it hits the ground, (*b*) plot the trajectory that the projectile would have had if the projectile does not move in the western direction. Given $v_0 = 300$ m/s and $\theta = 60°$.

Solution: *x* and *y* axes represent East and North directions. Consider vertical direction as *z*. As projectile is fired in north, the initial velocity v_0 can be resolved into *y* and *z* direction as:

$$v_{0y} = v_0 \cdot \cos\theta \text{ and } v_{0z} = v_0 \cdot \sin\theta$$

Time taken by projectile to reach highest point ($v_z = 0$) is

$$t_h = \frac{v_{0z}}{g}$$

Total flying time, $t_t = 2 \cdot t_h$

Due to wind, projectile has a constant velocity in the negative *x* direction, $v_x = -25$.

∴ Position of projectile is given by

$$x = x_0 + v_x t \text{ and } y = y_0 + v_{oy} t$$

$$z = z_0 + v_{0z} t - \frac{1}{2} g t^2$$

Start with initial point $(x_0, y_0, z_0) = (3000.0.0)$.

Complete MATLAB program is given below:

```
v0=300;g=9.81;theta=60;
x0=3000;vx=-25;
v0z=v0*sin(theta*pi/180);
v0y=v0*cos(theta*pi/180);
t=2*v0z/g;
tplot=linspace(0,t,100); % CREATING A TIME VECTOR WITH 100 ELEMENTS
z=v0z*tplot-0.5*g*tplot.^2;
y=v0y*tplot;
x=x0+vx*tplot;
 % CALCULATING X,Y,Z COORDINATES OF THE PROJECTILE AT EACH TIME
xnowind(1:length(y))=x0; % CONSTANT X-COORDINATE WHEN NO WIND
```

```
plot3(x,y,z,'k-',xnowind,y,z,'k-') % TWO 3-D LINE PLOTS
grid on
xlabel('x(m)');
ylabel('y(m)');
zlabel('z(m)')
```

Output is shown in Fig. ED3.26 (*a*).

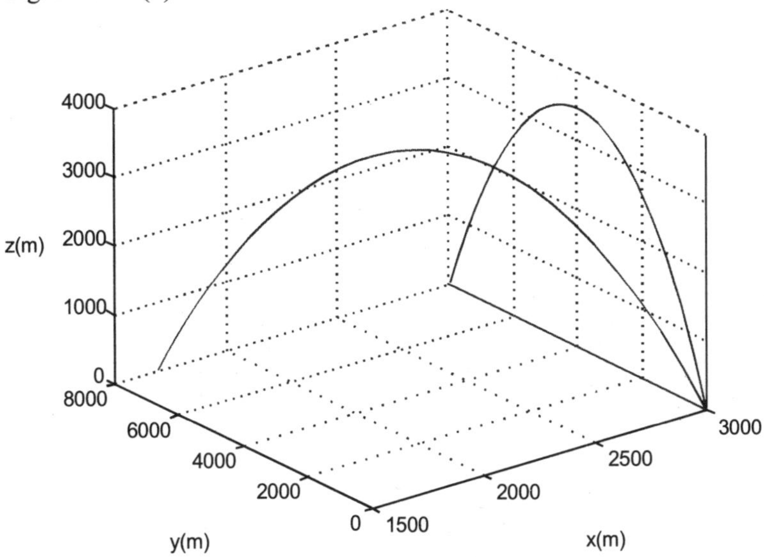

Fig. ED 3.26(a) MATLAB output

3.3 SUMMARY

In this Chapter, we have demonstrated the application of MATLAB for solving engineering mechanics problems. More example problems and solutions are given in Chapter 5.

CHAPTER 4

MATLAB FOR ENGINEERING APPLICATIONS

MATLAB enables to execute powerful features such as three-dimensional graphs and symbolic calculations very conveniently. This chapter gives some more solved examples using MATLAB in obtaining answers. It is similar to MATHCAD and other programming software. Some built-in functions must be properly used in order to get the outputs. Following programs are tested with MATLAB-R11 software.

4.1 NUMERICAL CALCULATIONS

In mechanics, MATLAB can be used for most of the time to compute the numerical expressions. In MATLAB some care should be taken in handling vectors and scalars that occur in expressions. This section helps to understand the evaluation procedure. Here along with basic line commands, the built-in functions for handling polynomials and matrices are also presented.

E4.1: Compute the following quantity using MATLAB in the Command Window:

$$\frac{17\left[\sqrt{5}-1\right]}{\left[15^2-13^2\right]}+\frac{5^7\log_{10}(e^3)}{\pi\sqrt{121}}+\ln(e^4)+\sqrt{11}$$

Solution: Following lines typed at the command prompt one after the other. Built-in functions like sqrt, exp, log, log10, pi are used as per requirement.

```
>> term1 = 17*(sqrt(5) -1)/(15^2-13^2);
>> term2 = 5^7*log10(exp(3))/(pi*sqrt(121));
>> term3 = log(exp(4));
>> term4 = sqrt(11);
>> result= term1+term2+term3+term4
```

Output is as follows:

```
result =
        2.9532e + 003
```

E4.2: Compute the following quantity using MATLAB in the Command Window:

$$B = \frac{\tan x + \sin 2x}{\cos x} + \log\left|x^5 - x^2\right| + \cosh x - 2\tanh x \quad \text{for } x = 5\pi/6.$$

Solution: Here the use of trigonometric built-in functions is explained.

```
>> x = 5*pi/6;
>> term1 = (tan(x)+sin(2*x))/cos(x);
>> term 2=log(abs(x^5- x^2)) + cosh(x) -2*tanh(x);
>> term = term1+ term2
```

Output is as follows:

```
>>term =
        11.3331
```

E4.3: Compute the following quantity using MATLAB in the Command Window:

$$x = a + \frac{ab}{c}\frac{(a+b)}{\sqrt{|ab|}} + c^a + \frac{\sqrt{14}\,b}{e^{3c}} + \ln(2) + \frac{\log_{10} c}{\log_{10}(a+b+c)} + 2\sinh a - 3\tanh b$$

for $a = 1, b = 2$ and $c = 1.8$.

Solution:

```
>> a=1; b=2; c=1.8;
>> x1= a + (a*b)*(a + b)/(c*sqrt (abs(a*b)));
>> x2=c^a + sqrt(14)*b/exp(3*c) +log(2);
>> x3=log10(c)/log10(a+b+c) +2*sinh(a) - 3*tanh(b);
>> x = x1+ x2 + x3
```

Output comes like this

```
>> x = 5.7170
```

E4.4: Use MATLAB to create

(*a*) a row and column vectors that has the elements: 11, –3, $e^{7.8}$, ln(59), tan(π/3), 5 $\log_{10}(26)$.

(*b*) a row vector with 20 equally spaced elements in which the first element is 5.

(*c*) a column vector with 15 equally spaced elements in which the first element is –2.

Solution:

(*a*)
```
»A = [11 - 3 exp(7.8) log(59) tan(pi/3) 5*log10(26)] % ROW-VECTOR
B= A' % COLUMN-VECTOR
```

Output is as follows:

```
A =
    1.0e+003 *
              0.0110 - 0.0030 2.4406 0.0041 0.0017 0.0071
B =
     1.0e + 003 *
     0.0110
    -0.0030
     2.4406
     0.0041
     0.0017
     0.0071
```

(*b*) » $B = [5:1:24]$ % Each separated by one

Output is as follows:

```
B =
    Columns 1 through 12
    5 6 7 8 9 10 11 12 13 14 15 16
    Columns 13 through 20
    17 18 19 20 21 22 23 24
    Length of B can be checked as follows:
    » c = length(B)
    Output is  20
```

(*c*) » len = 15;

```
     A = [-2:4:(len*4 - 4)]
```

Output is as follows:

```
A =
    Columns 1 through 12
    -2 2 6 10 14 18 22 26 30 34 38 42
    Columns 13 through 15
    46 50 54
```

E4.5: Enter the following matrix A in MATLAB and create:

$$A = \begin{bmatrix} 1 & 2 & 3 & 4 & 5 & 6 & 7 & 8 \\ 9 & 10 & 11 & 12 & 13 & 14 & 15 & 16 \\ 17 & 18 & 19 & 20 & 21 & 22 & 23 & 24 \\ 25 & 26 & 27 & 28 & 29 & 30 & 31 & 32 \\ 33 & 34 & 35 & 36 & 37 & 38 & 39 & 40 \end{bmatrix}$$

(*a*) a 4×5 matrix *B* from the 1st, 3rd, and the 5th rows, and the 1st, 2nd, 4th, and 8th columns of the matrix *A*.

(*b*) a 16 element-row vector *C* from the elements of the 5th row, and the 4th and 6th columns of the matrix *A*.

Solution: Here matrix *A* contains numbers in rows in increasing order of 1. Thus following command can be used:

```
» A = [1:8;9:16;17:24;25:32;33:40];
```

Output is something as follows:

```
A =

     1   2   3   4   5   6   7   8
     9  10  11  12  13  14  15  16
    17  18  19  20  21  22  23  24
    25  26  27  28  29  30  31  32
    33  34  35  36  37  38  39  40
```

(*a*)

```
» x = A(:,1:2);
y = A(:,4);
z = A(:,8);
b = [x y z]
```

Output is as follows:

```
b =

     1   2   4   8
     9  10  12  16
    17  18  20  24
    25  26  28  32
    33  34  36  40
```

Further following command is used to reshape it:

```
» B = reshape (b,4,5)

         1  33  26  20  16
   B =   9   2  34  28  24
        17  10   4  36  32
        25  18  12   8  40
```

(*b*)

```
» X = A(5,:);
x = reshape(X, 8, 1)
Y = A(:,4);
y = Y(1:4);
Z = A(:,6);
z = Z(1:4);
c = [x; y; z];
C = reshape(c,1,16)
```

```
C =
    Columns 1 through 12
    33 34 35 36 37 38 39 40 4 12 20 28
    Columns 13 through 16
    6 14 22 30
```

E4.6: Given the function $y=\left(x^{\sqrt{2}+0.02}+e^{x}\right)^{1.8}\ln x$. Determine the value of y for the following values of x: 2, 3, 8, 10, –1, –3, –5, –6.2. Solve the problem using MATLAB by first creating a vector **x**, and creating a vector **y**, using element-by-element calculations.

Solution:

```
» x = [2; 3; 8; 10; -1; -3; -5; -6.2];
» y = x. ^ (sqrt (2) +0.02) +exp(x)) . ^1.8.*log(x) % Vector (y) is vectorized
    using the rules in Table 2.18
» for i = 1: length(x)
   fprintf ('x = %f\ty = %f\n', x (i), y (i));
  end
x = 2.000000     y =           44.456815
x = 3.000000     y =          358.601333
x = 8.000000     y =      3775247.017492
x =10.000000     y =    151523614.007320
x =-1.000000     y =            1.775947
x =-3.000000     y =          -20.763918
x =-5.000000     y =         -113.745299
```

E4.7: Define a and b as scalars, a = 0.75, and b = 11.3, and x, y and z as the vectors, x = 2, 5, 1, 9, y = 0.2, 1.1, 1.8, 2 and z = –3, 2, 5, 4. Use these variables to calculate A given below using element-by-element computations for the vectors with MATLAB.

$$A=\frac{x^{1.1}y^{-2}z^{5}}{(a+b)^{b/3}}+a\frac{\left(\frac{z}{x}+\frac{y}{2}\right)}{z^{a}}$$

Solution: Here also first define all scalars and vectors. Then use inline function and vectorize.

```
» a = 0.75;
```

```
» b = 11.3;
» x = [2;5;1;9]; y = [0.2;1.1;1.8;2]; z = [-3;2;5;4];
» A = inline('x^1.1*y^- 2*z^5/(a + b)^(b/3) + a*(z/x + y/2)/z^a');
» vectorize(A)
```

This gives output as follows:

```
ans(a, b, x, y, z) =x.^1.1.*y.^-2.*z.^5./(a+b).^(b./3)+a.*(z./x+y./2)./z.^a
```

Use this function and paste it as again A to get its value

```
» A= (x.^1.1*y.^- 2*z.^5)/(a + b)^(b/3) + a.*(z./x + y./2)./z.^a
```

The final output is as follows:

```
A =

  - 0.7783 + 0.3257i
    0.4368
    1.4052
    0.6263
```

E4.8: Enter the following three matrices in MATLAB

$$A=\begin{bmatrix}1 & 2 & 3\\ -8 & 5 & 7\\ -8 & 4 & 6\end{bmatrix}, B=\begin{bmatrix}12 & -5 & 4\\ 7 & 11 & 6\\ 1 & 8 & 13\end{bmatrix}, C=\begin{bmatrix}7 & 13 & 4\\ -2 & 8 & -5\\ 9 & -6 & 11\end{bmatrix}$$

and show that

(*a*) $A+B=B+A$

(*b*) $A+(B+C)=(A+B)+C$

(*c*) $7(A+C)=7(A)+7(C)$

(*d*) $A*(B+C)=A*B+A*C$

Solution: Matrices of same sizes can be comfortably multiplied or added. Following statements provides the proof of basic matrix operations:

(*a*) A + B = B + A

```
» A + B

        13 - 3   7
ans =- 1 16  13
      - 7 12  19

» B + A

        13 - 3   7
ans =- 1 16  13
      - 7 12  19
```

(*b*) A + (B + C) = (A + B) + C

```
» A + (B + C)
        20  10 11
ans = - 324   8
         2   6 30
» (A + B) + C
        20  10 11
ans =- 3 24   8
         2   6 30
```

(*c*) 7(A + C) = 7A + 7C

```
 » 7*(A + C)
        56  105  49
ans =-70   91  14
         7 - 14 119
» 7*A+7*C
        56  105  49
ans =- 70 91  14
         7 - 14 119
```

(*d*) A(B + C) = AB + AC

```
» A*(B+C)
        59   52  82
ans =- 57 45 109
     - 72   24  84
» A*B+A*C
        59   52  82
ans = - 5745 109
     - 72   24  84
```

Following examples illustrate the use of polynomials.

E4.9: Consider the polynomials

$$p_1(s) = s^3 + 5s^2 + 3s + 10$$

$$p_2(s) = s^4 + 7s^3 + 5s^2 + 8s + 15$$

$$p_3(s) = s^5 + 15s^4 + 10s^3 + 6s^2 + 3s + 9$$

Determine $p_1(2)$, $p_2(2)$, and $p_3(3)$.

Solution: First define p_1, p_2 and p_3. Then use polyval built-in function.

```
» p1=[1 5 3 10]; p2=[1 7 5 8 15]; p3=[1 15 10 6 3 9];
»  polyval (p1,2)
```

Output is ans = 44

```
» polyval (p2,2)
```

Output is ans = 123

```
» polyval (p3,2)
```

Output is ans = 391

E4.10: The following polynomials are given:

$$p_1(x) = x^5 + 2x^4 - 3x^3 + 7x^2 - 8x + 7$$
$$p_2(x) = x^4 + 3x^3 - 5x^2 + 9x + 11$$
$$p_3(x) = x^3 - 2x^2 - 3x + 9$$
$$p_4(x) = x^2 - 5x + 13$$
$$p_5(x) = x + 5$$

Use MATLAB functions with polynomial coefficient vectors to evaluate the expressions at $x = 2$.

Solution: Define

```
» p1=[1 2 - 3 7 - 8 7]; p2=[1 3 - 5 9 11]; p3=[1 - 2 - 3 9];
      p4=[1 - 5 13]; p5=[1 5];
» polyval(p1,2)
```

Output is ans = 59

```
» polyval(p2,2)
```

Output is ans = 49

```
» polyval(p3,2)
```

Output is ans = 3

```
» polyval(p4,2)
```

Output is ans = 7

```
» polyval(p5,2)
```

Output is ans =7

E4.11: Determine the roots of the following polynomials:

(*a*) $p_1(x) = x^7 + 8x^6 + 5x^5 + 4x^4 + 3x^3 + 2x^2 + x + 1$

(*b*) $p_2(x) = x^6 - 7x^6 + 7x^5 + 15x^4 - 10x^3 - 8x^2 + 7x + 15$

(*c*) $p_3(x) = x^5 - 13x^4 + 10x^3 + 12x^2 + 8x - 15$

(*d*) $p_4(x) = x^4 + 7x^3 + 12x^2 - 25x + 8$

(*e*) $p_5(x) = x^3 + 15x^2 - 23x + 105$

(*f*) $p_6(x) = x^2 - 18x + 23$

(*g*) $p_7(x) = x + 7$

Solution: Define

```
»p1=[1 8 5 4 3 2 1 1]; p2=[1 - 7 7 15 - 10 - 8 7 15];
      p3=[1 - 13 10 12 8 - 15];
»p4=[1 7 12 - 25 8]; p5=[1 15 - 23 105]; p6=[1 - 18 23];
      p7=[1 7];
» a = roots(p1)
```

Output is

```
a =  -7.3898
     -0.6570 +0.3613i
     -0.6570 -0.3613i
      0.4268 +0.5473i
      0.4268 -0.5473i
     -0.0749 +0.7030i
     -0.0749 -0.7030i
» b = roots(p2)
```

Output is

```
b =   5.1625
      2.4906
      0.8961 +0.6912i
      0.8961 -0.6912i
     -1.2191
     -0.6131 +0.6094i
     -0.6131 -0.6094i
» c = roots(p3)
```

Output is

```
c =  12.0867
      1.4875
     -0.7057 +0.7061i
     -0.7057 -0.7061i
      0.8371
» d = roots(p4)
```

Output is

```
d =  -4.1331 +2.2411i
     -4.1331 -2.2411i
      0.8305
      0.4358
```

```
» e = roots(p5)
```

Output is

```
e =16.7477
    0.8738 +2.3465i
    0.8738 -2.3465i
» f = roots(p6)
```

Output is

```
f = 16.6158
     1.3842
» g = roots(p7)
```

Output is g =– 7

They will display the roots. Some may be real and some may be complex.

E4.12: An aluminium thin-walled sphere is used as a marker buoy. The sphere has a radius of 65 cm and a wall thickness of 10 mm. The density of aluminium is 2700 kg/m^3. The buoy is placed in the ocean where the density of the water is 1050 kg/m^3. Determine the height *H* between the top of the buoy and the surface of the water.

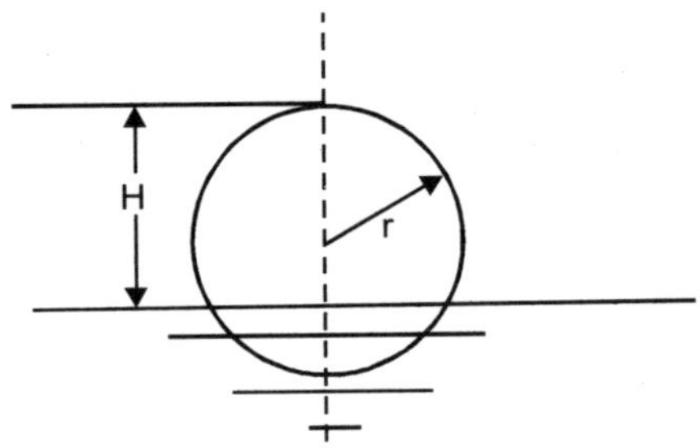

Fig. E4.12

Solution: According to Archimedes principle, the buoyancy force is equal to the weight of the fluid displaced by the body.

Weight of the sphere $W_s = \rho_{Al} g\, 4\pi(r_o^3 - r_i^3)/3$

Weight of water displaced $W_w = \rho_w \pi(2r_o - H)^2(r_o + H)g/3$

Setting the two weights equal to each other gives the following:

$$H^3 - 3r_o H^2 + 4r_o^3 - 4\rho_{Al}(r_o^3 - r_i^3)/\rho_w = 0$$

Last equation is a third degree polynomial in *H*. The root of the polynomial is found in MATLAB as follows:

```
%%%%%%%%%%%%%%%%%%
ro=0.65;rin=0.5;
rhoal=2700;rhow=1050;
a0=4*ro^3- 4*rhoal*(ro^3- rin^3)/rhow;
p=[1 - 3*ro 0 a0];
H=roots(p);
%%%%%%%%%%%%%%%%
```

When the script file is executed in the command window, three roots are obtained for H. The only answer that is physically possible is the first.

» $H=$

$$\begin{array}{r} 2.0544 \\ -0.0522+0.4601i \\ -0.0522-0.4601i \end{array}$$

E4.13: Consider the two matrices

$$A=\begin{bmatrix} 1 & 0 & 2 \\ 2 & 5 & 4 \\ -1 & 8 & 7 \end{bmatrix} \quad \text{and} \quad B=\begin{bmatrix} 7 & 8 & 2 \\ 3 & 5 & 9 \\ -1 & 3 & 1 \end{bmatrix}$$

Using MATLAB, determine the following:

(*a*) **A+B**

(*b*) **AB**

(*c*) $\mathbf{A}^2$

(*d*) $\mathbf{A}^T$

(*e*) $\mathbf{B}^{-1}$

(*f*) $\mathbf{B}^T\mathbf{A}^T$

(*g*) $\mathbf{A}^2+\mathbf{B}^2-\mathbf{AB}$

(*h*) determinant of **A**, determinant of **B** and determinant of **AB**.

Solution: Following commands are executed one after other. Here also matrices are square and of same size. Here inv function for finding inverse of square matrix is illustrated.

```
» A = [1 0 2;2 5 4;- 1 8 7]; B = [7 8 2;3 5 9;- 1 3 1];
» A + B
```

Output is:

```
ans =
      8    8    4
      5   10   13
    - 2   11    8
» A*B
ans =
      5   14    4
     25   53   53
     10   53   77
» A^2
```

```
ans =

   - 1   16   16
     8   57   52
     8   96   79

» A'

ans =

     1    2   -1

     0    5    8

     2    4    7

» inv(B)

ans =

    0.0991  0.0090 - 0.2793
    0.0541- 0.0405   0.2568
  - 0.0631  0.1306 - 0.0495

» B'*A'

ans =

     5   25   10
    14   53   53
     4   53   77

» A^2 + B^2- A*B

ans =

    65  104  100
    10   80   59
   - 1   53   28

» det(A)

ans =

        45

» det(B)

ans =

     - 222

» det(A*B)

ans =

     - 9990
```

E4.14: Use MATLAB to define the following matrices:

$$\mathbf{A}=\begin{bmatrix}2 & 1\\0 & 5\\7 & 4\end{bmatrix},\ \mathbf{B}=\begin{bmatrix}5 & 3\\-2 & -4\end{bmatrix},\ \mathbf{C}=\begin{bmatrix}2 & 3\\-5 & -2\\0 & 3\end{bmatrix} \text{ and } \mathbf{D}=[1\ 2]$$

Compute matrices and determinants if they exist.

(*a*) $(\mathbf{AC}^T)^{-1}$

(*b*) $|\mathbf{B}|$

(*c*) $|\mathbf{AC}^T|$

(*d*) $(\mathbf{C}^T\mathbf{A})^{-1}$

Solution: Following commands are executed to get the result. Here D is a vector while others are matrices.

```
» A = [2 1;0 5;7 4]; B = [5 3; - 2 - 4]; C = [2 3;- 5 - 2;0 3]; D = [1 2];
» inv(A*C¢)
```

Warning: Matrix is close to singular or badly scaled.

Results may be inaccurate. RCOND = 8.732891e–018.

```
ans =
   1.0e + 015 *
    1.0027   0.0286   -0.2865
    0.4011   0.0115   -0.1146
   -0.7353  -0.0210    0.2101
» det(B)
ans =
          - 14
» det(A*C')
ans =
      0
» inv(C'*A)
ans =
      0.0078    0.0359
     -0.0421    0.0062
```

E4.15: Determine the values of x, y, and z for the following set of linear algebraic equations:

$$x_2 - 3x_3 = -7$$

$$2x_1 + 3x_2 - x_3 = 9$$

$$4x_1 + 5x_2 - 2x_3 = 15$$

Solution: This set of equations can be written as: $AX = B$ and X is obtained through matrix inversion

```
» A = [0 1 - 3;2 3 - 1;4 5 - 2];
» B = [- 7;9;15];
» X = inv(A)*B;
» fprintf('solution is %f\n%f\%f\n', X);
```

Output is as follows:

```
solution is 1.666667        3.000000    3.333333
```

Sometimes it is advantageous to use *A*/*B* command instead of finding the inverse.

E4.16: Determine the eigenvalues and eigenvectors of the following matrices using MATLAB.

$$\mathbf{A} = \begin{bmatrix} 1 & -2 \\ 1 & 5 \end{bmatrix} \text{ and } \mathbf{B} = \begin{bmatrix} 1 & 5 \\ -2 & 7 \end{bmatrix}$$

Solution: Following commands are executed for computing eigenvalues and eigenvectors of matrices

```
» A = [1 - 2;1 5]; B = [1 5;- 2 7];
» [V1, D1] = eig(A)
» V1 =
     -0.9597   0.5054
      0.2811  -0.8629
  D1 =
        1.5858    0
        0     4.4142
» [V2, D2] = eig(B)
» V2 =
      0.2673 + 0.8018i    0.2673 -   0.8018i
      0 + 0.5345i           0 - 0.5345i
  D2 =
      4.0000 + 1.0000i          0
      0                    4.0000 -   1.0000i
```

This means eigenvalues of *A* are diagonal elements of *D*1 that is 1.5858 and 4.4142, while corresponding eigenvectors are two columns of matrix *V*1.

4.2 WRITING SCRIPT FILES AND FUNCTIONS

This section illustrates some examples of writing script files. User-defined functions are developed as script files for computing values at several points by repeated execution. All the script files must be stored under dot m (.m) extension with a file name. In command prompt, it is just sufficient to type the file name to execute the file data. Calling a user-defined function from another script file is quite common in MATLAB. Most of the works use script files only rather than command prompt operations.

E4.17: Write a simple script file to find (*a*) dot product (*b*) cross-product of 2 vectors: $a = \hat{j} - \hat{k}$ and $b = 3\hat{i} - 2\hat{j}$

Solution: Here the vectors are represented as row matrices. Then dot and cross commands can be comfortably used. First of all open a new *m* file from start menu and type the following commands:

```
%%%%%%%%%%% This is a script file %%%%%%%%%%%%%
a=[0 1 - 1];
b=[3 - 2 0];
d=dot(a,b);
c=cross(a,b);
fprintf('dot product of vectors is %f\n',d);
fprintf('cross product is\n');disp(c);
%%%%%%%%%%%%%%%%%%%%%%%%%%
```

Then save this file as vector.m and quit to command prompt. Type the following command

```
» vector
```

This gives the output of the program as follows:

dot product of vectors is – 2.000000

cross product is

– 2 – 3 – 3

E4.18: Write a function to find gradient of $f(x,y) = x^2 + y^2 - 2xy + 4$ at (*a*) (1,1) (*b*) (1,– 2) and (*c*) (0,– 3). Use the function name from command prompt.

Solution: This is something like built-in functions but first of all the gradient function is to be created. So open a new file and type the following:

```
%%%%%%%%%%%%%%%%%%%%%%%%%%%%%%%%
function y = grad(X)
y=[2*X(1)- 2*X(2);2*X(2)- 2*X(1)];  % vector of partial derivatives (gradient)
%%%%%%%%%%%%%%%%%%%%%%%%%%%%%%%%%%%%%%%%%%%%%%%%%
```

Thus every function should begin with the word 'function' and the name of function 'grad' is optional and should be saved as *m*–file under the same name. The variables *y* and *X* are local or global (common in calling program also). Now this is saved as grad.m and the file menu is closed. To get the output, pass *X* value as vector through the command prompt.

```
» grad([1 1])
ans =
     0
     0
» grad([1 - 2])
ans =
     6
   - 6
```

```
» grad([0 - 3])
ans =
        6
       -6
```

E4.19: Write MATLAB functions $f = x^2 - 3x + 1$ and $g = e^x - 4x + 6$ and find the result $f(127)/g(5)$ from a script file.

Solution: Create first two function f.m and g.m separately as follows:

```
%%%%  f.m  %%%%%%
function p= f (x)
p=x^2 -  3*x+1;
%%%%%%%%%
%%%%  g.m  %%%%%%
function q=g(x)
q=exp(x) - 4*x+6;
%%%%%%%%%%%%
```

Then write another script file result.m as follows:

```
%%%% result.m  %%%%%
r=f (127)/g(5);
disp(r);
%%%%%%%%%%%%%
```

After doing this, just type 'result' at the command prompt and see the output as follows:

```
» result
» 117.1686
```

Thus in many instances several files are to be called from one script file.

4.3 GRAPHICS

It is very convenient to obtain graph of a function in MATLAB. These graphs allow us to understand insights in parametric variations. Easiest way to create a plot of a function is to use the range variables. Following examples illustrate plotting in two dimensions and three dimensions.

E4.20: Plot the function $y = |x| \cos(x)$ for $-200 \le x \le 200$.

Solution: First define the range variable and then obtain corresponding values of y. Following statements are executed at the command prompt.

```
» x = [- 200 : 1 : 200];
» y = abs(x) .* cos(x);
» plot (x,y)
```

Things like title, x-axis and y-axis, grid can be given after this.

```
»title('|x| cos(x) for -200 \leq x \leq 200')
»xlabel('x')
»ylabel('y')
```

Output is shown in Fig. E4.20.

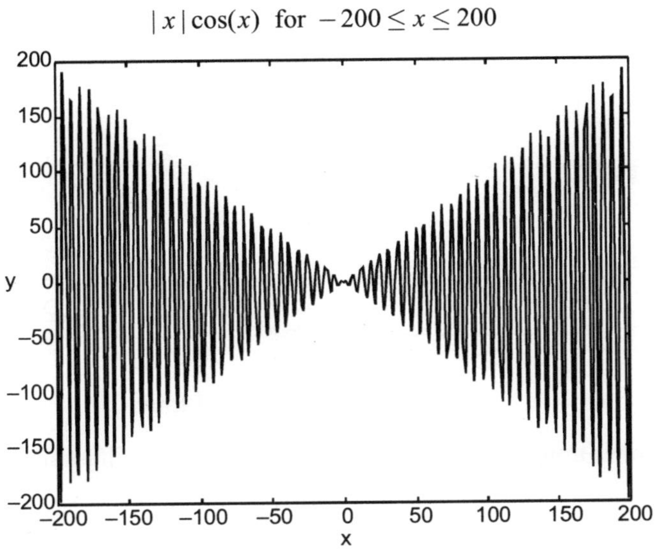

Fig. E4.20

E4.21: Plot the following functions on the same plot for $0 \le x \le 2\pi$ using the plot function:

(*a*) $\sin^2(x)$

(*b*) $\cos^2(x)$

(*c*) $\cos(x)$

Solution: Following program is typed and executed.

```
»x = [0 : .01 : 2*pi];
»ys2 = (sin(x)).^2;
»yc2 = (cos(x)).^2;
»yc = cos(x);
»plot (x, ys2, x, yc2,'—', x, yc, '- o');
»legend('sin^2(x)','cos^2(x)','cos(x)',4)   % 4 refers to quadrant number
   (right bottom)
»xlabel('x')
```

The plot obtained is shown in Fig. E4.21.

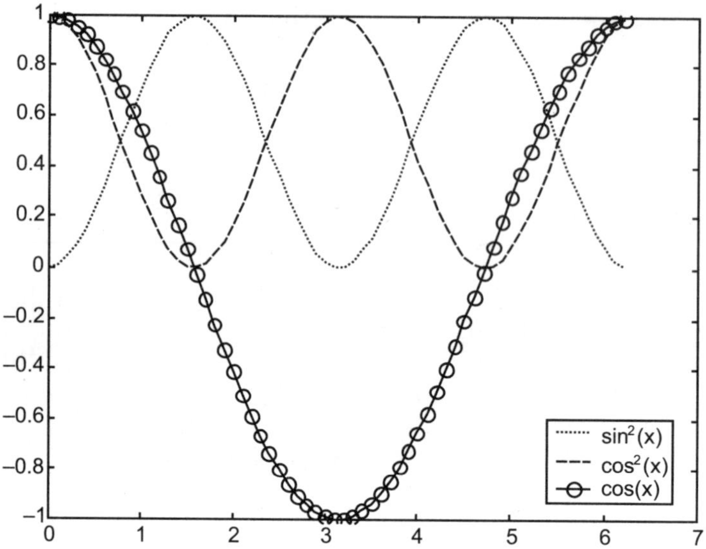

Fig. E4.21 MATLAB output

Sometimes, plots are drawn from index values after computation. This is illustrated with the following example.

E4.22: Plot a graph of the function $y = 45\ \sin(0.4t)$ for $t \in [0, 3]$.

Solution: Here t is varied first and values are substituted in y. Complete program is presented below:

```
» i=1;
» for t=0:0.2:3
    y(i)=45*sin(0.4*t);
     i=i+1;
    end
» T=[0:0.2:3];
» plot(T,y,'-p');
» xlabel('t'); ylabel('y(t)');
» grid on;
```

Output is shown in Fig. E4.22.

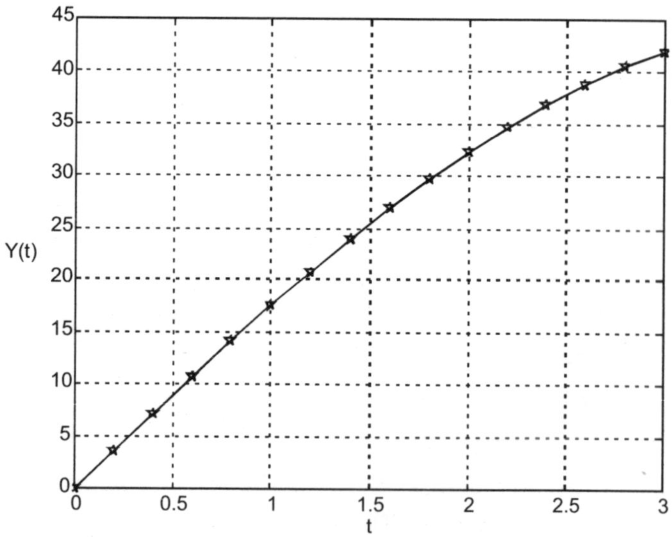

Fig. E4.22 MATLAB output

E4.23: Consider the function $z = 0.56\ \cos(xy)$. Draw a surface plot showing variation of z with x and y. Given $x \in [0, 10]$ and $y \in [0,100]$.

Solution:

```
x = 0:1:10;
y = 0:5:100;
[x, y]=meshgrid (x, y);
z = 0.56.*cos(x.*y);
mesh(x, y, z)
surf(x, y, z)
xlabel('x');ylabel('y');zlabel('z')
title('\bfPlot of variation of amplitude in time and frequency domain');
```

Output is shown in Fig. E4.23.

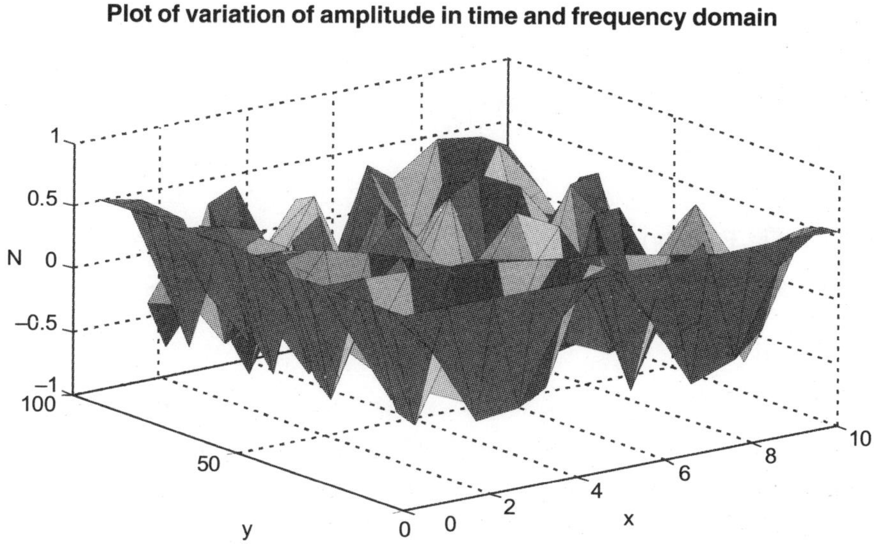

Fig. E4.23 MATLAB output

E4.24: Figure E4.24 shows two boats: boat *A* travels south at a speed of 10 mph, and boat *B* travels east at a speed of 19 mph. The ships are positioned at 8 am are also shown in figure. Write a MATLAB program to plot the distance between the ships as a function of time for the next 5 hours.

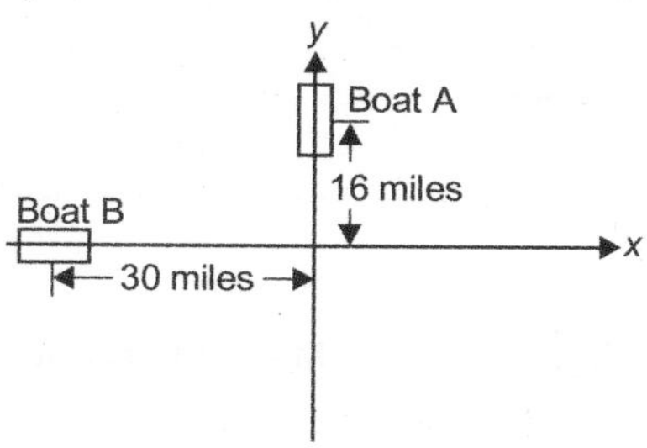

Fig. E4.24

Solution: From basic physics, distance is $d=\sqrt{x^2+y^2}$, where (x, y) are coordinates of boats which can be written as:

$$x = -30 + 10t \text{ and } y = 16 - 19t$$

MATLAB program is given below:

```
%%%%%%%%%%%%%%%%%%%%%%%
t=0:0.2:5;
x=-30+10.*t;
y=16-19.*t;
d=sqrt(x.^2+y.^2);
plot(t,d,'- p');
xlabel('Time (h)');
ylabel('Distance between ships(mile)');
%%%%%%%%%%%%%%%%%%%
```

Output is as shown in Fig. E4.24(*a*).

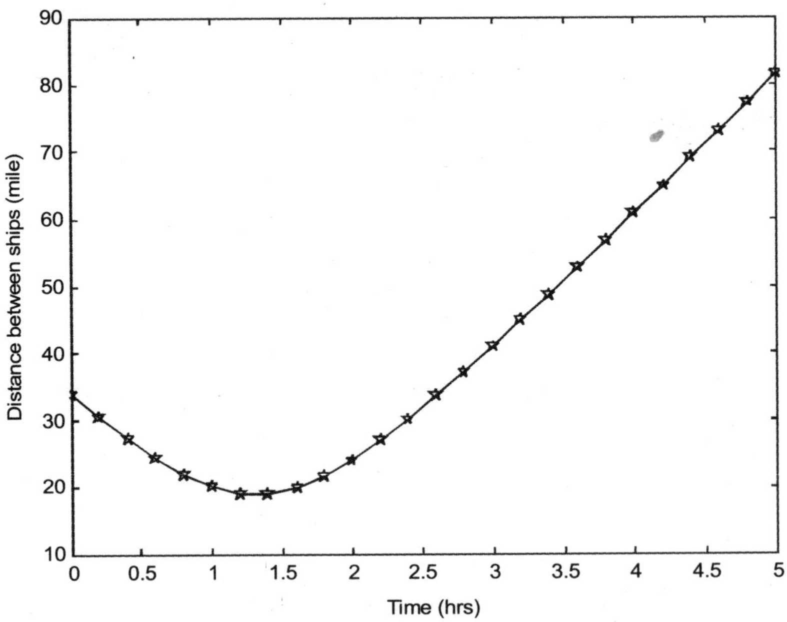

Fig. E 4.24 (a) MATLAB output

4.4 SYMBOLIC CALCULATIONS

Symbolic maths means operations on expression containing undefined variables rather than numbers. Maple engine is part of the Symbolic Math toolbox. It requires symbolic math toolbox. Following examples illustrate Maple expressions that involve symbolic mathematics. Symbolic calculus and solution of one or more equations is illustrated. In Mechanics problems, solving an equation for finding maximum or minimum values is done with symbolic calculations.

E4.25: Consider the given symbolic expressions defined below:

$$S1 = \text{'}2/(x-5)\text{'};\ S2 = \text{'}x \wedge 5 + 9 * x - 15\text{'};\ S3 = \text{'}(x \wedge 3 + 2 * x + 9) * (x * x - 5)\text{'};$$

Perform the following symbolic operations using MATLAB.

(*a*) $S1S2/S3$ (*b*) $S1/S2S3$ (*c*) $S1/(S2)^2$ (*d*) $S1S3/S2$ (*e*) $(S2)^2/(S1S3)$

Solution: Following steps are typed at the command prompt.

```
syms x S1 S2 S3 a b c d e  % Shortcut for constructing symbolic objects
»S1=2/(x - 5);
»S2 =x ^ 5 + 9 * x - 15;
»S3 = (x ^ 3 + 2 * x +9) * (x * x - 5);
»a = S1*S2/S3
»b = S1/(S2*S3)
»c = S1/S2^2
»d = S1*S3/S2
»e = S2^2/(S1*S3)
```

Output is as follows:

```
a =
   2/(x - 5)*(x^5 + 9*x- 15)/(x^3 + 2*x + 9)/(x^2 - 5)
b =
   2/(x - 5)/(x^5 + 9*x - 15)/(x^3 + 2*x + 9)/(x^2 - 5)
c =
   2/(x - 5)/(x^5 + 9*x - 15)^2
d =
   2/(x - 5)/(x^5 + 9*x - 15)*(x^3 + 2*x + 9)*(x^2 - 5)
e =
   1/2*(x^5 + 9*x - 15)^2*(x - 5)/(x^3 + 2*x + 9)/(x^2 - 5)
```

Following example illustrates the method of solving a single equation.

E4.26: Solve the following equations using symbolic mathematics.

(*a*) $x^2+9=0$

(*b*) $x^2+5x-8=0$

(*c*) $x^3+11x^2-7x+8=0$

(*d*) $x^4+11x^3+7x^2-19x+28=0$

(*e*) $x^7-8x^5+7x^4+5x^3-8x+9=0$

Solution: First define the equation and then use solve function.

(*a*)
```
»f = 'x^2 + 9 = 0'; solve (f, 'x')
ans =
        [  3*i]
        [ -3*i]
```

(*b*)
```
»f = 'x^2 + 5*x - 8 = 0'; solve (f, 'x')
ans =
    [ - 5/2+1/2*57^(1/2)]
    [ - 5/2 - 1/2*57^(1/2)]
```

(*c*)
```
» f = 'x^3 +11*x^2 - 7*x + 8 = 0'; solve(f, 'x')
»ans =
    [ - 1/6*(14284+12*144321^(1/2))^(1/3)-284/3/(14284+12*144321^
      (1/2))^(1/3)-11/3]
    [1/12*(14284+12*144321^(1/2))^(1/3)+142/3/(14284+12*144321^
      (1/2))^(1/3) -11/3+1/2*i*3^(1/2)*
    (- 1/6*(14284+12*144321^(1/2))^(1/3)+284/3/(14284+12*144321^
      (1/2))^(1/3))]
    [1/12*(14284+12*144321^(1/2))^(1/3)+142/3/(14284+12*144321^
      (1/2))^(1/3) -11/3-1/2*i*3^(1/2)*
    (- 1/6*(14284+12*144321^(1/2))^(1/3)+284/3/(14284+12*144321^
      (1/2))^(1/3))]
```

This is difficult to read. Thus we can write as eval(ans) to evaluate this solution.

```
» eval(ans)
ans =
    -11.6592
     0.3296 -0.7599i
     0.3296 +0.7599i
```

(*d*)
```
»f = 'x^4 +11*x^3 + 7*x^2-19*x + 28 = 0'; solve (f, 'x'); eval (ans)
ans =
      0.7152 +0.8217i
      0.7152 -0.8217i
      -2.3378
      -10.0927
```

(e)
```
»f = 'x^7 - 8*x^5 + 7*x^4 + 5*x^3 - 8*x + 9'; solve(f, 'x'); eval(ans)
ans =
      -3.1362
      -0.8007 -0.6944i
      -0.8007 +0.6944i
       0.6139 -0.6513i
       0.6139 +0.6513i
       1.7549 -0.3308i
       1.7549 +0.3308i
```

E4.27: Determine the values of *x*, *y*, and *z* for the following set of linear algebraic equations:

$$2x + y - 3z = 11$$
$$4x - 2y + 3z = 8$$
$$-2x + 2y - z = -6$$

Solution: This can be attempted as solution of simultaneous equations.

```
»syms x y z
»% ANOTHER WAY OF WRITING EQUATIONS AS FUNCTION
»eqn1 = 2*x + y - 3*z -11;
»eqn2 = 4*x - 2*y + 3*z - 8;
»eqn3 = -2*x + 2*y - z + 6;
» [x, y, z] = solve (eqn1, eqn2, eqn3)
```

Output is as follows:

```
x =
     3
y =
    -1
z =
    -2
```

E4.28: Figure E4.28 shows a scale with two springs.

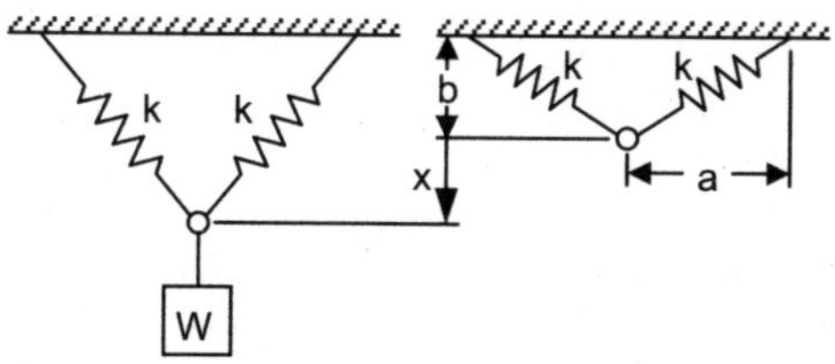

Fig. E4.28

The two springs are unstretched initially and will stretch when a mass is attached to the ring and the ring will displace downwards a distance of x. The weight W of the object is given by

$$W = \frac{2k}{\ell}(\ell - \ell_0)(b+x)$$

where ℓ_0 = initial length of a spring = $\sqrt{a^2+b^2}$

and ℓ = the stretched length of the spring = $\sqrt{a^2+(b+x)^2}$.

If k = spring constant,

write a MATLAB program to determine the distance x when W=350 N. Given: $a = 0.16$ m, $b = 0.045$ m, and the spring constant $k = 3000$ N/m.

Solution: Substituting the expressions for ℓ in terms of x into W, we get

$$W - 2k\left(1 - \frac{\ell_0}{\sqrt{a^2+(b+x)^2}}\right)(b+x) = 0,$$

$$\Rightarrow \qquad 350 - 6000\left(1 - \frac{0.1662}{\sqrt{0.0256+(0.045+x)^2}}\right)(0.045+x) = 0,$$

where $\ell_0 = \sqrt{a^2+b^2} = 0.1662$.

Here we can use the 'solve' command.

MATLAB program for this problem is given below:

```
%%%%%%%%%%%%%
syms x
a=0.16;b=0.045;W=350;k=3000;
lo=sqrt(a^2+b^2);
eqn=350-6000*(1-0.1662/sqrt(0.0256+(0.045+x)^2))*(0.045+x);
» x=solve(eqn)
x =
[ -.92911489530951193030693377153214e-1-.11894920892604858985051553592798*i]
[ -.92911489530951193030693377153214e-1+.11894920892604858985051553592798*i]
[ .13864416954998267806469927987985]
```

There is only one real solution $x = 0.13864416954$ m.

E4.29: Determine the solutions of the following first-order ordinary differential equations using MATLAB's symbolic mathematics.

(*a*) $y' = 8x^2 + 5$ with initial condition $y(2) = 0.5$.

(*b*) $y' = 5x \sin^2(y)$ with initial condition $y(0) = \pi/5$.

(*c*) $y' = 7x \cos^2(y)$ with initial condition $y(0) = 2$.

(*d*) $y' = -5x + y$ with initial condition $y(0) = 3$.

(*e*) $y' = 3y + e^{-5x}$ with initial condition $y(0) = 2$.

Solution: First-order differential equations are solved using dsolve command.

(*a*)
```
» y = dsolve('Dy = 8*x^2 + 5','y(2)=0.5','x')
```
Output is as follows:
```
y =
   8/3*x^3+5*x-185/6
```

(*b*)
```
» y=dsolve('Dy=5*x*sin(y)^2','y(0)=pi/5', 'x')
```
Output contains 2 solutions as follows:
```
y =
   [2*atan(5/2*x^2-1/(5-2*5^(1/2))^(1/2)+1/2*(25*x^4-20*x^2/
         5-2*5^(1/2))^(1/2)+4/(5-2*5^(1/2))+4)^(1/2))]
   [2*atan(5/2*x^2-1/(5-2*5^(1/2))^(1/2)-1/2*(25*x^4-20*x^2/
         (5-2*5^(1/2))^(1/2)+4/(5-2*5^(1/2))+4)^(1/2))]
```

(*c*)
```
» y=dsolve('Dy=7*x*cos(y)^2','y(0)=2','x')
```
Output is as follows:
```
y =
   atan(7/2*x^2+tan(2))
```

(*d*)
```
» y=dsolve('Dy=-5*x+y','y(0)=3', 'x')
y =
   5*x+5-2*exp(x)
```

(*e*)
```
» y=dsolve('Dy=3*y+exp(-5*x)',' y(0)=2', 'x')
```
Output is as follows:
```
y =
   -1/8*exp(-5*x)+17/8*exp(3*x)
```

E4.30: For the following differential equations, use MATLAB to find *x* (*t*) when (*a*) all the initial conditions are zero, (*b*) *x* (*t*) when *x* (0) = 1 and $\dot{x}(0) = -1$.

(*a*) $\dfrac{d^2x}{dt^2} + 10\dfrac{dx}{dt} + 5x = 11$ (*b*) $\dfrac{d^2x}{dt^2} - 7\dfrac{dx}{dt} - 3x = 5$

(*c*) $\dfrac{d^2x}{dt^2} + 3\dfrac{dx}{dt} + 7x = -15$ (*d*) $\dfrac{d^2x}{dt^2} + \dfrac{dx}{dt} + 7x = 26$

Solution: These second order differential equations are solved symbolically.

(*a*)
```
»x=dsolve('D2x=11-10*Dx-5*x','x(0)=0','Dx(0)=0')
y=dsolve('D2x=11-10*Dx-5*x','x(0)=0','Dx(0)=-1')
```
Output is as follows:
```
x =
   11/5+(-11/10-11/20*5^(1/2))*exp((-5+2*5^(1/2))*t)-11/100*
        (-5+2*5^(1/2))*5^(1/2)*exp(-(5+2*5^(1/2))*t)
y =
   11/5+(-11/10-3/5*5^(1/2))*exp((-5+2*5^(1/2))*t)-1/50*
        (-30+11*5^(1/2))*5^(1/2)*exp(-(5+2*5^(1/2))*t)
```

(*b*)
```
»x=dsolve('D2x=5+7*Dx+3*x', 'x(0)=0','Dx(0)=0')
y=dsolve('D2x=5+7*Dx+3*x', 'x(0)=0' ,'Dx(0)=-1')
```
Output is as follows:
```
x =
   -5/3+(5/6+35/366*61^(1/2))*exp(-1/2*(-7+61^(1/2))*t)+5/366*
       (-7+61^(1/2))*61^(1/2)*exp(1/2*(7+61^(1/2))*t)
y =
   -5/3+(5/6+41/366*61^(1/2))*exp(-1/2*(-7+61^(1/2))*t)+1/366*
       (-41+5*61^(1/2))*61^(1/2)*exp(1/2*(7+61^(1/2))*t)
```

(*c*)
```
»x=dsolve('D2x=-15-3*Dx-7*x','x(0)=0','Dx(0)=0')
y=dsolve('D2x=-15-3*Dx-7*x','x(0)=0','Dx(0)=-1')
```
Output is as follows:
```
x =
   -15/7+15/7*exp(-3/2*t)*cos(1/2*19^(1/2)*t)+45/133*19^
        (1/2)*exp(-3/2*t)*sin(1/2*19^(1/2)*t)
y =
   -15/7+15/7*exp(-3/2*t)*cos(1/2*19^(1/2)*t)+31/133*19^
        (1/2)*exp(-3/2*t)*sin(1/2*19^(1/2)*t)
```

(*d*)
```
»x=dsolve('D2x=26-Dx-7*x','x(0)=0','Dx(0)=0')
y=dsolve('D2x=26-Dx-7*x','x(0)=0','Dx(0)=-1')
```
Output is as follows:
```
x =
   26/7-26/7*exp(-1/2*t)*cos(3/2*3^(1/2)*t)-26/63*3^(1/2)*exp
     (-1/2*t)*sin(3/2*3^(1/2)*t)
y =
   26/7-40/63*3^(1/2)*exp(-1/2*t)*sin(3/2*3^(1/2)*t)-26/7*exp
     (-1/2*t)*cos(3/2*3^(1/2)*t)
```

They give solution of *x* in terms of default variable time *t*.

Thus here we have to specify the differential equation in a *string*, using *Dy* for *y* (*t*) and *y* for *y*′(*t*). The last argument *'t'* is the name of the independent variable. If MATLAB can't find a solution it will return an empty symbol. If MATLAB finds several solutions it returns a vector of solutions. Sometimes MATLAB can't find an explicit solution, but returns the *solution in **implicit form***.

E.g., dsolve('*Dy*=1/(*y* – exp(*y*))', '*t*') returns

$$t - 1/2*y\text{\^{}}2 + \exp(y) + C1 = 0$$

Unfortunately MATLAB cannot handle initial conditions in this case. The solution will contain a constant *C*1. We can substitute values for the constant using **subs(sol, '*C1*',value)**. E.g., to set *C*1 to 5 and plot this solution for *t* = –2 to 2 use ezplot(subs(sol, 'C1', 5) , [–2 2]).

Sometimes it is better to solve differential equations numerically.

E4.31: Figure E4.31 shows a water tank (shaped as an inverted frustum cone with a circular hole at the bottom on the side).

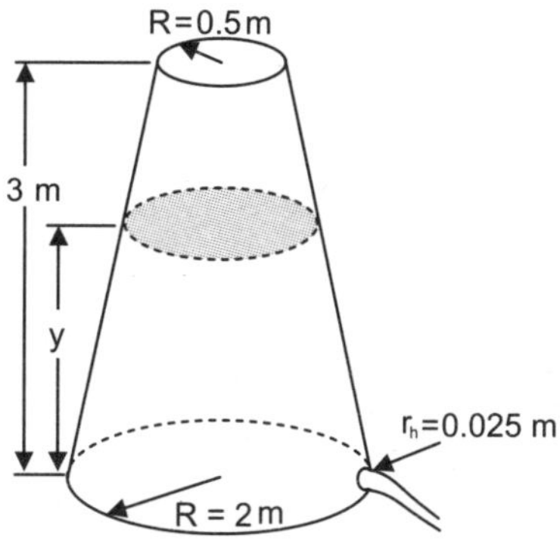

Fig. E4.31 Water tank

The velocity of water discharged through the hole is given by $v = \sqrt{2gy}$ where y = height of the water and g = acceleration due to gravity (9.81 m/s²). The rate of discharge of water in the tank as the water drains out through the hole is given by: $\frac{dy}{dt} = -\frac{\sqrt{2gy}r_h^2}{(2-0.5y)^2}$ where y = height of water and r_h = radius of the hole. Write a MATLAB program to solve and plot the differential equation. Assume, that the initial height of the water is 2.5 m.

Solution: Differential equation is solved using numerical method as follows:

```
%%%%%%%%%%%%%%%%
y0=2.5; % Initial level
tspan = [0:1:5000]; % time - span
[t y] = ode45('flow',tspan,y0); % Solving the ODE
for i = 1:length(tspan)
```

```
if y(i) > =1e-5 T=t(i); end
end
plot(t, y);
xlabel('t(s)');ylabel('level of water (m)');grid on;
axis([0 T 0 y0])
%%%%%%%%%%%%%%%%%%%%%%
```

The function file with the differential equation named flow.m is listed below:

```
function dydt = flow(y,t)
rh=0.025;g=9.81;
dydt=-abs(sqrt(2*g*y))*rh^2/(2 - 0.5*y)^2;
```

Output is shown in Fig. E4.31(*a*) below:

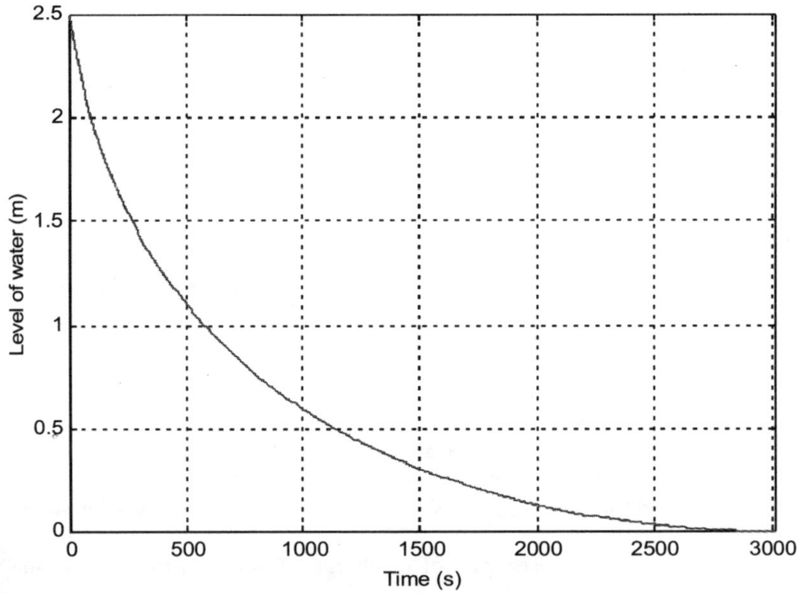

Fig. E4.31(a) MATLAB output

E4.32: An airplane uses a parachute (see Fig. E4.32) and other means of braking as it slow down on the runway after landing. The acceleration of the airplane is given by $a = -0.005\, v^2 - 4$ m/s^2

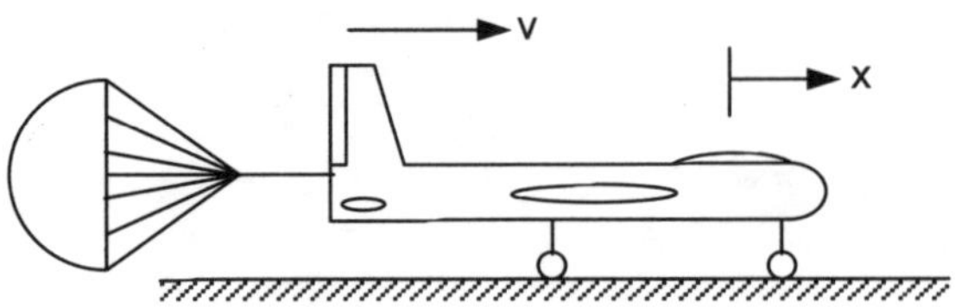

Fig. E4.32

Considering the airplane with a velocity of 500 km/h opens its parachute and starts decelerating at $t = 0$ second, write a MATLAB program to solve the differential equation and plot the velocity from $t = 0$ second until the airplane stops.

Solution: The deceleration of the airplane is : $a = -0.005\ v^2 - 4$

or $$\frac{dv}{dt} = -0.005\ v^2 - 4$$

Further it can be written as $\dfrac{dv}{0.005v^2 + 4} + dt = 0$

The last equation is a first-order ODE that needs to be solved with the initial condition:

$v = 500$ km/h at $t = 0$.

Numerical solution of differential equation is shown in the following program written in a script file (*m*-file), which should be executed by typing the file name in command prompt.

MATLAB Program:

```
%%%%%%%%%%%%%%%%%%%%%%%%%%%
v0=500;
v0mps=v0*1000/3600; % changing velocity into m/s
tspan= [0:0.2:100];
% a vector that specifies the interval of the solution
[t v]=ode45('para',tspan,v0);% Solving the ODE
plot(t,v)
xlabel('t(s)');ylabel('velocity(m/s)')
axis([0 100 0 v0])
%%%%%%%%%%%%%%%%%%%%%%%%%%%%%%
The function file with the differential equation named para.m is listed below:
function dtdv=para(v, t)
dtdv=-(0.005*v^2+4);
```

The output is shown in Fig. E4.32(*a*).

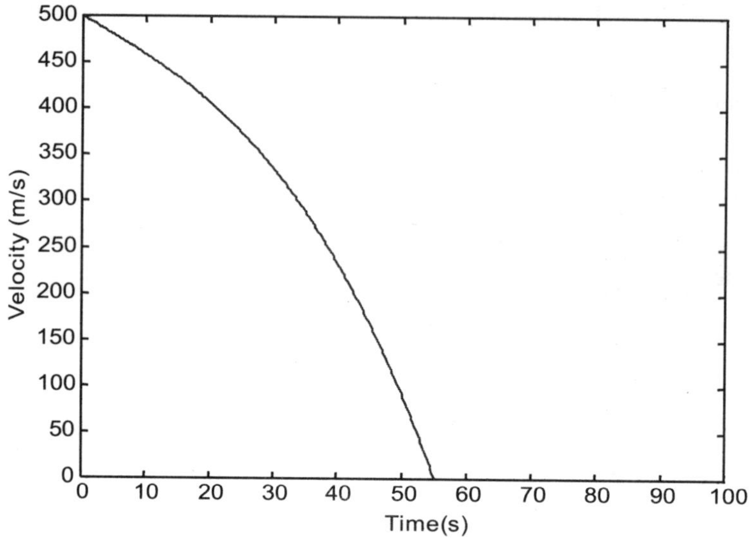

Fig. E4.32 (a) MATLAB output

4.5 DIFFERENTIATION AND INTEGRATION

Another set of important symbolic operations is differentiation and integration. Maple command 'diff' is used for differentiation. Symbolic integration is performed with int command. These two functions are very useful in Mechanics problems.

E4.33: Obtain the first and second derivatives of the following functions using MATLAB's symbolic mathematics.

(*a*) $F(x) = x^5 - 8x^4 + 5x^3 - 7x^2 + 11x - 9$

(*b*) $F(x) = (x^3 + 3x - 8)(x^2 + 21)$

(*c*) $F(x) = (3x^3 - 8x^2 + 5x + 9)/(x + 2)$

(*d*) $F(x) = (x^5 - 3x^4 + 5x^3 + 8x^2 - 13)^2$

(*e*) $F(x) = (x^2 + 8x - 11)/(x^7 - 7x^6 + 5x^3 + 9x - 17)$

Solution:

(*a*)
```
» f ='x^5-8*x^4+5*x^3-7*x^2+11*x-9';
» df=diff(f,'x') % Here x is symbolic variable
» ddf=diff( f,'x',2)
```

Output is as follows:

```
df =
    5*x^4-32*x^3+15*x^2-14*x+11
```

```
ddf =
   20*x^3-96*x^2+30*x-14
```

(*b*)
```
» f ='(x^3 +3*x-8)/(x^2+21)';
df=diff(f,'x') % Here x is symbolic variable
ddf=diff(f,'x',2)
```

Output is as follows:

```
df = (3*x^2+3)/(x^2+21)-2*(x^3+3*x-8)/(x^2+21)^2*x
ddf =
    6*x/(x^2+21)-4*(3*x^2+3)/(x^2+21)^2*x+8*(x^3+3*x-8)
/(x^2+21)^3*x^2-2*(x^3+3*x-8) (x^2+21)^2
```

(*c*)
```
» f ='(3*x^3 -8*x^2+5*x+9)/(x+2)';
df=diff(f,'x') % Here x is symbolic variable
ddf=diff(f,'x',2)
```

Output is as follows:

```
»df =
    (9*x^2-16*x+5)/(x+2)-(3*x^3-8*x^2+5*x+9)/(x+2)^2
»ddf =
    (18*x-16)/(x+2)-2*(9*x^2-16*x+5)/(x+2)^2+2*
 (3*x^3-8*x^2+5*x+9)/(x+2)^3
```

(*d*)
```
» f ='(x^5-3*x^4+5*x^3+8*x^2-13)^2';
df=diff(f,'x') % Here x is symbolic variable
ddf=diff(f,'x',2)
```

Output is as follows:

```
df =
    2*(x^5-3*x^4+5*x^3+8*x^2-13)*(5*x^4-12*x^3+15*x^2+16*x)
ddf =
    2*(5*x^4-12*x^3+15*x^2+16*x)^2+2*(x^5-3*x^4+5*x^3
 +8*x^2-13)*(20*x^3-36*x^2+30*x+16)
```

(*e*)
```
» f ='(x^2+8*x-11)/(x^7-7*x^6+5*x^3+9*x-17)';
df=diff(f,'x')
ddf=diff(f,'x',2)
```

Output is as follows:

```
df =
(2*x+8)/(x^7-7*x^6+5*x^3+9*x-17)-(x^2+8*x-11)/
 (x^7-7*x^6+5*x^3+9*x-17)^2*(7*x^6 42*x^5+15*x^2+9)
ddf =
2/(x^7-7*x^6+5*x^3+9*x-17)-2*(2*x+8)/(x^7-7*x^6+5*x^3
   +9*x-17)^2*(7*x^6 42*x^5+15*x^2+9)+2*(x^2+8*x-11)/
   (x^7-7*x^6+5*x^3+9*x-17)^3*(7*x^6
```

```
42*x^5+15*x^2+9)^2-(x^2+8*x-11)/(x^7-7*x^6+5*x^3+9*x
-17)^2*(42*x^5 210*x^4+30*x)
```

E4.34: Determine the values of the following integrals using MATLAB's symbolic functions.

(*a*) $\int\left(5x^7 - x^5 + 3x^3 - 8x^2 + 7\right)dx$

(*b*) $\int \sqrt{x}\cos x$

(*c*) $\int x^{2/3}\sin^2 2x$

(*d*) $\int_{0.2}^{1.8} x^2 \sin x \, dx$

(*e*) $\int_{-1}^{-0.2} |x| \, dx$

Solution:

(*a*) This integral is computed using following command:

```
int('5*x^7-x^5+3*x^3-8*x^2+7')
```

Output is as follows:

```
» 5/8*x^8-1/6*x^6+3/4*x^4-8/3*x^3+7*x
```

(*b*)

```
» int('sqrt(x)*cos(x)')
```

Output is

```
sin(x)*x^(1/2)-1/2*2^(1/2)*pi^(1/2)*FresnelS(2^(1/2)/pi^(1/2)*x^(1/2))
```

(*c*)

```
» int('x^(2/3)*sin(2*x)^2')
```

Output is

```
ans =
    12/11*x^(11/3)*hypergeom([1, 11/6],[2, 3/2, 17/6],-4*x^2)
```

(*d*)

```
» int('x^2*sin(x)',0.2,1.8)
ans =
    -31/25*cos(9/5)+18/5*sin(9/5)-49/25*cos(1/5)-2/5*sin(1/5)
» eval(ans)
ans =
    1.7872
```

(*e*)

```
» int('abs(x)',-1,-0.2)
ans = 12/25
```

E4.35: Use MATLAB to calculate the following integral: $\int_0^5 \frac{1}{0.8x^2+0.5x+2}dx$

Solution: Following statement is typed at the command prompt:

```
» int('1/(0.8*x^2+0.5*x+2)',0,5)
```

Output will come like this

```
ans =
        .877422374644841693246125127907 06
```

E4.36: Use MATLAB to calculate the following integral: $\int_0^{10} \cos^2(0.5x)\sin^4(0.5x)dx$

Solution: Following statement is typed at the command prompt:

```
» int('cos(0.5*x)^2*sin(0.5*x)^4',0,10)
```

Output will come like this

```
ans =
        .603179124545753798823939522153 36
```

E4.37: The variation of gravitational acceleration g with altitude y is given by:

$$g = \frac{R^2}{(R+y)^2} g_o ,$$

where R= 6371 km is radius of the earth and g_o= 9.81 m/s^2 is gravitational acceleration at sea level.

The change in the gravitational potential energy ΔU of an object that is raised up from the earth is given by:

$$\Delta U = \int_0^y mgdy$$

Determine the change in the potential energy of a satellite with a mass of 500 kg that is raised from the surface of the earth to a height of 800 km.

Solution: The integral is computed by substituting all given values in meters as follows:

```
» dU=int('500*(6371e3)^2/(6371e3 + y)^2*9.81', 0, 800e3)

dU =

     3486236787.0589875888997350439269
```

In some engineering dynamic applications like equations of motion in vibrations, Laplace transforms are quite useful. They convert differential equation into algebraic form. MATLAB employs lapace command.

E4.38: Find the Laplace transform of the following function using MATLAB:

$$f(t) = 7t^3\cos(5t+60°)$$

Solution: Following MATLAB script is used to get the Laplace transform:

```
» syms t % tell MATLAB that "t" is a symbol.
» f = 7 * t^3*cos(5*t + (pi/3)); % define the function.
» laplace(f )
ans =
      -84/(s^2+25)^3*s^2+21/(s^2+25)^2+336*(1/2*s-5/2*3^(1/2))/
(s^2+25)^4*s^3-168*(1/2*s-5/2*3^(1 2))/(s^2+25)^3*s
» pretty(laplace(f)) % the pretty function prints symbolic output
% in a format that resembles typeset mathematics.
```

$$\frac{-84s^2}{(s^2+25^3)}+\frac{21}{(s^2+25)^2}+\frac{336\left(\frac{1}{2}s-\frac{5}{2}(3)^{\frac{1}{2}}\right)\times s^3}{(s^2+25)^4}-\frac{168\left(\frac{1}{2}s-\frac{5}{2}\times(3)^{\frac{1}{2}}\right)\times s}{(s^2+25)^3}$$

E4.39: Use MATLAB program to find the transforms of the following functions.

(*a*) $f(t) = -7t\,e^{-5t}$

(*b*) $f(t) = -3\cos 5t$

(*c*) $f(t) = t\sin 7t$

(*d*) $f(t) = 5\,e^{-2t}\cos 5t$

(*e*) $f(t) = 3\sin(5t + 45°)$

(*f*) $f(t) = 5\,e^{-3t}\cos(t - 45°)$

Solution: Following command prompt scripts are self-understanding.

(*a*)
```
»syms t x
» f = -7*t*exp(-5*t);
» laplace(f, x)
ans =
      -7/(x+5)^2
```

(*b*)
```
»syms t x
»f = -3*cos(5*t);
» laplace( f, x)
ans =
      -3*x/(x^2+25)
```

(*c*)
```
»syms t x
» f = t*sin(7*t);
» laplace( f, x)
ans =
      1/(x^2+49)*sin(2*atan(7/x))
```

(*d*)
```
»syms t x
»f = 5*exp(-2*t)*cos(5*t);
» laplace( f, x)
ans =
     5*(x+2)/((x+2)^2+25)
```

(*e*)
```
»syms t x
»f = 3*sin(5*t + (pi/4));
» laplace( f, x)
ans =
     3*(1/2*x*2^(1/2)+5/2*2^(1/2))/(x^2+25)
```

(*f*)
```
»syms t x
»f = 5*exp(-3*t)* cos(t-(pi/4));
» laplace( f, x)
ans =
     5*(1/2*(x+3)*2^(1/2)+1/2*2^(1/2))/(x+3)^+2+1)
```

4.6 SUMMARY

The MATLAB features such as the three-dimensional graphs and symbolic calculations are illustrated with several numerical examples.

CHAPTER 5

MATLAB—A COMPUTATIONAL TOOL FOR MECHANICS

This chapter gives additional examples of using MATLAB for Mechanics. In all the problems, formulations are explained first and then MATLAB is used to solve the equations or to plot the figures or to get the optimum data. Objective of presenting this chapter is to provide the insights of using MATLAB in classical mechanics.

5.1 STATICS

Problems in statics are given starting from equilibrium of current forces to evaluation of frictional forces in order. Equations of equilibrium are used to compute the resultant forces acting on the body or the magnitudes of unknown forces.

ES5.1: Determine the resultant or equivalent force applied to the bracket for the system shown in Fig. ES5.1

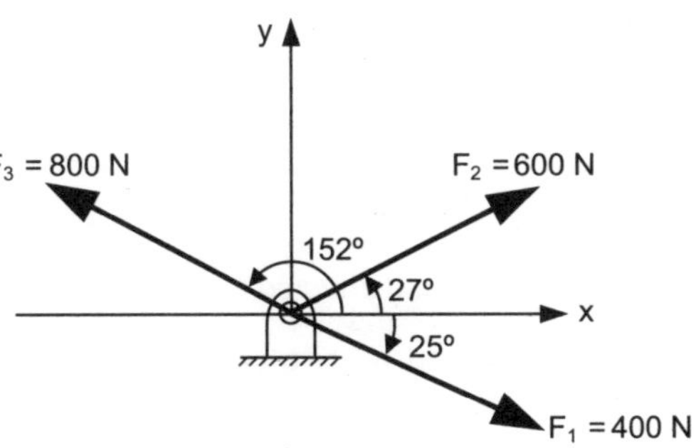

Fig. ES5.1

Solution: A force is a vector (physical quantity that has both magnitude and direction). In a cartesian coordinate system a 2-*D* vector *F* can be written as:

$$F = F_x\,\hat{i} + F_y\,\hat{j} = F\cos\theta\,\hat{i} + F\sin\theta\,\hat{j}$$

Here F is magnitude and θ is its angle relative to the x-axis. F_x and F_y are the components of F in the directions of x and y axis respectively and $\hat{i}, \hat{j}$, are the unit vectors in these directions.

F can be written in magnitude and phase form:

$$F = \sqrt{F_x^2 + F_y^2} \quad \text{and} \quad \theta = \tan^{-1}\frac{F_y}{F_x}$$

MATLAB solution given below follows three steps:

1. Write each force as a vector with two elements, where the first element is the x component of the vector and the second element is the y component.
2. Determine the vector form of the equivalent force by adding the vectors.
3. Determine the magnitude and direction of the equivalent force.

Complete MATLAB program is given below:

```
%%%%%%%%%%%%%%%%%%%%%%%%%%%%%%%
F1m=400;F2m=600;F3m=800;
th1=-25*pi/180;th2=27*pi/180;th3=152*pi/180;
F1=F1m*[cos(th1) sin(th1)]; %DEFINITION OF VECTOR
F2=F2m*[cos(th2) sin(th2)]; %DEFINITION OF VECTOR
F3=F3m*[cos(th3) sin(th3)]; %DEFINITION OF VECTOR
Fnet=F1+F2+F3;
Fnet_m=sqrt(Fnet(1)^2+Fnet(2)^2);
Fnet_d=atan(Fnet(2)/Fnet(1))*180/pi;
fprintf('\nEquivalent force has a magnitude of %f and directed %f
degrees (CCW) relative to x-axis\n',Fnet_m,Fnet_d);
%%%%%%%%%%%%%%%%%%%%%%%%%%%%%%%%
```

This gives output as follows:

»

Equivalent force has a magnitude of 515.520346 and directed 68.281221degrees (CCW) relative to x-axis

ES5.2: Figure ES5.2 shows a load W supported by two cables AC and BC.

Use the following three sets of values:

(i) $\alpha = 30^\circ, \beta = 80^\circ, W = 1.5$ kN

(ii) $\alpha = 60^\circ, \beta = 40^\circ, W = 2.5$ kN

(iii) $\alpha = 35^\circ, \beta = 65^\circ, W = 1.2$ kN

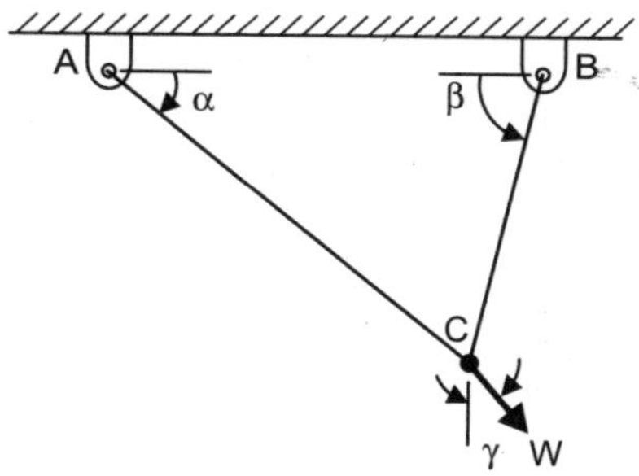

Fig. ES5.2

(*a*) Determine the tension in *AC* and *BC* as a function of *W* and γ.

(*b*) Write a MATLAB program to plot the tensions T_{AC} and T_{BC} for values of γ ranging from $(\beta - 90^\circ)$ to $(90^\circ - \alpha)$.

(*c*) Determine from the plots the value of γ for which the tension in the cables T_{AC} and T_{BC} is as small as possible and the respective values of T_{AC} and T_{BC}.

Solution: Free-body diagram is shown in Fig. ES5.2 (*a*) showing all the forces in the system.

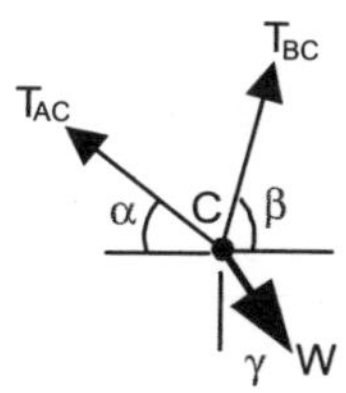

Fig. ES5.2 (a) Free-body diagram

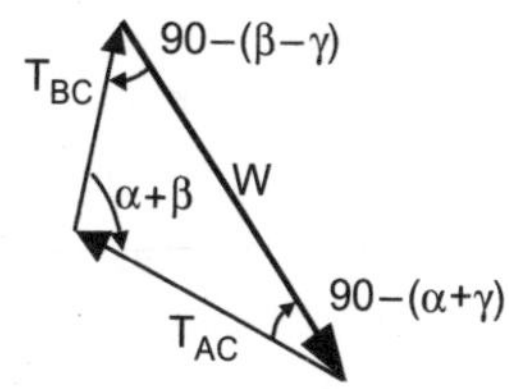

Fig. ES5.2 (b) Force-triangle

A force-triangle can be constructed as shown in Fig. ES5.2 (*b*).

Applying law of sines, we get:

$$\frac{W}{\sin(\alpha+\beta)} = \frac{T_{AC}}{\cos(\beta-\gamma)} = \frac{T_{BC}}{\cos(\alpha+\gamma)}$$

Hence $$T_{AC} = \frac{W}{\sin(\alpha+\beta)}\cos(\beta-\gamma)$$

and $$T_{BC} = \frac{W}{\sin(\alpha+\beta)}\cos(\beta+\gamma)$$

Complete MATLAB script is given below:

```
%%%%%%%%%%%%%%%%%%%%%%%%%%%%%%
W=1500; alpha=30;beta=80;a=alpha*pi/180;b=beta*pi/180;
t= (b-a)/2;
```

```
t1=beta-90;
t2=90-alpha;
gama=[t1:5:t2];
ga=gama*pi/180;
Tac=W.*cos(b-ga)./sin(a+b);
Tbc=W.*cos(a+ga)./sin(a+b);
plot(gama,Tac,'-*',gama,Tbc,'-p');
xlabel('\gamma(degrees)');
ylabel('Tension in (N)');
legend('Tension in AC','Tension in BC');
title('Variation of tensions \alpha=30 deg, \beta=80 deg., W=1.5 kN');
%%%%% Finding value of gamma for which the tension in 2 cables is
as small as possible %%%%
f=abs(Tac-Tbc);
[fmin,i]=min(f);
Ang_min=gama(i);
Tsm=Tac(i);
fprintf('\n');
fprintf('(b) angle for which tension is as small as possible
gama=%g degrees\n',Ang_min);
fprintf('(c) The corresponding tension is %g N\n', Tsm);
%%%%%%%%%%%%%%%%%%%%%%%%%%%%%%%%%%%%
```

Output is shown in Fig. ES5.2 (c).

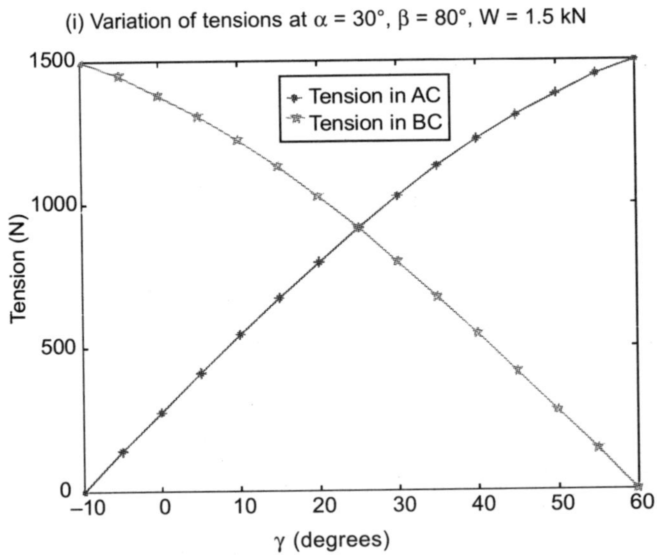

Fig. ES5.2 (c) MATLAB output

Following additional output will also come at the command prompt.

(*a*) Angle for which tension is as small as possible gama=25 degrees

(*b*) The corresponding tension is 915.581 N

Running the program for other two cases of α, *W* and β, we get following outputs (see Fig. ES5.2 (*d*) and (*e*)).

(*i*) (*a*) angle for which tension is as small as possible gama = –10 degrees

(*b*) The corresponding tension is 1631.76 N

(*ii*) (*a*) angle for which tension is as small as possible gama = 15 degrees

(*b*) The corresponding tension is 783.244 N

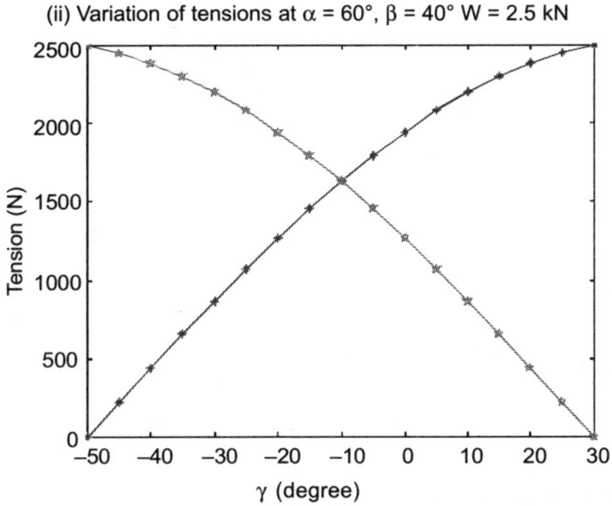

Fig. ES5.2 (d) MATLAB output

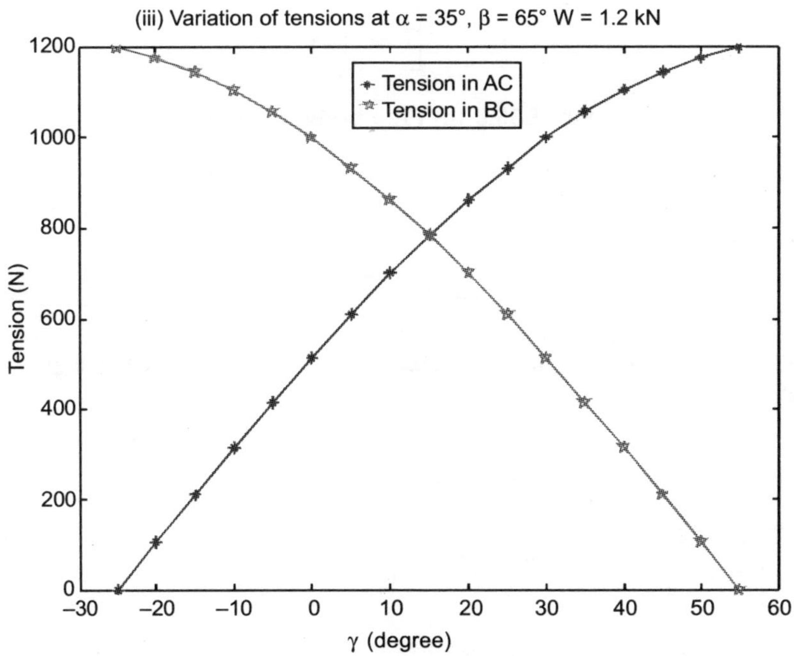

Fig. ES5.2 (e) MATLAB output

ES5.3: A collar P sliding freely on the horizontal frictionless rod as shown in Fig. ES5.3, is attached with a spring (spring constant k).

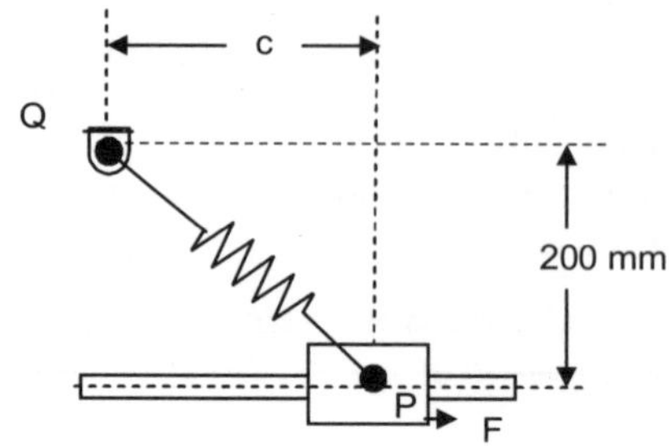

Fig. ES5.3

The spring is undeformed when the collar is directly below the support Q. Express the force F required to maintain the equilibrium of the system in terms of k and the distance c. Plot F as a function of c for values of c from 0 to 400 mm when (*i*) $k = 1$ N/mm, (*ii*) $k = 2$ N/mm and (*iii*) $k = 3$ N/mm

Solution: Initially when $c = 0$, the spring is not stretched. Thus the stretch of the spring Δ is given by:

$$\Delta = PQ - 200 = \sqrt{200^2 + c^2} - 200$$

Hence force in spring $F_s = k\Delta$

The free-body diagram is shown in Fig. ES5.3 (*a*).

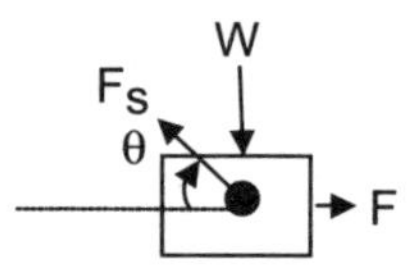

Fig. ES5.3 (a) Free-body diagram

For equilibrium of the collar:

$$\Sigma F_x = 0 \Rightarrow F - F_s \cos\theta = 0$$

$$\Rightarrow \quad F = F_s \frac{c}{\sqrt{200^2 + c^2}} = k\frac{c(\sqrt{200^2 + c^2} - 200)}{\sqrt{200^2 + c^2}}$$

Complete MATLAB program for studying the variation of F as a function c and k is given below:

```
%%%%%%%%%%%%%%%%%%%%%%%%
k=[1;2;3];
c=0:40:400;
num=(sqrt(200^2+c.^2)-200).*c;
den=sqrt(200^2+c.^2);
F1=k(1)*num./den;
F2=k(2)*num./den;
F3=k(3)*num./den;
```

```
plot(c,F1,'-*',c,F2,'-p',c,F3,'-o')
xlabel('c(mm)');
ylabel('Force on collar (in N)');
legend('k=1 N/mm', 'k=2 N/mm', 'k=3 N/mm')
grid on;
%%%%%%%%%%%%%%%%%%%%%%%%
```

Output is shown in Fig. ES5.3 (*b*).

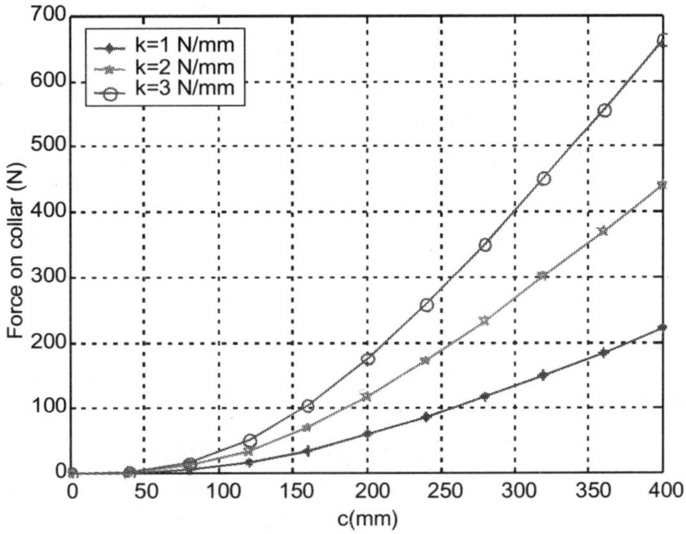

Fig. ES5.3 (b) MATLAB output

ES5.4: Figure ES5.4 shows an acrobat walking on a tight rope attached to supports at *A* and *B*. The friction between his shoes and the rope is high enough to prevent him from slipping. The weight of the rope and elastic deformation may be neglected. Write a MATLAB program to determine the deflection *y* and the tension in *AC* and *BC* of the rope for values of *x* ranging from 0.2 m to 50 m using 0.5 m increments. Determine also

(*a*) the maximum deflection of the rope

(*b*) the maximum tension in the rope

(*c*) the minimum values of the tension in portions *AC* and *BC* of the rope.

The length of the tight rope = 90.6 m

The combined mass of the acrobat and his balancing pole is 80 kg.

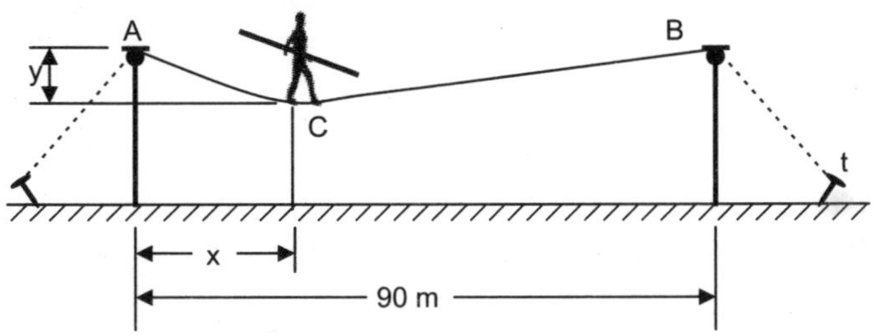

Fig. ES5.4

Solution: First of all from the geometry (see Fig.ES5.4 (*a*)), let us obtain a relation between *x* and *y* as follows:

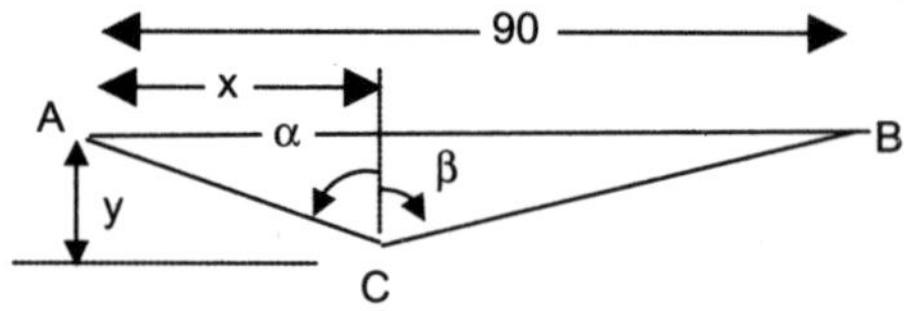

Fig. ES5.4 (a) Geometry at an instant

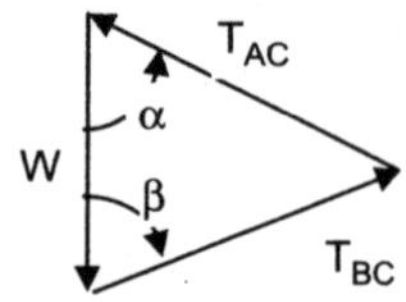

Fig. ES5.4 (b) Force-triangle

From triangle *ABC*: $\tan\alpha = x/y$ and $\tan\beta=(90-x)/y$.

Also rope-length $L = AC + BC$

$$\Rightarrow \qquad L=\sqrt{x^2+y^2}+\sqrt{(90-x)^2+y^2}$$

$$\Rightarrow \qquad \text{deflection } y=\frac{1}{2L}\sqrt{(L^2+180x-90^2)^2-4L^2x^2}$$

Now force triangle is shown in Fig. ES5.4 (*b*).

From law of sines, $\dfrac{T_{AC}}{\sin\beta}=\dfrac{T_{BC}}{\sin\alpha}=\dfrac{W}{\sin(180°-(\alpha+\beta))}$

Thus $\qquad T_{AC}=W\dfrac{\sin\beta}{\sin(\alpha+\beta)}$ and $T_{BC}=W\dfrac{\sin\alpha}{\sin(\alpha+\beta)}$

Complete MATLAB script for finding the tensions as a function of *x* is given below:

```
%%%%%%%%%%%%%%%%%%
L=90.6;
m=80;g=9.81;
x=[0.2:0.5:50];
y=1/(2*L)*sqrt((L^2+180*x-90^2).^2-4*L^2*x.^2);
alfa=atan(x./y);
beta=atan((90-x)./y);
TAC=m*g*sin(beta)./sin(alfa+beta);
TBC=m*g*sin(alfa)./sin(alfa+beta);
[y_max,i]=max(y);
TAC_max=max(TAC);
TBC_max=max(TBC);
TAC_min=min(TAC);
TBC_min=min(TBC);
if TAC_max>TBC_max
T=TAC_max;
```

```
else
T=TBC_max;
end
z=[x;y;TAC;TBC];
fprintf(' x(m) y(m) TAC(N) TBC(N)\n');
fprintf(' %f\t%f\t%f\t%f\n',z);
fprintf('(a) Maximum deflection of rope: %f m \n',y_max);
fprintf('(b) Maximum tension in rope: %g N \n',T);
fprintf('(c)Minimum tensions in AC=%gN and in BC=%gN \n',
TAC_min, TBC_min);
%%%%%%%%%%%%%%%%%%%%%%%%%%
```

Output is as follows:

x(m)	y(m)	TAC(N)	TBC(N)
0.200000	0.771176	808.961407	203.088407
0.700000	1.087578	926.046798	501.230956
1.200000	1.328284	1043.536796	699.630115
1.700000	1.529460	1151.220886	855.959251
2.200000	1.705157	1249.767379	987.987179
2.700000	1.862597	1340.612080	1103.759200
3.200000	2.006085	1424.994651	1207.682423
3.700000	2.138431	1503.892371	1302.467964
....	....	....	....
....	....	....	....
43.700000	5.202662	3415.147955	3412.541951
44.200000	5.203994	3415.520590	3413.917306
44.700000	5.204691	3415.477241	3414.876090
45.200000	5.204755	3415.017892	3415.418654
45.700000	5.204184	3414.142375	3415.545197
46.200000	5.202979	3412.850369	3415.255765
46.700000	5.201139	3411.141403	3414.550252
47.200000	5.198664	3409.014848	3413.428400
47.700000	5.195552	3406.469924	3411.889800
48.200000	5.191803	3403.505693	3409.933888
48.700000	5.187415	3400.121057	3407.559945
49.200000	5.182387	3396.314762	3404.767099
49.700000	5.176716	3392.085389	3401.554317

(*a*) Maximum deflection of rope: 5.204755 m

(*b*) Maximum tension in rope: 3415.55 N

(*c*) Minimum tensions in $AC = 808.961$ N and in $BC = 203.088$N

ES5.5: A barge is pulled by two tugboats as shown in Fig. ES5.5. To move the barge along the water properly, the tug-boats must exert a resultant force of 3 kN along the direction of motion of the barge. Supposing that tug-2 can move anywhere such that $0 \leq \beta \leq 90°$, determine angle β at which a minimum tension in the rope connected to tug-2 is obtained.

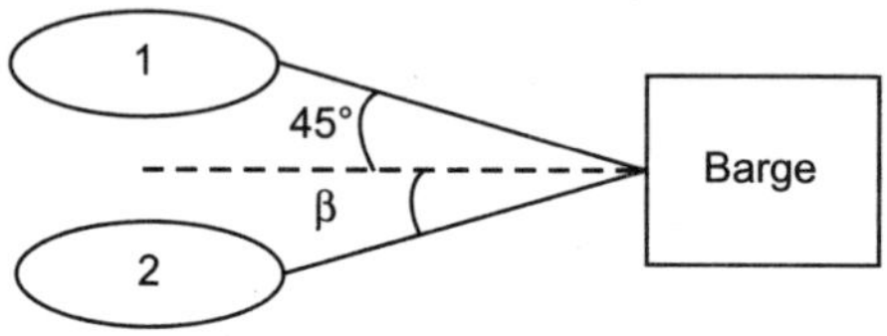

Fig. ES5.5

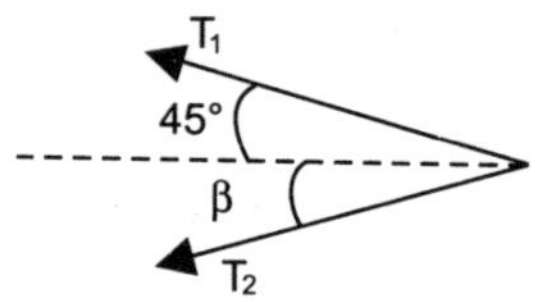

Fig. ES5.5 (a) Free-body at the barge

Solution: This problem deals with finding maximum and minimum force T_2 in the rope between the tug-2 and the barge when the angle β varies from 0 to 90 degrees. First of all draw the free-body diagram and then apply the equilibrium equations. Free-body diagram is shown in Fig. ES5.5(*a*).

Resolving the forces along the direction of motion of barge:

$$R = T_1 \cos \pi/4 + T_2 \cos \beta$$

$$\Rightarrow \quad 3000 = T_1 \cos \pi/4 + T_2 \cos \beta \quad \text{...(1)}$$

Resolving forces perpendicularly

$$T_1 \sin \pi/4 = T_2 \sin \beta \quad \text{...(2)}$$

Solving equations (1) and (2):

$$T_2 = \frac{3000}{(\sin \beta + \cos \beta)}$$

Following MATLAB script is used:

```
%%%%%%%%%%%%%%%%%%%%%%%%%
beta=[0:1:90];
b=beta*pi/180;
T2=3000./(sin(b)+cos(b));
[Tmin,i]=min(T2);
[Tmax,j]=max(T2);
plot(beta,T2);
xlabel('\beta(degree)');ylabel('T_2(N)');grid on;
fprintf('\nMaximum tension:%f N at angle %f degrees and min
tension:%f N at an angle %f deg\n',Tmax, beta(j), Tmin, beta(i));
%%%%%%%%%%%%%%%%%%%%%%%%%%
```

Output is shown below: (Fig. ES5.5 (*b*)) along with following command prompt result:

»

Maximum tension:3000.000000 N at angle 0.000000 degree and min tension:2121.320344 N at an angle 45.000000 degree.

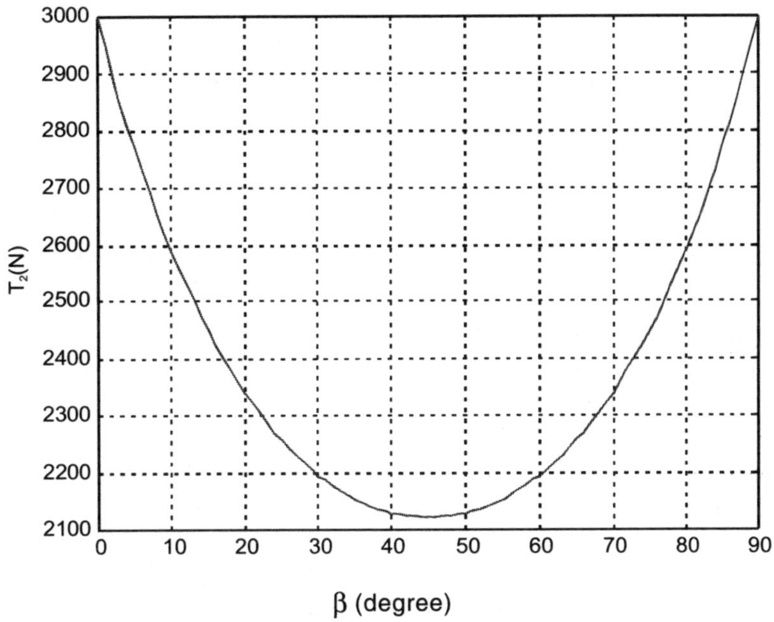

Fig. ES5.5 (b) MATLAB output

ES5.6: A 50 kg crate is held in equilibrium as it rests on a frictionless inclined plane making an angle θ with the horizontal as shown in Fig. ES5.6. Compute the tension in the cable *T* and the normal force at the surface for θ at 2° increments, increments increasing from 0 to 90°. Is there some value of θ for which $T = N$? If cable is replaced by a spring of stiffness 100 N/m, plot the deflection of the cable as a function of θ.

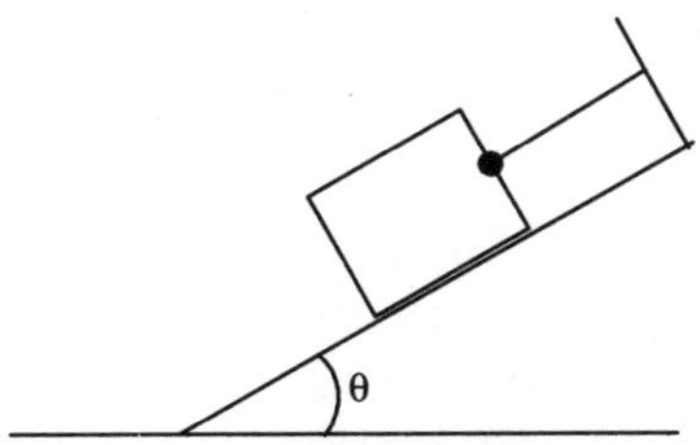

Fig. ES5.6

Solution: This problem deals with equilibrium of particle. It requires first to draw the free-body diagram as shown in Fig. ES5.6(*a*).

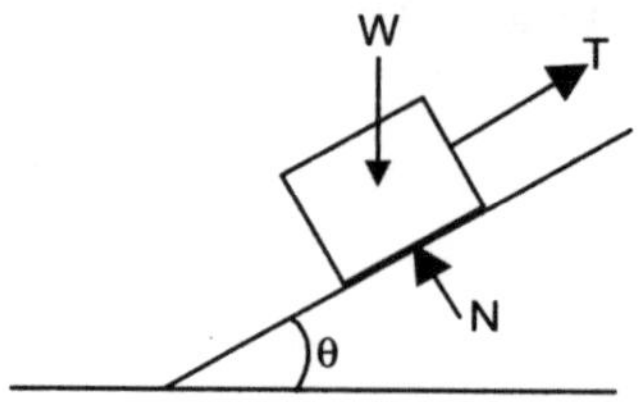

Fig. ES5.6 (a) Free-body diagram

Resolving forces parallel to and perpendicular to the inclined plane, we obtain:

$$T - W\sin\theta = 0$$

$$N - W\cos\theta = 0$$

For equilibrium, the deflection of spring is $x = T/k$

Complete MATLAB program is given below:

```
%%%%%%%%%%%%%%%%%%%%
g=9.81;m=50;k=100;
theta=[0:2:90];
th=theta*pi/180;
T=m*g.*sin(th);
N=m*g.*cos(th);
x=T./k;
v=[theta;T;N];
fprintf('theta(deg)T(N)N(N)\n');
fprintf('%f\t%f\t%f\n',v);
plot(theta,x);
xlabel('\Theta(degree)');
ylabel('Spring deflection (m)');
grid on;
%%%%%%%%%%%%%%%%%%%%
```

Output is as follows:

Theta(deg)	T(N)	N(N)
0.000000	0.000000	490.500000
2.000000	17.118203	490.201201
4.000000	34.215550	489.305167
6.000000	51.271211	487.812990
....		
....		
86.000000	489.305167	34.215550
88.000000	490.201201	17.118203
90.000000	490.500000	0.000000

Obviously at $\theta = 45°$, $T = N$.

Figure ES5.6 (*b*) shows the variation of spring deflection as a function of angle of inclined plane.

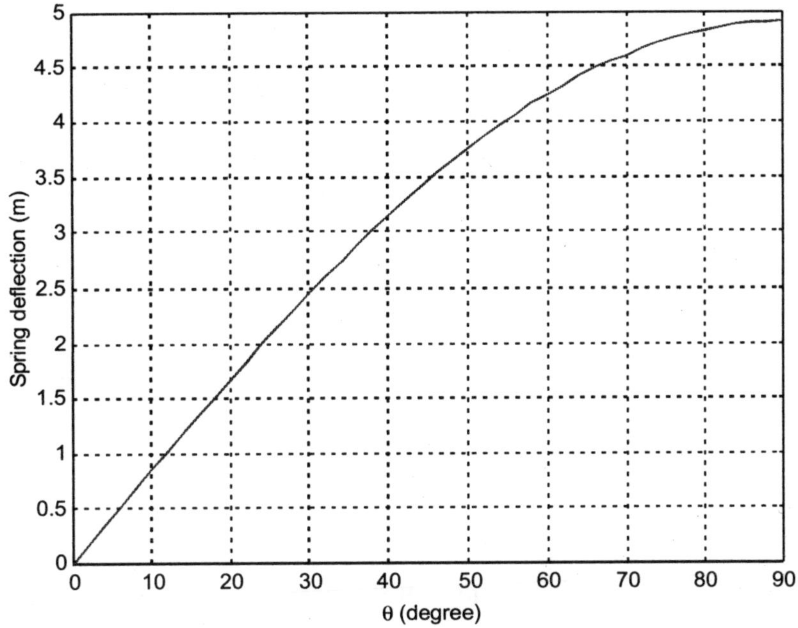

Fig. ES5.6 (b) MATLAB output

ES5.7: Write a generalized MATLAB script to find the perpendicular distance between the line of a force $F = 6\hat{i} + 10\hat{j} + 18\hat{k}$ acting at point *A*(1,2,3) and the line *OB* shown in Fig. ES5.7 having direction angles (60°, 100° and acute angle).

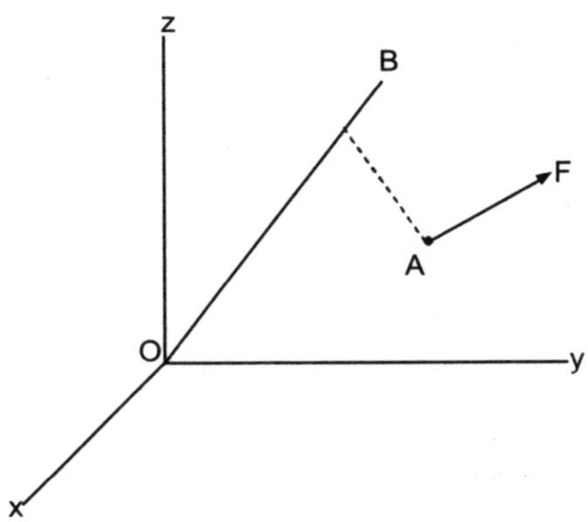

Fig. ES5.7

Solution: The moment of a force *F* about an axis *p* is defined as the projection *OB* on *p* of the moment M_o as shown in Fig. ES5.7(*a*).

One of the most powerful theorems of statics is the principle of moments, or Varignon's theorem. It states that: "The moment caused by the resultant force (of some system of forces) about some arbitrary point is equal to the sum of the moments due to all of the component forces of the system."

Denoting λ as the unit vector of axis p, the moment M_p of force F about an axis p can be expressed as *scalar product* (dot product)

$$M_p = \lambda \cdot M_O$$

where M_O is moment of force about point O with respect to x-y-z coordinate frame, given by

$$M_O = r \times F$$

Also the force F can be resolved on to the axis p as

$$F_1 = F \cdot \lambda$$

Hence the magnitude of perpendicular force on p-axis is:

$$F_p = \sqrt{F_x^2 + F_y^2 + F_z^2 - F_1^2}$$

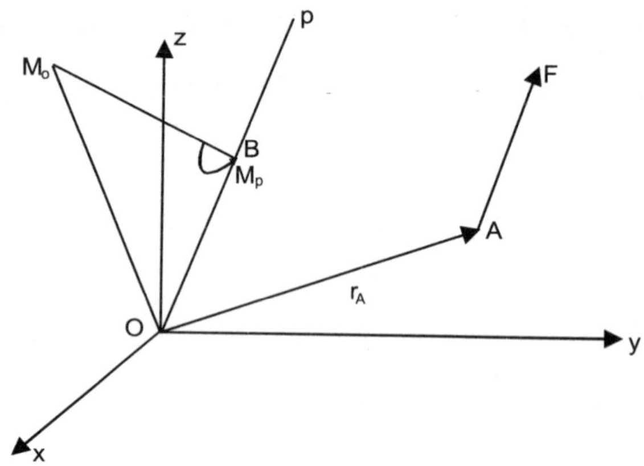

Fig. ES5.7 (a) Moment of a force about another axis

Thus the perpendicular distance $d = \left|\dfrac{M_p}{F_p}\right|$

Complete MATLAB script for obtaining *d* is presented below:

```
%%%%%%%%%%%%%%%%%%%%%%%%%%%%
syms Fx Fy Fz x y z l m n
F=[Fx Fy Fz];
r=[x y z];
dc=[l m n];  % direction cosines
Mo=cross(r,F);
Mp=dot(dc,Mo);
F1=dot(F,dc);
Fp=sqrt(Fx^2+Fy^2+Fz^2-F1^2);
d=abs(Mp/Fp);
% finding direction cosines from the angles
dc1=cos(60*pi/180);dc2=cos(100*pi/180);dc3=abs(sqrt(1-dc1^2-dc2^2));
% substitution of numerical values
dn=subs(d,{l,m,n,Fx,Fy,Fz,x,y,z},{dc1,dc2,dc3,6,10,18, 1,2,3});
fprintf('\nPerpendicular distance is %g m\n', dn)
%%%%%%%%%%%%%%%%%%%%%%%%%%%%%%
```

Output is as follows:

Perpendicular distance is 0.0954009 m.

ES5.8: A tension T is applied to the cable attached to the top A of the rigid mast and secured to the ground at B as shown in Fig.ES5.8.

(*a*) Obtain a general expression for the moment of T about the base O (i.e., M_O) as a function of x_B and z_B

(*b*) Write a MATLAB program to plot the magnitude of M_O and its components about the x and z axes as a function of x_B for z_B=12 m.

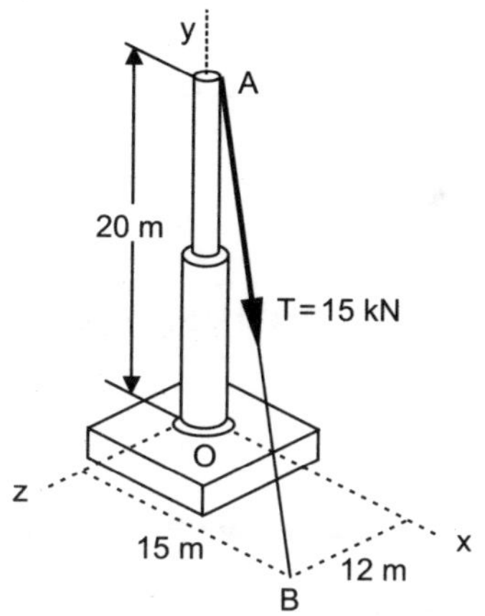

Fig. ES5.8 Tension applied to mast

Solution: First we express T as a cartesian vector and then find the moment by taking a cross-product.

$$T = 15\left[\frac{x_B\hat{i} - 20\hat{j} + z_B\hat{k}}{\sqrt{x_B^2 + 20^2 + z_B^2}}\right]$$

Moment of force about O is $M_O = r_{OA} \times T = 20\,\hat{j} \times 15\left[\frac{x_B\hat{i} - 20\hat{j} + z_B\hat{k}}{\sqrt{x_B^2 + 20^2 + z_B^2}}\right] = 300\left[\frac{-x_B\hat{k} + z_B\hat{i}}{\sqrt{x_B^2 + 20^2 + z_B^2}}\right]$

Hence the components of M_O in x and z directions are:

$$M_x = \left[\frac{300 z_B}{\sqrt{x_B^2 + 20^2 + z_B^2}}\right] \quad \text{and} \quad M_z = \left[\frac{-300 x_B}{\sqrt{x_B^2 + 20^2 + z_B^2}}\right]$$

Resultant moment is $M = \sqrt{M_x^2 + M_z^2} = 300\sqrt{\dfrac{x_B^2 + z_B^2}{x_B^2 + 20^2 + z_B^2}}$

After substituting z_B=12 m, moments can be plotted as a function of x_B.

Following MATLAB script is used to show the result.

```
%%%%%%%%%%%%%%%%%%%%
xB=-20:2:20;
d=inline('sqrt(xB.^2+20^2+12^2)'); % denominator
```

```
Mx=300*12./d(xB);
Mz=-300*xB./d(xB);
M=sqrt(Mx.^2+Mz.^2);
plot(xB,Mx,'-*',xB,Mz,'-p',xB,M,'-o');
xlabel('x_B(m)');ylabel('M(kNm)');
legend('M_x','M_z','M',3);
title('Plot of Moment versus x_B');grid on;
%%%%%%%%%%%%%%%%%%%%%%
```

Output is shown in Fig. ES5.8 (*a*).

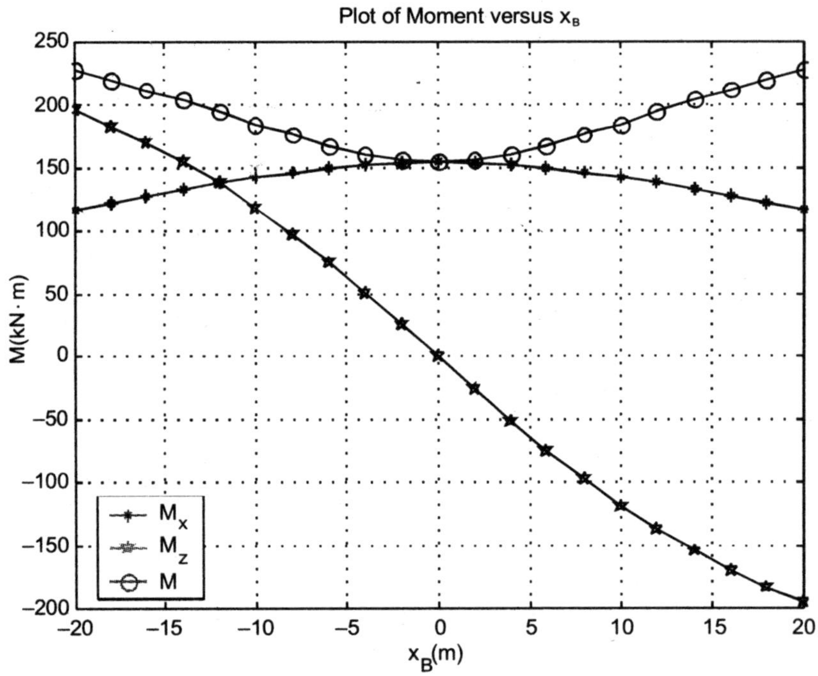

Fig. ES5.8 (a) MATLAB output

ES5.9: Write a MATLAB program to determine the single equivalent force *R* and the point where the line of action of *R* intersects the *yz* plane for the system shown in Fig. ES5.9. Take $F_1 = \hat{i} + \hat{j}$,

$r_1 = \hat{i} + \hat{j} - \hat{k}$, $M_1 = \hat{j} - 2\hat{k}$, $F_2 = -\hat{i} + \hat{k}$, $r_2 = 2\hat{i} - \hat{j} + \hat{k}$, $M_2 = \hat{j}$ and $F_3 = \hat{i} + 4\hat{j} - \hat{k}$,

$r_3 = 2\hat{i} - \hat{j} - \hat{k}$, $M_3 = \hat{i} - \hat{k}$.

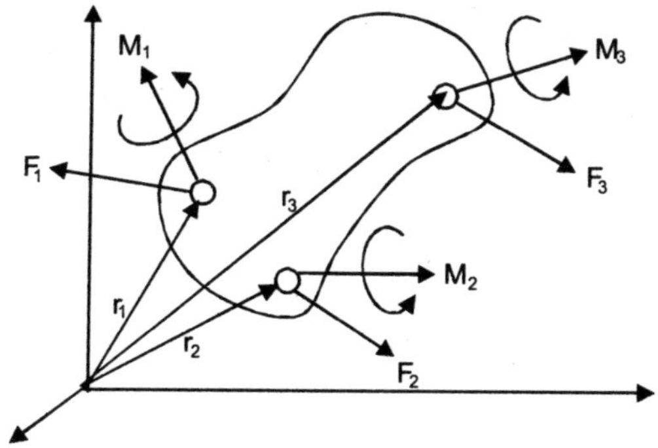

Fig.ES5.9

Solution: The system of forces and couples as shown can be represented as:

$$F_p = F_{xp}\hat{i} + F_{yp}\hat{j} + F_{zp}\hat{k}$$
$$M_p = M_{xp}\hat{i} + M_{yp}\hat{j} + M_{zp}\hat{k}$$
$$r = x_p\hat{i} + y_p\hat{j} + z_p\hat{k}$$

where $p = 1, 2, 3$

The net moment and force are thus given by:

$$M = M_1 + M_2 + M_3 + r_1 \times F_1 + r_2 \times F_2 + r_3 \times F_3$$
$$F = F_1 + F_2 + F_3$$

If the system can be reduced to a single equivalent force R, then

$$F \cdot M = p = 0$$

Since $R = F$, the point where the line of action of R intersects the yz-plane is given by:

$$r_{yz} \times R = M, \text{ where } r_{yz} = y\,\hat{j} + z\,\hat{k}$$

This can be solved for y and z coordinates.

When dot product is not zero, the resultant force along with moment acts at certain point.

Complete MATLAB program is given below:

```
%%%%%%%%%%%%%%%%%%%%%%%%
F1=[1 1 0];F2=[-1 0 1];F3=[1 4 -1];M1=[0 1 -2];M2=[0 1 0];M3=[1 0 -1];
r1=[1 1 -1];r2=[2 -1 1];r3=[2 -1 -1];
F=F1+F2+F3;
M=M1+M2+M3+cross(r1,F1)+cross(r2,F2)+cross(r3,F3);
p=dot(F,M);
if abs(p)<1e-10
p=0;
```

```
end
if p~=0
fprintf('\nThe system can not be reduced to a single resultant\n')
R=sqrt(F*F');%RESULTANT FORCE
lamda=F/R;
q=dot(F,M);
pitch=q/R^2;
Mp=pitch*F;
a1=Mp(1)-M(1); a2=Mp(2)-M(2); a3=Mp(3)-M(3);
eqn1='F(3)*y-F(2)*z+a1';
eqn2='F(1)*z+a2';
eqn3='-F(1)*y+a3';
y,z]=solve(eqn2,eqn3);
y=eval(y);z=eval(z);
fprintf('\nResultant=%gN with pitch=%g and intersects the yz
plane at y=%g m and z=%fm\n',R,pitch, y,z);
elseif p==0
eq1='F(3)*y-F(2)*z-M(1)';
eq2='F(1)*z-M(2)';
eq3='-F(1)*y-M(3)';
y,z]=solve(eq2,eq3);
y=eval(y);z=eval(z);
fprintf('\nResultant=%gN and intersects the yz plane at y=%g m
and z=%f m\n',R,y,z);
end
%%%%%%%%%%%%%%%%%%%%%%%%%%%%%%%%%%%%
```

Output comes as follows at the command prompt:

» The system can not be reduced to a single resultant

Resultant = 5.09902 N with pitch = 0.0384615 and intersects the *yz* plane at $y = -5$ m and $z = -1.192308$ m

ES5.10: A beam *AB* is subjected to several vertical forces as shown in Fig. ES5.10. Using MATLAB, determine the magnitude of the resultant of the forces and the distance x_C, where the line of action of resultant intersects *AB*.

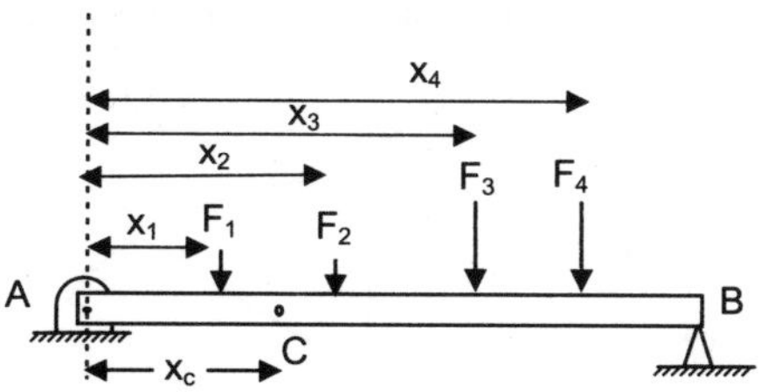

Fig. ES5.10

Take $F_1 = 10$ N, $F_2 = 20$ N, $F_3 = 30$ N, $F_4 = 40$ N and $X_1 = 0.3$ m, $X_2 = 1.2$ m, $X_3 = 2$ m, $X_4 = 2.2$ m.

Solution: The resultant force is given by $R = \Sigma F = -F_1\,\hat{j} - F_2\,\hat{j} - F_3\,\hat{j} - F_4\,\hat{j} = -(F_1 + F_2 + F_3 + F_4)\,\hat{j}$

Moment about point A, $\Sigma M_A = \Sigma r \times F = (x_1\hat{i}) \times (-F_1\hat{j}) + (x_2\hat{i}) \times (-F_2\hat{j}) + (x_3\hat{i}) \times (-F_3\hat{j}) + (x_4\hat{i}) \times (-F_4\hat{j})$

$$= -(x_1F_1 + x_2F_2 + x_3F_3 + x_4F_4)\hat{k}$$

The resultant of the given system of forces is equal to R and its point of application must be such that the moment of R about A is equal to ΣM_A.

Thus $$r \times R = \Sigma M_A$$

$$x_C\hat{i} \times (-R\hat{j}) = -(x_1F_1 + x_2F_2 + x_3F_3 + x_4F_4)\hat{k}$$

$$\Rightarrow \quad -x_C R\hat{k} = -(x_1F_1 + x_2F_2 + x_3F_3 + x_4F_4)\hat{k}$$

$$\Rightarrow \quad x_C = -(x_1F_1 + x_2F_2 + x_3F_3 + x_4F_4)/R$$

MATLAB program script to compute R and x_c is given below:

```
%%%%%%%%%%%%%%%%%%%%
F=[10 20 30 40];
X=[0.3 1.2 2 2.2];
sumF=0;sumM=0;
for i=1:4
sumF=sumF+F(i);
sumM=sumM+X(i)*F(i);
end
xC=sumM/sumF;
fprintf('\nResultant of forces is R=%g N acting at a distance %g m from support
A\n',sumF,xC);
%%%%%%%%%%%%%%%%%%%%
```

Output is as follows:

Resultant of forces is $R = 100$ N acting at a distance 1.75 m from support A.

ES5.11: Figure ES5.11 shows a uniform circular plate supported by three vertical wires that are equally spaced around its edge.

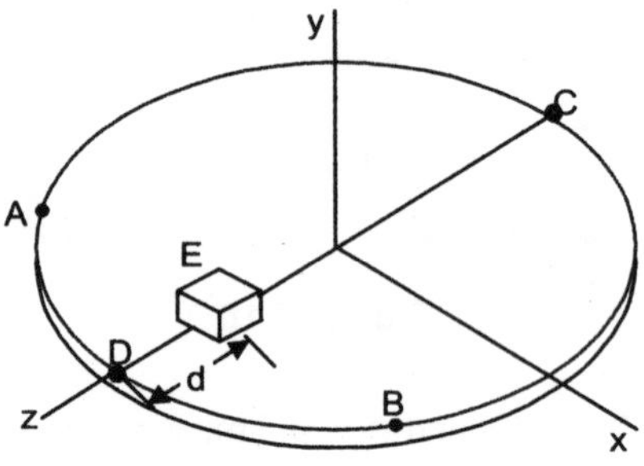

Fig. ES5.11

A small block E is placed on the plate at D and is then slowly moved along the diameter CD until it reaches C. Write a MATLAB program to plot the tension in the wires A and C as a function of d, where d is the distance of the block from D. Find the value of d for which the tension in wires A and C is minimum. Take radius of the circular plate $= 310$ mm, mass of the plate $= 30$ kg and mass of the block $E = 3.5$ kg.

Solution: The free-body diagram of the system is shown in Fig. ES5.11(a).

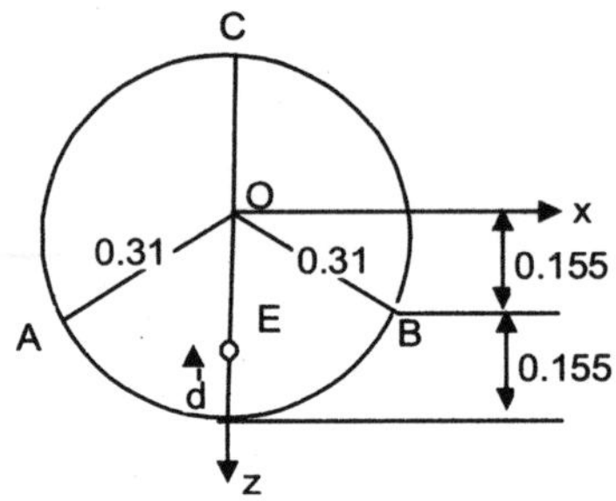

Fig. ES5.11 (a) Free-body diagram

Due to symmetry $T_A = T_B$. From equilibrium, we have:

$$\Sigma F_y = 0 \Rightarrow 2T_A + T_C - (30 + 3.5) \times 9.81 = 0$$

$$\Sigma M_{AB} = 0 \Rightarrow T_C(0.45) - 30(9.81) \times 0.155 + 3.5(9.81) \times (0.155 - d) = 0$$

Hence
$$T_C = 89.5435 + 76.3d$$

$$T_A = 119.545 - 38.15d$$

Following MATLAB script plots the tension in the wires A and C.

```
%%%%%%%%%%%%%%%%%%%%%%%%%%%
d = 0:0.02:0.62;
TA=119.545-38.15*d;
TC=89.5435+76.3*d;
fA=abs(TA);fC=abs(TC);
```

```
[fAmin,I]=min(fA);[fCmin,J]=min(fC);
dminA=d(I);dminC=d(J);
TAmin=TA(I);TCmin=TC(J);
fprintf('Value of d for which tension in wire A is minimum is %g m\n',dminA);
fprintf('Corresponding tension is %g N\n',TAmin);
fprintf('Value of d for which tension in wire C is minimum is %g m\n',dminC);
fprintf('Corresponding tension is %g N\n',TCmin);
plot(d,TA,'-*',d,TC,'-p');
xlabel('d(m)');
ylabel('Wire tension (N)');
legend('T_A','T_C',2);
grid on;
%%%%%%%%%%%%%%%%%%%%%%%%%%%%%
```

Output is shown in Fig. ES5.11 (*b*), along with following command prompt output.

»

Value of *d* for which tension in wire *A* is minimum is 0.62 m.

Corresponding tension is 95.892 N.

Value of *d* for which tension in wire *C* is minimum is 0 m.

Corresponding tension is 89.5435 N.

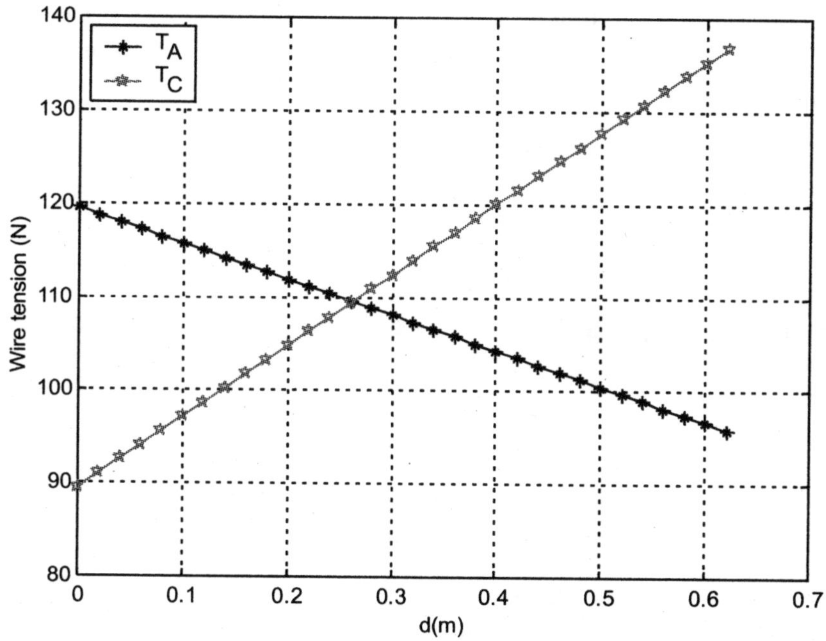

Fig. ES5.11 (b) MATLAB output

ES5.12: Figure ES5.12 shows a slender rod *AB* attached to blocks *A* and *B*.

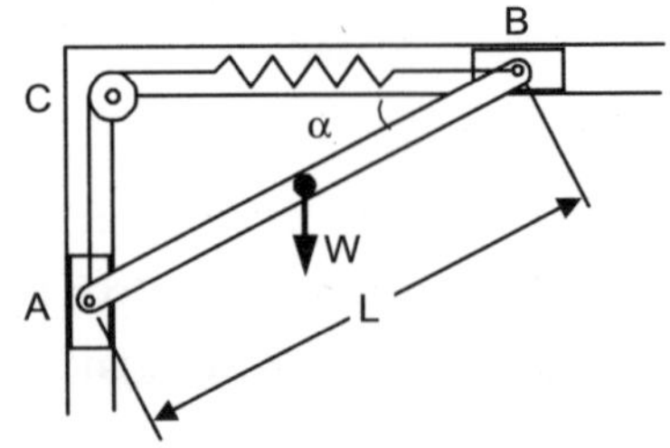

Fig. ES5.12

The weight of the rod *AB* is *W* and the blocks *A* and *B* move freely in the guides. The spring constant is *k* and the spring is unstretched when *AB* is horizontal. Write a MATLAB program to determine the three values of the angle α corresponding to equilibrium for values of *W* from 1 to 10 N in 2 N increments and $\ell = 25$ cm, and $k = 1.5$ N/cm.

Solution: This is a problem from equilibrium of rigid bodies. First of all, the free-body diagram of the rod *AB* is drawn as shown in Fig. ES5.12 (*a*).

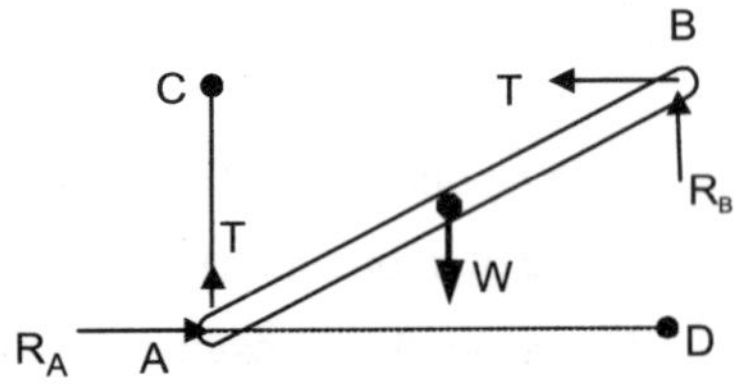

Fig. ES5.12
(a) Free-body diagram

As there is no deflection in spring when $\alpha = 0^o$, we can write the relation between deflection *x* and α as:

$$X = AC + CB - \ell = \ell\,(\sin\alpha + \cos\alpha - 1)$$

The tension in the spring $T = kx = k\ell\,(\sin\alpha + \cos\alpha - 1)$

Moment about a point *D* is given by

$$\Sigma M_D = T(\ell\sin\alpha) - T(\ell\cos\alpha) + W(0.5\ell\cos\alpha)$$

For equilibrium $\Sigma M_D = T(\ell\sin\alpha) - T(\ell\cos\alpha) + W(0.5\ell\cos\alpha) = 0$

Substituting for *T* gives:

$$k\ell^2(\sin\alpha + \cos\alpha - 1)(\sin\alpha - \cos\alpha) + 0.5\,W\ell\cos\alpha = 0$$

or

$$\frac{W}{2k\ell} = (\sin\alpha + \cos\alpha - 1)(1 - \tan\alpha)$$

This equation is solved for α at different values of *W*.

Complete MATLAB script is given below:

```
k=150; L=0.25;
for W=1:2:10
LHS=W/(2*k*L);
eqn='sin(x)+cos(x)-1)*(1-tan(x))-LHS';
x=eval(solve(eqn)); % evaluate numerical values
z1=isreal(x(1));z2=isreal(x(2));z3=isreal(x(3)); % checking for imaginary
values of solution
      fprintf('When W=%g N the equilibrium solution is',W);
      if z1==1
         th1=x(1)*180/pi
      elseif fprintf('No solution\n');
      end
```

```
      if z2==1
      th2=x(2)*180/pi
   elseif fprintf('No solution\n');
   end
      if z3==1
      th3=x(3)*180/pi
   elseif fprintf('No solution\n');
   end
   end
```

Output is as follows:

```
When W = 1 N the equilibrium solution is
   th1 =
        0.7798
   th2 =
        -134.8422
   th3 =
        44.0624
   When W = 3 N the equilibrium solution is
   th1 =
        2.4471
   th2 =
        -134.5292
   th3 =
        42.0821
   When W = 5 N the equilibrium solution is
   th1 =
        4.2948
   th2 =
        -134.2197
   th3 =
        39.9249
   When W = 7 N the equilibrium solution is
   th1 =
        6.3916
   th2 =
        -133.9135
   th3 =
        37.5219
   When W = 9 N the equilibrium solution is
   th1 =
        8.8662
   th2 =
        -133.6106
   th3 =
        34.7444
```

ES5.13: Figure ES5.13 shows a two-rod mechanism where the rods *AC* and *BD* are connected by a slider block *D*. Write a MATLAB program to compute and plot the couple M_A required to hold the rods in equilibrium for values of a from 0 to 120°. Also plot the magnitude of the force *F* exerted by rod *AC* on the slider-block for values of a ranging from 0 to 120°. Neglect the effect of friction.

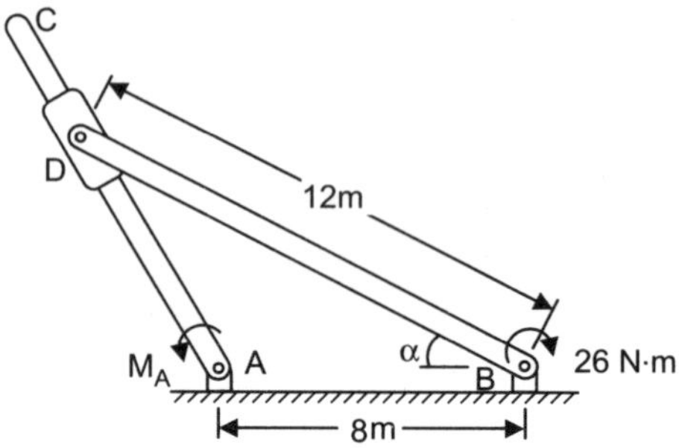

Fig. ES5.13

Solution: Let us draw the free-bodies of the bodies *AD* and *BD* as shown in Fig. ES5.13(*a*):

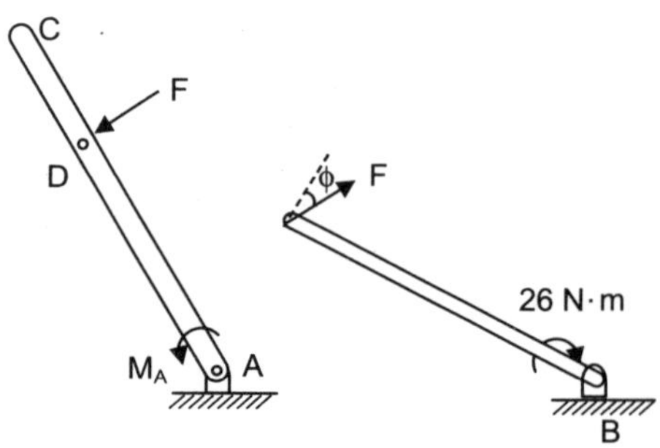

Fig. ES5.13 (a) Free-body diagrams

From free-body of *BD*:

$$(F\cos\phi)\times 12 = M_B = 26 \quad \text{or} \quad F = \frac{26}{12\cos\phi} \qquad \text{... (1)}$$

$$\text{From free-body of } AD\text{: } M_A = F(AD) = \frac{26}{12\cos\phi}AD \qquad \text{... (2)}$$

Also from the geometry in the triangle *ABD* (see Fig. ES5.13(*b*)) and using the law of cosines,
we have:

$$AD^2 = 12^2 + 8^2 - 2(12)(8)\cos\alpha$$

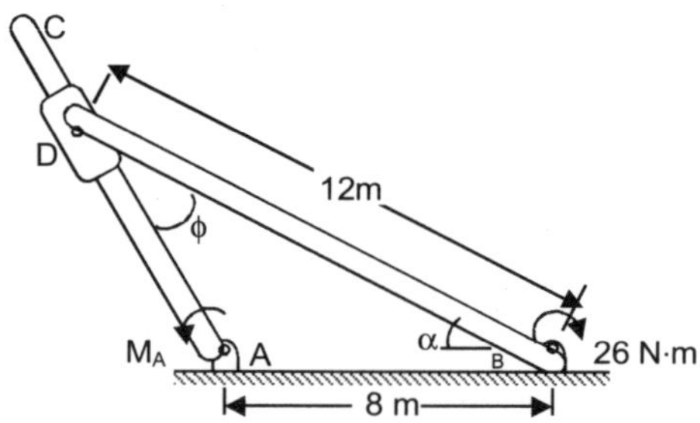

Fig. ES5.13 (b) Geometry

And from the law of sines:

$$\sin\phi = 8\times\frac{\sin\alpha}{AD}, \text{ where } \phi = \angle ADB$$

Substituting AD and ϕ in eqs.(1) and (2), it is possible to study the variation of M_A and F as a function of angle α.

Complete MATLAB script for this is given below:

```
%%%%%%%%%%%%%%%%%%%%%%%
alpha=[0:10:120];% range of alpha in degrees
al=alpha*pi/180;
AD=(12^2+8^2-2*12*8*cos(al)).^0.5;
sphi=8.*sin(al)./AD;
cphi=(1-sphi.^2).^0.5;
F=26./(12.*cphi); M=F.*AD;
subplot(2,1,1);
plot(alpha,F);
xlabel('\alpha(degree)'); ylabel('F(N)'); grid on;
subplot(2,1,2)
plot(alpha,M)
xlabel('\alpha(degree)'); ylabel('M_A(Nm)');grid on;
%%%%%%%%%%%%%%%%%%%%%%%%
```

Output is shown in Fig. ES5.13(*c*).

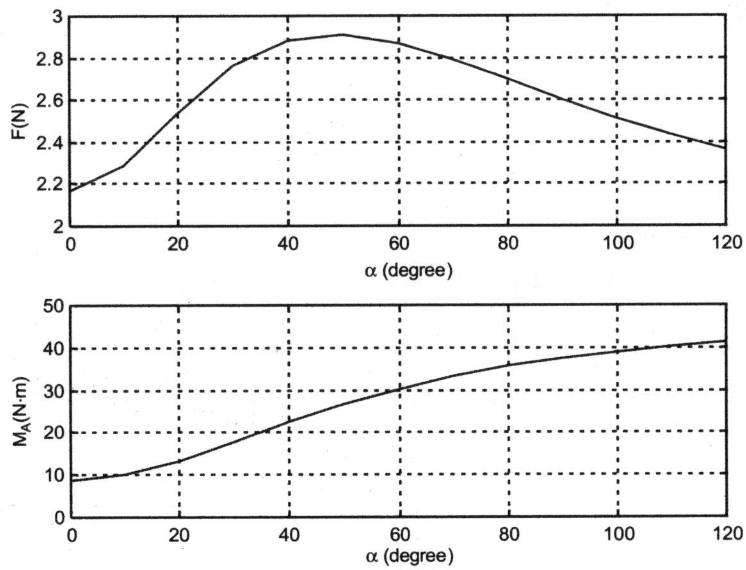

Fig. ES5.13 (c) MATLAB output

ES5.14: T he magnitude of the force P applied to the piston of an engine system during one revolution of crank AB is shown in Fig. ES5.14. Plot the magnitude of the couple M required to hold the system in equilibrium as a function of θ for $0 \le \theta \le 2\pi$.

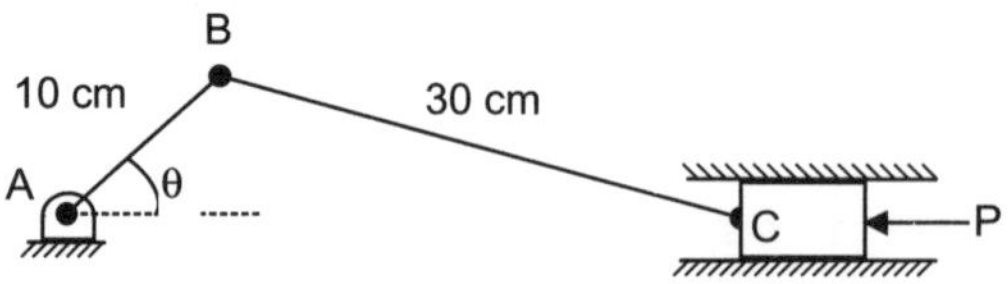

(i) Engine mechanism

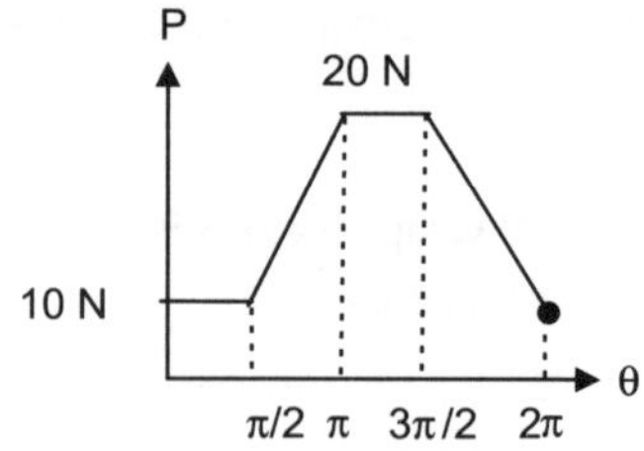

(ii) Variation of P

Fig. ES5.14 (i) and (ii)

Solution: This is a problem of static force analysis. The various forces acting are normal reaction *N* at *C* and gas force *P*. First of all free-body diagram of the linkage is drawn as shown in Fig. ES5.14 (*a*).

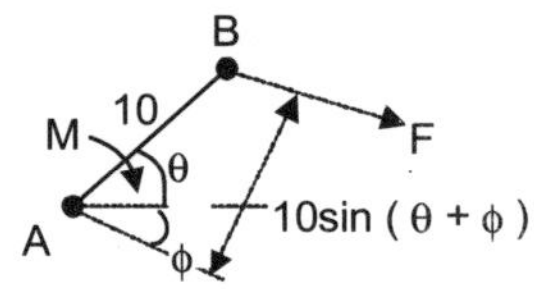

(i)

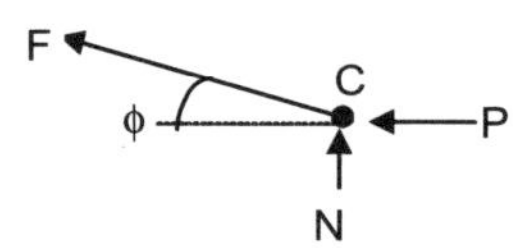

(ii)

Fig. ES5.14 (a) Free-body diagram of the links AB and C

From Fig. ES5.14 (*a*)(*i*) for the body *AB*, we can write:

Magnitude of moment $M = 10\sin(\theta+\phi)\times F = 10F\sin(\theta+\phi)$ ... (1)

From Fig. ES5.14 (*a*) (*ii*) for body *C*,

Force along *BC* can be written as:

$$F\cos\phi = P \quad \text{... (2)}$$

Using equations (1) and (2), the moment *M* required to hold the system in equilibrium is given by:

$$M = \frac{10P\sin(\theta+\phi)}{\cos\phi}$$

The angle ϕ can be written in terms of θ from geometry as follows:

$$10\sin\theta = 30\sin\phi$$

Now, the complete MATLAB program for plotting the couple M can be written as follows:

```
%%%%%%%%%%%%%%%%%%%%%%%
i=0;
for theta=0:1:360
th=theta*pi/180;
i=i+1;
ang(i)=theta;
phi=asin(sin(th)/3);
if theta>=0 && theta<90
```

```
P(i)=10;
elseif theta>=90 && theta<180
P(i)=(20/pi) *th; % EQUATION OF STRAIGHT LINE between pi/2 and pi
elseif theta>=180 && theta<270
P(i)=20;
elseif theta>=270
P(i)=-(20/pi)*th+50; % EQUATION OF STRAIGHT LINE between 3pi/2
and 2pi
end
M(i)=10*P(i)*sin(th+phi)/cos(phi);
end
plot(ang,M),
xlabel('\theta(degree)');
ylabel('Crank effort M(N·cm)')
axis([0 360 -2.1 1.2])
title('\bfVariation of moment on crank-shaft in N·m')
grid on;
%%%%%%%%%%%%%%%%%%%%%%%%%
```

Output of the program is shown in Fig. ES5.14 (*b*).

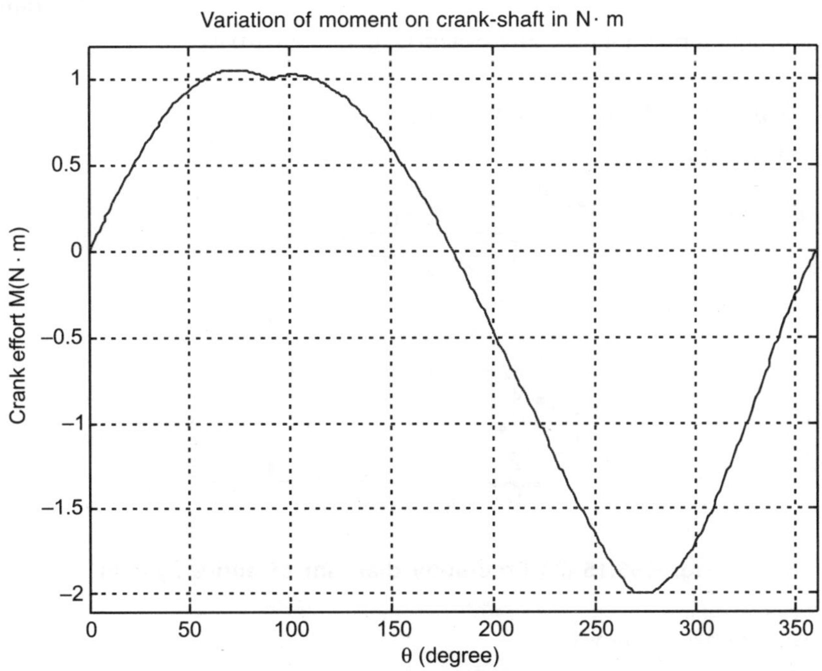

Fig. ES5.14 (b) MATLAB output

ES5.15: Figure ES5.15 shows a rod *CD* attached to collar *D* and passes through another collar welded to the end *B* of lever *AB*. As an initial step in the design of lever *AB*, use MATLAB to plot the magnitude of couple

M required to hold the system in equilibrium as a function of θ for $10^\circ \le \theta \le 90^\circ$. Also determine the value of θ at which M becomes minimum and corresponding value of M.

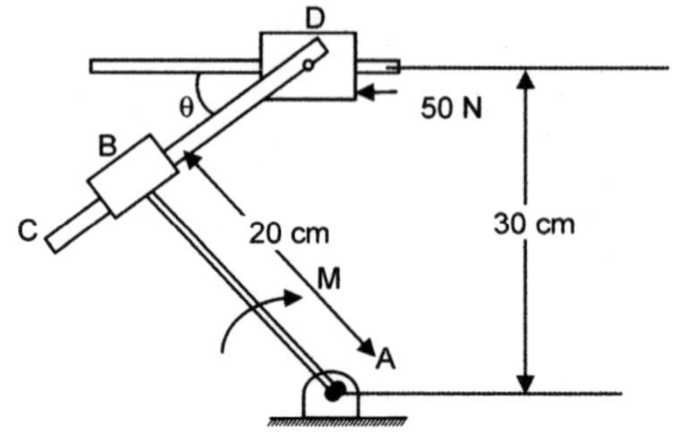

Fig. ES5.15

Solution: Various forces acting on the system are force in rod CD, external force P and holding moment M on AB. First of all consider the free-body diagram of rod CD (see Fig. ES5.15 (*a*)):

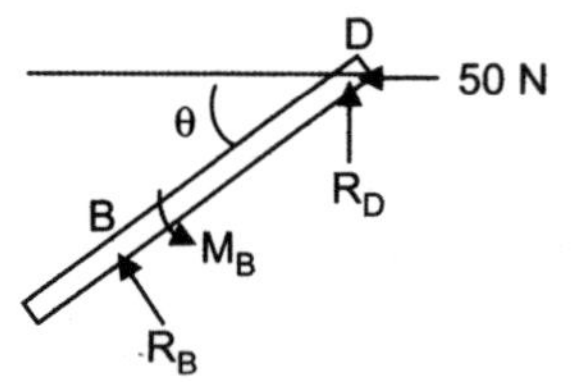

Fig. ES5.15 (a) Free body diagram of rod CD

Resolving forces along CD and equating to zero, we get:

$$R_D \sin\theta - 50\cos\theta = 0 \quad \Rightarrow \quad R_D = \frac{50}{\tan\theta}$$

Writing equations of equilibrium for the entire system (Fig. ES5.15 (*b*)) we obtain:

$$\Sigma M_A = 0 \Rightarrow M - R_D(d) - 50(0.3) = 0$$

or $$M = R_D(d) + 15$$

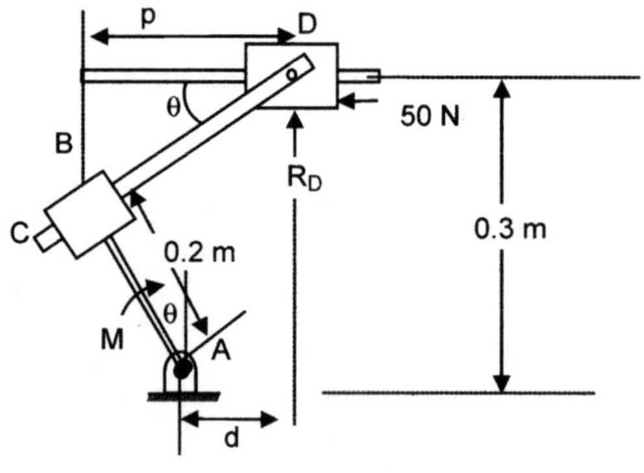

Fig. ES5.15 (b) Free-body diagram of entire system

From geometry, we obtain d as follows:

At an instant θ,

$$d = p - 0.2\sin\theta, \text{ where } p = \frac{(0.3 - 0.2\cos\theta)}{\tan\theta} \quad \text{(see Fig. ES5.15(}b\text{))}$$

$$d = p - 0.2\sin\theta, \text{ where } p = \frac{(0.3 - 0.2\cos\theta)}{\tan\theta} \text{ (see Fig. ES5.15(}b\text{))}$$

Complete MATLAB script for plotting *M* as a function of angle θ is given below:

```
%%%%%%%%%%%%%%%%%%%%%
theta=10:5:90;
th=theta*pi/180;
p=(0.3-0.2.*cos(th))/tan(th);
d=p-0.2.*sin(th);
RD=50./tan(th);
M=RD.*d+15;
[M_min,I]=min(M);
fprintf('\nMinimum couple is %g Nm and corresponding angle is %g
degrees\n',M_min,theta(I));
plot(theta, M,'-p');
xlabel('\theta(degree)');
ylabel('Holding couple (N·m)');
grid on;
%%%%%%%%%%%%%%%%%%%%%
```

Output is shown in Fig. ES5.15(c) along with the following command prompt result.

» Minimum couple is 5.15192 N·m and corresponding angle is 10 degree.

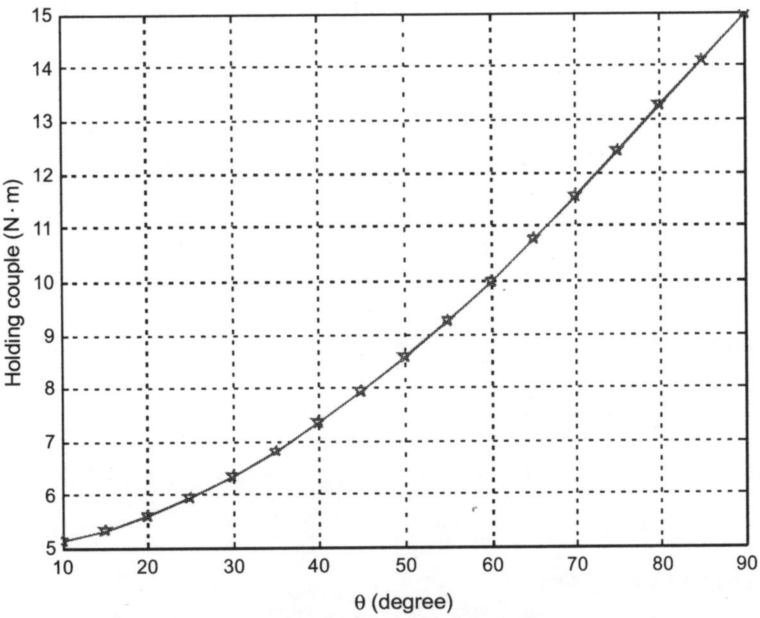

Fig. ES5.15 (c) MATLAB output

ES5.16: A 100 N force is pushing a 50 kg block as shown in Fig. ES5.16. The angle of inclination of the plane is α which varies from 0 to 45° in increments of 5°. The coefficient of static friction between the block and the incline is $\mu_s = 0.75$ and the coefficient of kinetic friction is $\mu_k = 0.65$. Will the block slide on the plane? If it does, will it slide up or down the plane? What is the friction force between the block and the plane?

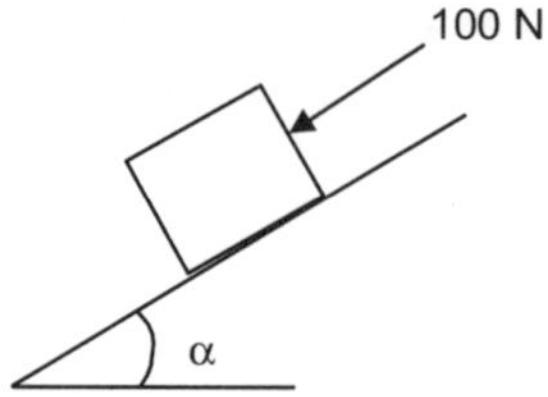

Fig. ES5.16

Solution: Various forces acting on the block are the weight W, applied force P=100 N, normal reaction N and frictional force F.

For equilibrium:

$$\Sigma F_x = 0 \Rightarrow 100 + W\sin\alpha - F = 0 \quad \text{...(1)}$$

$$\Sigma F_y = 0 \Rightarrow W\cos\alpha - N = 0 \quad \text{...(2)}$$

Motion begins (non-equilibrium) when F found from eq.(1) exceeds $\mu_s N$. Otherwise block will be at rest.

If block slides, then the relation $F = \mu_k N$ holds good.

Complete MATLAB program for finding the positions where block remains at rest is given below:

```
%%%%%%%%%%%%%%%%%%%%%%%%%%%%%
m=50;g=9.81;muk=0.65;mus=0.75;
W=m*g;
P=100;
for alfa=0:5:45;
al=alfa*pi/180;
N=W.*cos(al);
Fs=mus*N;
F=P+W.*sin(al);
I=0;
for alfa=0:5:45
I=I+1;
if Fs(I)<F(I)
fprintf('Block will slide for angle %g degrees\n',alfa);
Fk=muk*N;
fprintf('Frictional force at the instant is %g N\n',Fk);
else
fprintf('Block will not slide for angle %g degrees\n',alfa);
end
end
%%%%%%%%%%%%%%%%%%%%%%%%%%%%%
```

Output is as follows:

Block will not slide at angle 0 degrees.
Block will not slide at angle 5 degrees.
Block will not slide at angle 10 degrees.
Block will not slide at angle 15 degrees.
Block will not slide at angle 20 degrees.
Block will not slide at angle 25 degrees.
Block will slide at angle 30 degrees.
Frictional force at the instant is 276.111 N
Block will slide at angle 35 degrees.
Frictional force at the instant is 261.166 N
Block will slide at angle 40 degrees.
Frictional force at the instant is 244.234 N
Block will slide at angle 45 degrees.
Frictional force at the instant is 225.443 N

ES5.17: Figure ES5.17 shows the position of the rod *AB* controlled by the block which is slowly moved to the left by the force *P*. The weight of the rod *AB* is 30 N and that of the block is 5 N. The coefficient of kinetic friction between all surfaces contact is 0.3. Write a MATLAB program to plot the magnitude *P* of the force as a function of *x* for values of *x* from 55 into 5 cm. Find the maximum value of *P* and the corresponding values of *x*.

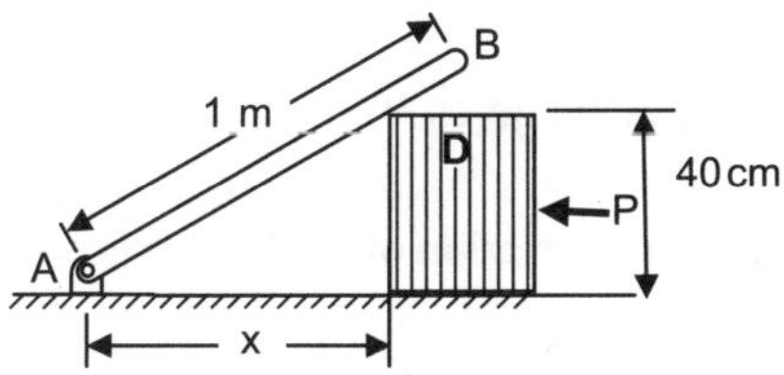

Fig. ES5.17

Solution: First of all free-body diagram of the rod is drawn as shown in Fig. ES5.17(*a*).

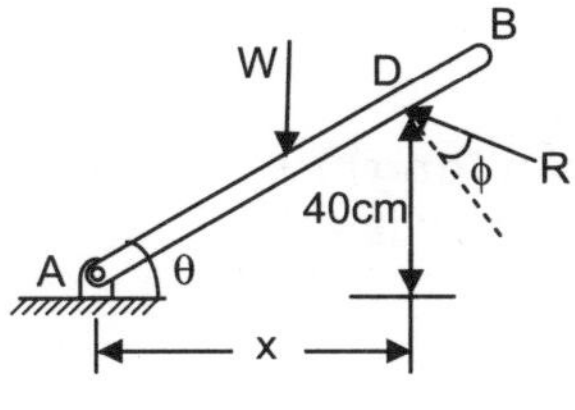

Fig. ES5.17 (a) Free-body diagram of rod AB

From geometry, $\tan\theta = 40/x$.

By definition, friction angle $\phi = \tan^{-1}(\mu)$

At *D*, we have $R_x = R\sin(\theta + \phi)$ ...(1)

$R_y = R\cos(\theta + \phi)$ ...(2)

Summing moments about *A* gives:

$$\Sigma M_A = 0 \Rightarrow R_x(40) + R_y x - W_{AB}(50\cos\theta) = 0 \quad ...(3)$$

Now consider the free-body diagram of block D (Fig. ES5.17(b)):

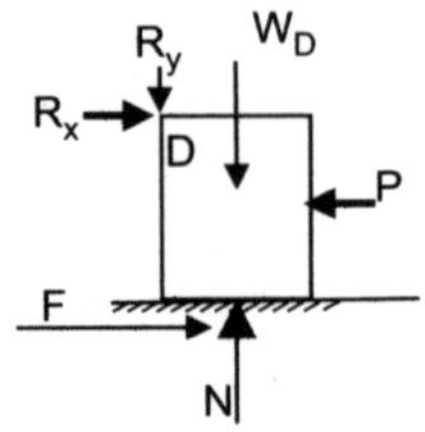

Fig. ES5.17 (b) Free-body diagram of block

Summing forces on the block yields:

$$\Sigma F_x = 0 \Rightarrow R_x + \mu N - P = 0 \Rightarrow P = R_x + \mu N \quad \text{...(4)}$$

$$\Sigma F_y = 0 \Rightarrow N - R_y - W_D = 0 \Rightarrow N = R_y + W_D \quad \text{...(5)}$$

From equations (1) and (2) we can eliminate R and relate R_x and R_y and then substituting in eq. (3) we obtain,

$$R_x = \frac{50 W_{AB} \cos\theta}{40 + \dfrac{x}{\tan(\theta+\phi)}} \quad \text{...(6)}$$

$$R_y = R_x \frac{\cos(\theta+\phi)}{\sin(\theta+\phi)} \quad \text{...(7)}$$

Substituting R_x and R_y in equations (4) and (5) gives P.

Complete MATLAB script for plotting the variation of P with x is given below:

```
%%%%%%%%%%%%%%%%
WAB=30;WD=5;
mu=0.3;
phi=atan(mu);
x=55:-1:5;
th=atan(40./x);
ang=th+phi;
Rx=50*WAB*cos(th)./(40+x./tan(ang));
Ry=Rx.*cos(ang)/sin(ang);
N=Ry+WD;
P=Rx+mu*N;
[Pmax,I]=max(P);
fprintf('Maximum value of P is %f N which occurs when x=%f
cm\n',Pmax,x(I));
plot(x,P);
xlabel('x(cm)');
ylabel('Force P(N)');
grid on;
%%%%%%%%%%%%%%%%%%%%%%
```

Output is shown below (Fig.ES5.17(*c*)) along with command prompt result.

» Maximum value of P is 21.603668 N which occurs when $x = 37.000000$ cm.

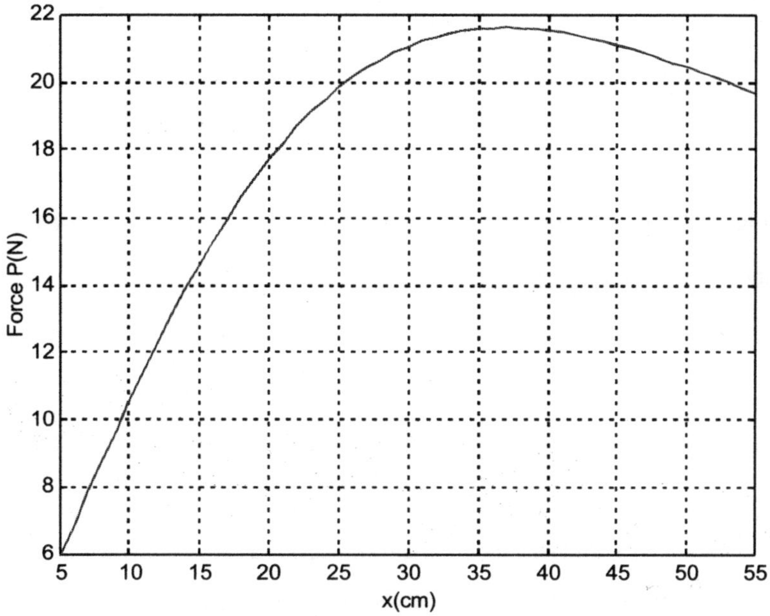

Fig. ES5.17 (c) MATLAB output

Note: In the above problem if friction is ignored, $\phi = 0$ and methodology will remain same.

ES5.18: Figure ES5.18 shows a truss supporting a ramp (shown as a dashed line).

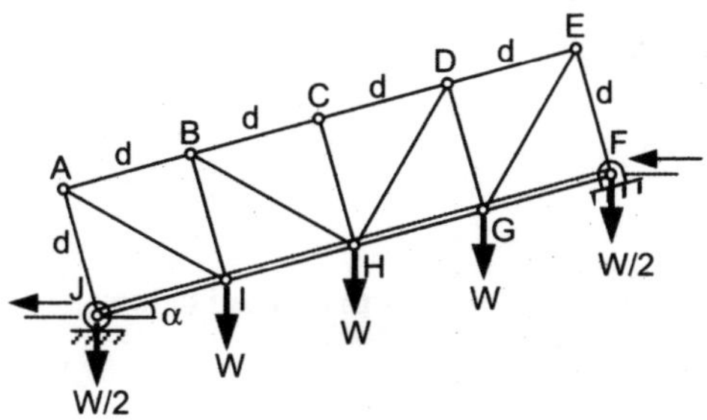

Fig. ES5.18

The ramp extends from a fixed approach level at joint F to a fixed exit level at joint J. The loads shown in figure represent the weight of the ramp.

(*a*) Determine the forces in members CD, BH, and IH as a function of a and W.

(*b*) Write a MATLAB program to plot the non-dimensional forces in CD, BH and IH as a function of α for the values of α ranging between 0° and 45°. Take $W = 1$ newton.

Solution: First of all the reaction force at J from the free-body diagram for the entire truss is determined. Fig. ED5.18(*a*) shows the free-body diagram.

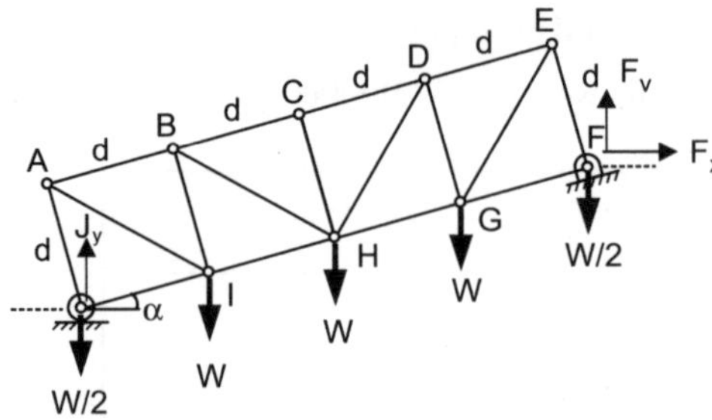

Fig. ES5.18 (a) Free-body diagram

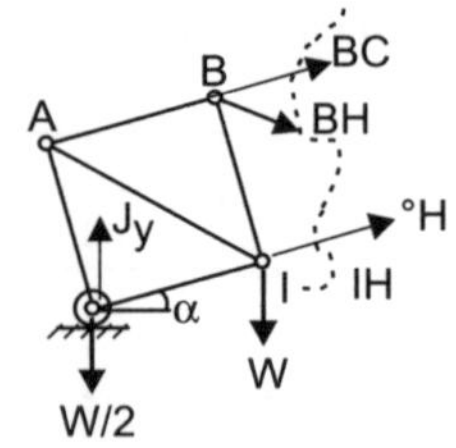

Fig. ES5.18 (b) Free-body diagram

Taking moments about the point F, we get:

$$\Sigma M_F = 0 \Rightarrow \left(\frac{W}{2} - J_y\right) 4d\cos\alpha + W(3d\cos\alpha) + W(2d\cos\alpha) + W(d\cos\alpha) = 0 \quad ...(1)$$

This gives $J_y = 2W$.

Now a free-body diagram by sectioning the truss through the members BC, BH and IH is drawn in Fig. ES5.18 (*b*).

The forces in the three members can be found by summing moments about points B, H and I. It can be observed that the forces in the members BC and CD are identical since BI is a zero-force member.

$$\Sigma M_B = 0 \Rightarrow IH(d) - Wd\sin\alpha - 1.5\,Wd\cos\alpha + 1.5\,Wd\sin\alpha = 0 \quad ...(2)$$

$$\Sigma M_H = 0 \Rightarrow -BC(d) + Wd\cos\alpha - 1.5\,W(2d\cos\alpha) = 0 \quad ...(3)$$

$$\Sigma M_I = 0 \Rightarrow -BH(d\sin 45°) - BC(d) - 1.5\,Wd\cos\alpha = 0 \quad ...(4)$$

These three equations can be solved for the forces in the three members in terms of α and W as follows:

$$IH = 1.5W\cos\alpha - 0.5W\sin\alpha$$

$$BC = -2W\cos\alpha = CD$$

$$BH = \frac{W}{\sqrt{2}}\cos\alpha.$$

Complete MATLAB script for plotting the forces is presented below:

```
%%%%%%%%%%%%%%
W=1;
alfa=0:5:45;
al=alfa*pi/180;
IH=1.5*W.*cos(al)-0.5*W.*sin(al);
CD=-2*W.*cos(al);
BH=(W/2^0.5).*cos(al);
plot(alfa,CD,'-*',alfa,BH,'-p',alfa,IH,'-o');
xlabel('\alpha(degree)');
```

```
ylabel('Forces in members (N)');
grid on;
legend('CD','BH','IH');
title('\bfForces in truss-members by Method of sections');
%%%%%%%%%%%%%%%%%%%%
```

Output is shown in Fig. ES5.18(*c*).

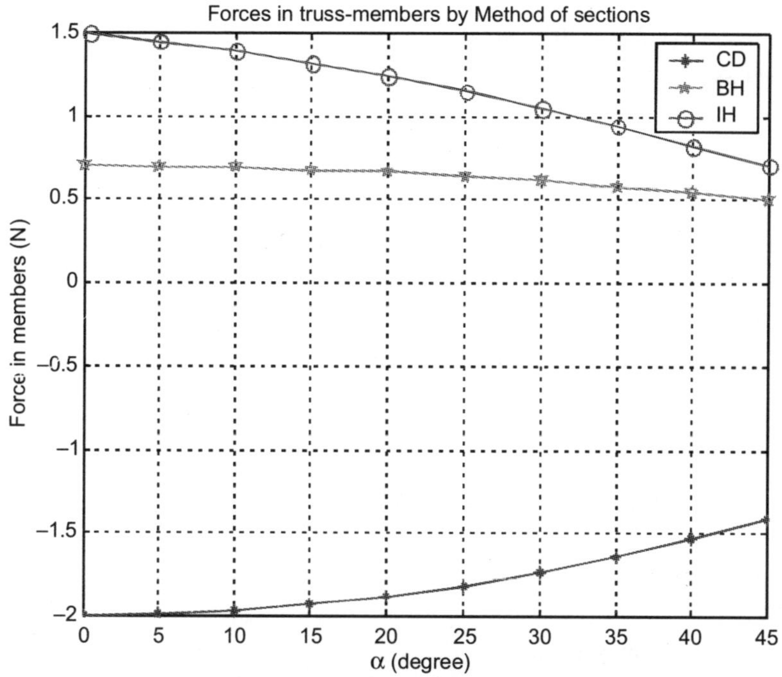

Fig. ES5.18 (c) MATLAB output

ES5.19: For the truss shown in Fig.ES5.19, determine the forces in the members F_1 to F_7 by writing the equilibrium equations and solve them using MATLAB.

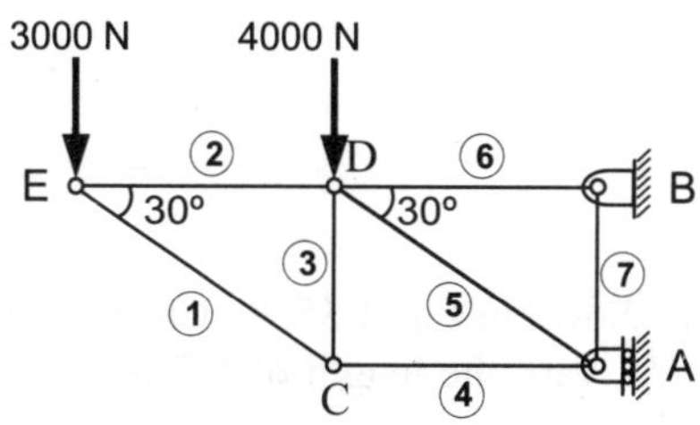

Fig. ES5.19

Solution: Starting from joint E, the equilibrium equations in terms of member forces can be obtained as follows:

At joint E:

$$\Sigma F_y = 0 \Rightarrow F_1 \sin 30^\circ = -3000 \quad ...(1)$$

$$\Sigma F_x = 0 \Rightarrow F_1 \cos 30^\circ + F_2 = 0 \quad ...(2)$$

At joint C:

$$\Sigma F_y = 0 \Rightarrow F_3 + F_1 \sin 30^\circ = 0 \quad ...(3)$$

$$\Sigma F_x = 0 \Rightarrow F_4 - F_1 \cos 30^\circ = 0 \quad ...(4)$$

At joint D:

$$\Sigma F_y = 0 \Rightarrow -F_3 - F_5 \sin 30^\circ = 4000 \quad ...(5)$$

$$\Sigma F_x = 0 \Rightarrow F_6 + F_5 \cos 30^\circ - F_2 = 0 \quad ...(6)$$

At joint A:

$$\Sigma F_y = 0 \Rightarrow F_5 \sin 30^\circ + F_7 = 0 \quad ...(7)$$

Equations (1) to (7) can be solved using following MATLAB script:

```
%%%%%%%%%%%%%%%%%%%
th=30*pi/180;c=cos(th);s=sin(th);
A=[s 0 0 0 0 0 0;c 1 0 0 0 0 0;s 0 1 0 0 0 0;-c 0 0 1 0 0 0;
  0 0 -1 0 -s 0 0;0 -1 0 0 c 1 0;0 0 0 0 s 0 1] ;
B=[-3000;0;0;0;4000;0;0];
F=inv(A)*B;
fprintf('Forces in members 1 to 7 in Newtons are\n');
disp(F)
%%%%%%%%%%%%%%%%%%%
```

Output is as follows:

```
Forces in members 1 to 7 in Newtons are
  1.0e+004*
   -0.6000
    0.5196
    0.3000
   -0.5196
   -1.4000
    1.7321
    0.7000
```

ES5.20: Consider the cantilever truss shown in Fig. ES5.20. Calculate the forces in each member if a wind load of 3000 N is added at the joint D. Vary the length of CE from 2 to 10 m and plot the corresponding forces in member CD.

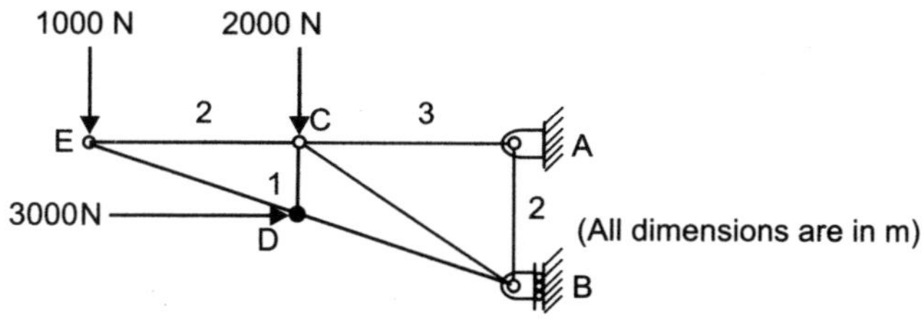

Fig. ES5.20

Solution:

Following MATLAB program can be readily understood with detailed explanations.

```
%%%%%%%%%%%%%%%%%
%LAC, LAB, LCD, LCE Lengths of different truss members
% FAC, FAB, FCD, FCE, FBC, FDB, FDE  Forces in different truss members
% THETA-1 angle between CE and DE
% THETA-2 angle between AC and BC
% THETA-3 angle between BC and BD
LAC=3;LAB=2;LCD=1;LCE=2;
%COMPUTATION OF ANGLES
theta1=atan(LCD/LCE);theta2=atan(LAB/LAC);
% CONSIDER EQUILIBRIUM OF JOINT E
FED=-1000/sin(theta1);
FCE=-FED*cos(theta1);
% CONSIDER EQUILIBRIUM OF JOINT D
FDB=(FED*cos(theta1)-3000)/cos(theta1);
FCD=FDB*sin(theta1)-FED*sin(theta1);
% CONSIDER EQUILIBRIUM OF JOINT C
FCB=(2000-FCD)/sin(theta2);
FCA=-FCB*cos(theta2)+FCE;
% CONSIDER EQUILIBRIUM OF JOINT B
FBA=-FCB*sin(theta2)-FDB*sin(theta1);
% SUPPORT REACTIONS
RAX=FCA;RAY=FBA;
RBX=FDB*cos(theta1)+FCB*cos(theta2);RBY=0;
% DISPLAY THE RESULTS
fprintf('Forces in members are EC=%f N, ED=%f N\n DB=%f N, CD=%f N\n
  CB=%f N,CA=%f N and AB=%f N\n',FCE,FED,FDB,FCD,FCB,FCA,FBA);
% VARYING THE LENGTH OF MEMBER AB
LCE=2:1:10;
% COMPUTATION OF ANGLES
theta1=atan(LCD./LCE);
% CONSIDER EQUILIBRIUM OF JOINT E
```

```
FED=-1000./sin(theta1);
FCE=-FED.*cos(theta1);
% CONSIDER EQUILIBRIUM OF JOINT  D
FDB=(FED.*cos(theta1)-3000)./cos(theta1);
FCD=FDB.*sin(theta1)-FED.*sin(theta1);
plot(LCE,FCD,'-p');
xlabel('Length of bay-CE (meter)');
ylabel('Force in member CD(N)');
grid on;
%%%%%%%%%%%%%%%%%%%%%%%%%%%%%%
```

Output is shown in Fig. ES5.20 (*a*) along with following command prompt result:

»

Forces in members are $EC = 2000.000000$ N, $ED = -2236.067977$ N
$DB = -5590.169944$ N, $CD = -1500.000000$ N
$CB = 6309.714732$ N, $CA = -3250.000000$ N and $AB = -1000.000000$ N

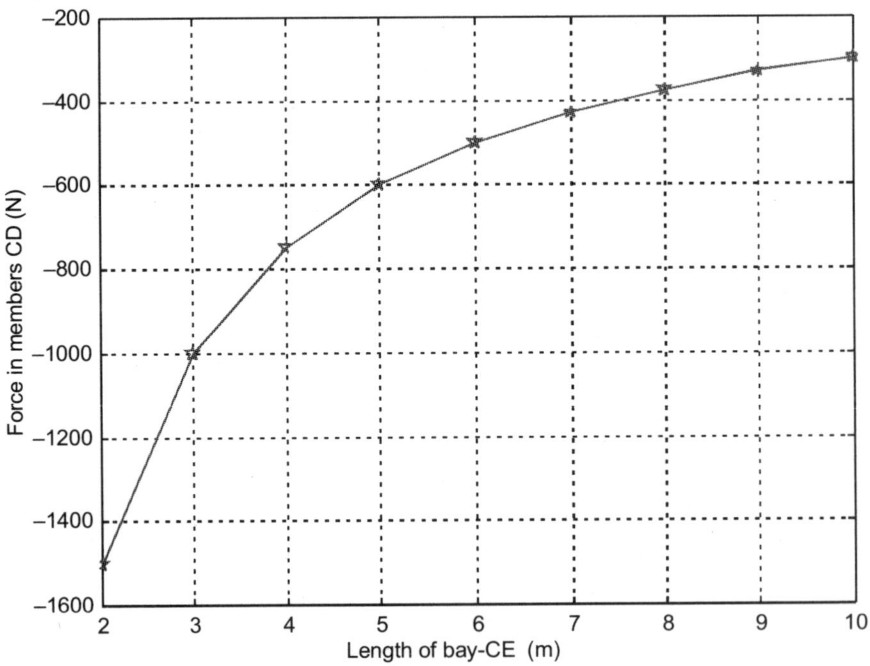

Fig. ES5.20 (a) MATLAB output

ES5.21: Figure ES5.21 shows the Fink truss. A single point load of 4 kN is to be applied to the top chord of the truss at one of the joints 1, 2, ..., 9. Write a MATLAB program to calculate the force in member *MN* as the load is successively applied at joints 1, 2, ..., 9.

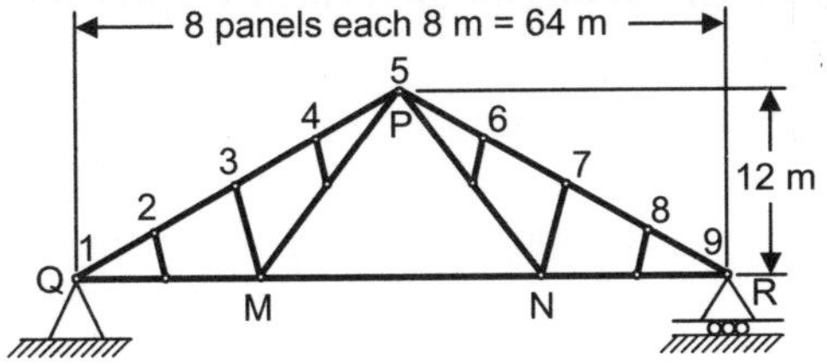

Fig. ES5.21

Solution: Since only one load is applied at a time, we assume that it is applied at some position x as shown in Fig. ES5.21(a).

Applying the equation of equilibrium, we get:

$$\Sigma F_y = 0 \Rightarrow Q + R - F = 0 \qquad ...(1)$$

$$\Sigma M_Q = 0 \Rightarrow 64R - xF = 0 \qquad ...(2)$$

Hence $$R = xF/64 \text{ and } Q = \left(1 - \frac{x}{64}\right) \qquad ...(3)$$

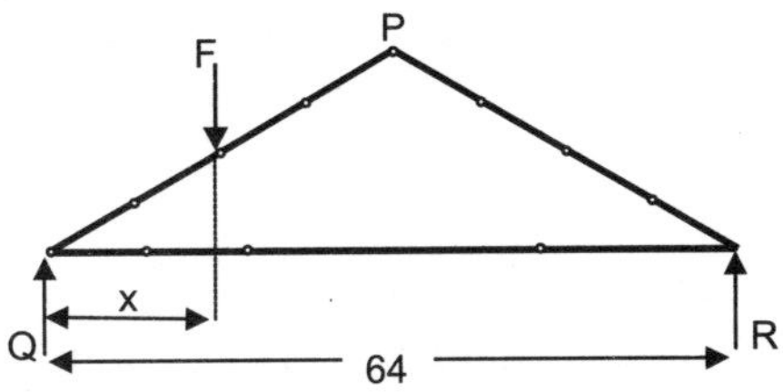

Fig. ES5.21 (a)

Now for computing force in MN, consider left half and right half of the truss separately as follows:

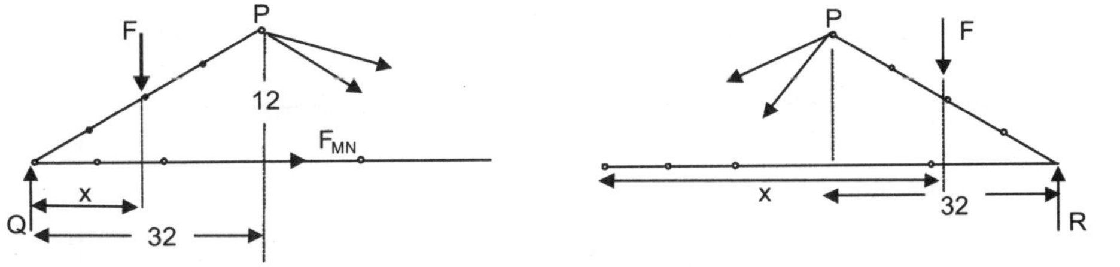

(i) **(ii)**

Fig. ES5.21(b) Free-bodies of two halves

For $0 \le x \le 32$ (Fig. ES5.21(b)(i)):

$$\Sigma M_P = 0 \Rightarrow 12F_{MN} - 32F\left(1 - \frac{x}{64}\right) + F(32 - x) = 0$$

$$\Rightarrow F_{MN} = \frac{32F}{12}\left(1 - \frac{x}{64}\right) - \frac{F}{12}(32 - x) = \frac{xF}{24} \qquad ...(4)$$

For $32 < x \le 64$ (Fig.ES5.21 (b) (ii)):

$$\Sigma M_P = 0 \Rightarrow 32\left(\frac{xF}{64}\right) - 12F_{MN} - F(x - 32) = 0$$

$$\Rightarrow F_{MN} = \frac{F}{12}\left(\frac{x}{2} + 32 - x\right) = \frac{F}{12}\left(32 - \frac{x}{2}\right) \qquad ...(5)$$

Now a complete MATLAB program for plotting the forces in MN is presented below:

```
%%%%%%%%%%%%%%%%%%%%%%
i=0;
F=4000;
for x=0:8:64
i=i+1;
if x<=32
FMN(i)=F*x/24;
c(i)=x;
else
FMN(i)=(F/12)*(32-0.5*x);
c(i)=x;
end
end
plot(c,FMN,'-p')
xlabel('Load application point x(m)');
ylabel('Force in Member MN (N)');
grid on;
%%%%%%%%%%%%%%%%%%%%%%%
```

Output is shown in Fig. ES5.21(*c*).

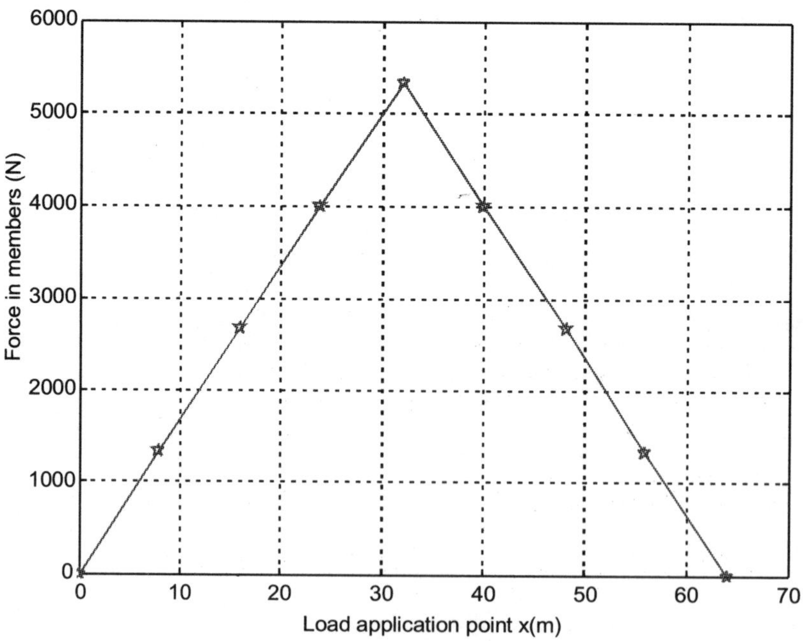

Fig. ES5.21 (c) MATLAB output

ES5.22: Figure ES5.22 shows the loading on a beam. At $x = 0$, the load is increasing at the rate of 80 N/m.

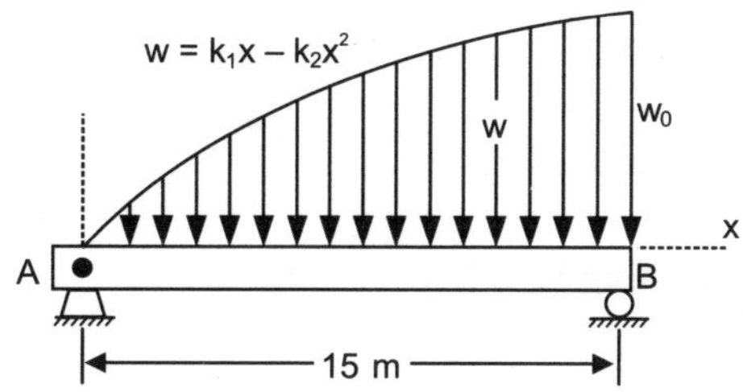

Fig. ES5.22

Write a MATLAB program to plot

(*a*) the distributed load $w(x)$ for $w_0 = 0, 100, 200, 300$ and 400 N/m.

(*b*) the reactions at the two supports as a function of w_0 for $0 \le w_0 \le 400$ N/m.

Solution: The loading function w and its derivatives dw/dx are:

$$w = k_1x - k_2x^2$$

$$dw/dx = k_1 - 2k_2x$$

Constant k_1 and k_2 are obtained from the boundary conditions:

$$w(x = 15) = w_0 = k_1(15) - k_2(15)^2 \Rightarrow 15k_1 - 225k_2 = w_0 \quad ...(1)$$

$$dw/dx(x = 0) = 80 = k_1 - 2k_2(0) \rightarrow k_1 = 80 \quad ...(2)$$

Solving equations (1) and (2) we obtain

$$k_1 = 80 \text{ and } k_2 = \left(\frac{80}{15} - \frac{w_0}{225}\right)$$

Hence $$w = 80x - \left(\frac{80}{15} - \frac{w_0}{225}\right)x^2$$

This function can now be plotted for the specified values of w_0. The reactions at the supports are determined from the free-body diagram shown in Fig. ES5.22(*a*). The load has been discretized into infinitesimal force dF acting at a distance x from A.

The total force due to the distributed load is

$$\int dF = \int w dx$$

and its moment about A is

$$\int x dF = \int x w dx.$$

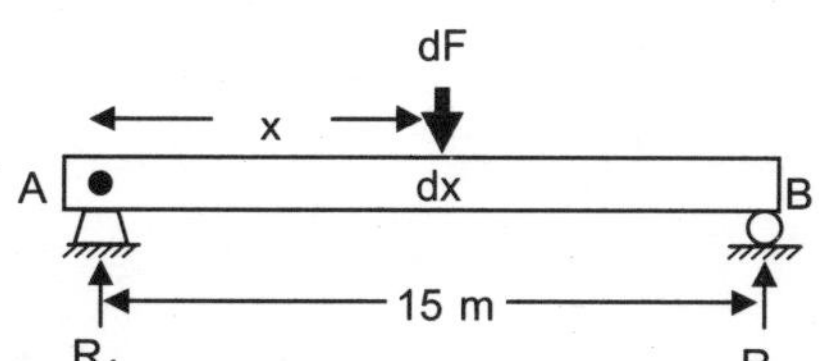

Fig. ES5.22 (a) Free-body diagram

Now $\Sigma M_A = 0 \Rightarrow 15\,R_B - \int_0^{15} xwdx = 0 \Rightarrow R_B = \frac{1}{15}\int_0^{15} x\left(80x - \left(\frac{80}{15} - \frac{w_0}{225}\right)x^2\right)dx$

$$\Sigma F_y = 0 \Rightarrow R_A + R_B = \int_0^{15} wdx \Rightarrow R_A = \int_0^{15}\left(80x - \left(\frac{80}{15} - \frac{w_0}{225}\right)x^2\right)dx - R_B$$

Integrals are evaluated from the following MATLAB script:

```
%%%%%%%%%%%%%%%%%%%%%
syms w RB RA
RB=int('x*(80*x-(80/15-w/225)*x^2)/15','x',0,15)
RA=int('(80*x-(80/15-w/225)*x^2)','x',0,15)-RB
%%%%%%%%%%%%%%%%%%%%%
```

This gives output as follows:

RB =

1500 + 15/4**w*

RA =

1500 + 5/4**w*

Now load and reactions are plotted using following MATLAB script:

```
%%%%%%%%%%%%%%%%%%%%%
w=inline('80*x-(80/15-wo/225)*x^2');
w=vectorize(w); % ADDS '.' TO ENSURE TERM BY TERM RATER THAN MATRIX  OPERA.
x=0:1:15;
subplot(2,1,1);
plot(x,w(0,x),'-*',x,w(100,x),'-p', x,w(200,x),'-',x,w(300,x),
 '- +',x,w(400,x));
xlabel('x(m)');
ylabel('w(N/m)');
legend('w0=0','w0=100','w0=200','w0=300','w0=400',2);
grid on;
subplot(2,1,2);
w0=0:20:400;
RA=1500+5/4.*w0;
RB=1500+15/4.*w0;
plot(w0,RA,'-*',w0,RB,'-p');
xlabel('w_0(N/m)');
ylabel('Support reactions (N)');
legend('R_A','R_B',2);
grid on;
%%%%%%%%%%%%%%%%%%%%%
```

Output is shown in Fig. ES5.22 (*b*).

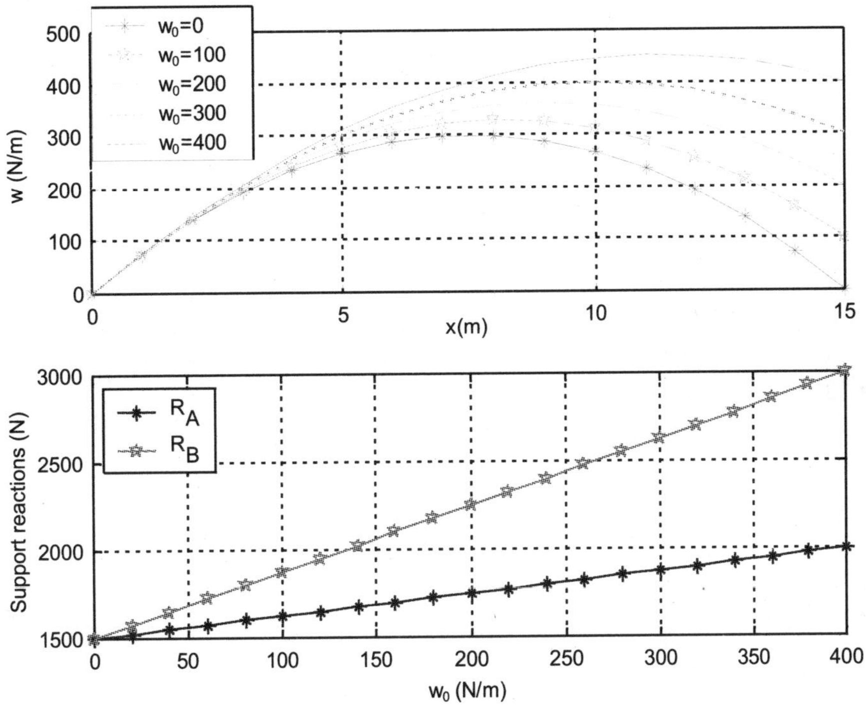

Fig. ES5.22 (b) MATLAB output

ES5.23: Figure ES5.23 shows a beam subjected to distributed and varying loads. Write a MATLAB program to plot the magnitude of the vertical reactions at supports *A* and *B* as functions of distance *d* over $0 \leq d \leq 4$m.

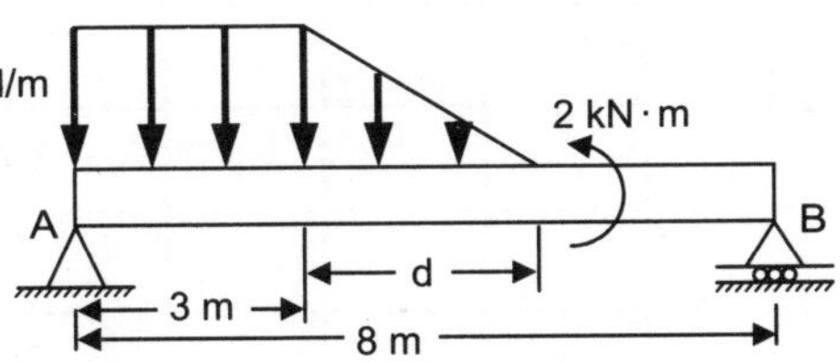

Fig. ES5.23

Solution: Figure ES5.23(*a*) shows the free-body diagram of the beam.

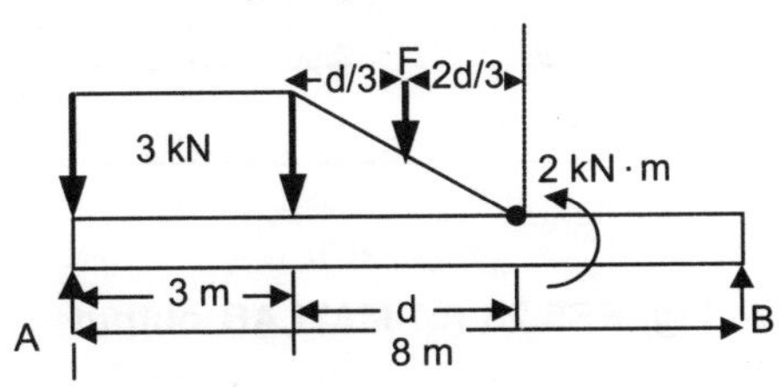

Fig. ES5.23 (a) Free-body diagram

Taking moments about the points A and B:

$$\Sigma M_A = 0 \quad \Rightarrow \quad 8B + 2 - F(3 + d/3) - 3(1.5) = 0 \quad \Rightarrow \quad B = \frac{2.5 + F(3 + d/3)}{8}$$

$$\Sigma M_B = 0 \quad \Rightarrow \quad -8A + 3(5 + 1.5) + F(5 - d + 2d/3) + 2 = 0 \quad \Rightarrow \quad A = \frac{21.5 + F(5 - d/3)}{8}$$

Complete MATLAB script to plot A and B as a function of d is given below:

```
%%%%%%%%%%%%%%%%%
d=[0:0.5:4];
F=1*d; % UNIFORMLY VARYING LOAD
A=(21.5+F.*(5-d./3))/8;
B=(2.5+F.*(3+d./3))/8;
plot(d,A,'-p',d,B,'-*');
xlabel('d(m)');
ylabel('Reaction (kN)');
grid on;
legend('R_A','R_B',2);
%%%%%%%%%%%%%%%%
```

Figure ES5.23 (b) shows the output of the script.

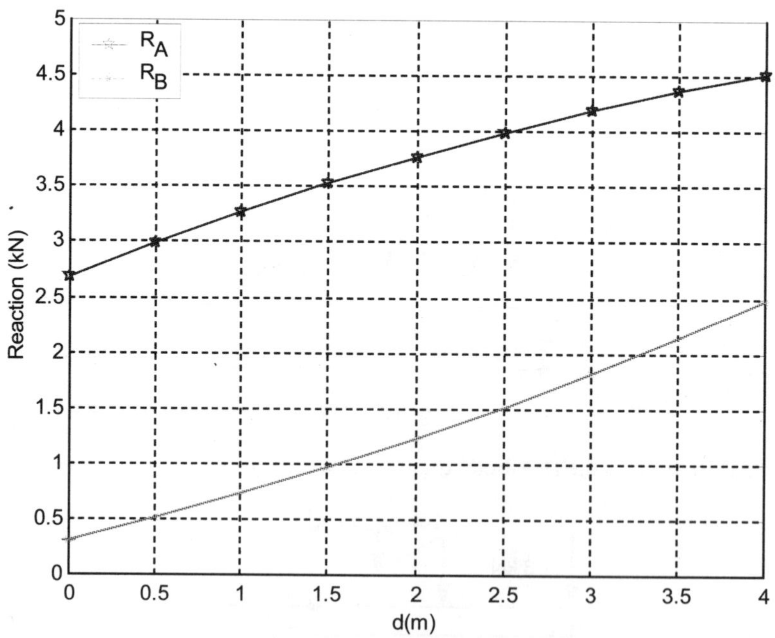

Fig. ES5.23 (b) MATLAB output

ES5.24: Figure ES5.24 shows a cantilever loaded with uniformly distributed load.

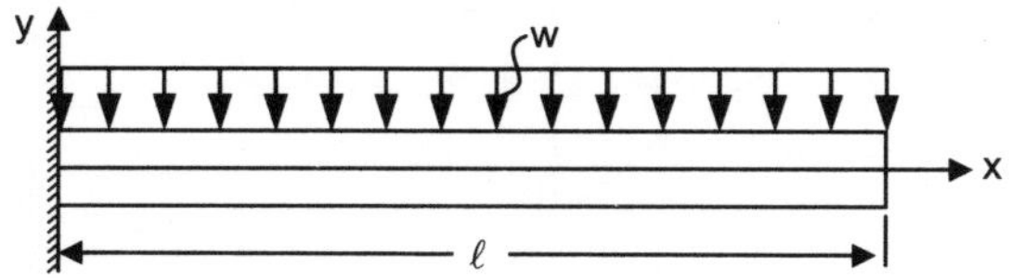

Fig. ES5.24

The deflection y at any point located at a distance x with a uniformly distributed load w is given by:

$$y = -\frac{wx^2}{24EI}\left(6\ell^2 - 4x\ell + x^2\right),$$

where E = the Young's modulus of elasticity,

I = the area moment of inertia and

ℓ = the length of the beam.

Write a MATLAB program to plot the deflection y of the beam as a function of x.

Take $\ell = 10$ m, $E = 70 \times 10^9$ Pa, $I = 10 \times 10^{-6}$ m^4, and $w = 1000$ N/m.

Also plot the deflection of a simply supported beam under same conditions by considering the deflection relation as:

$$y = \frac{wx}{24EI}(2\ell x^2 - x^3 - \ell^3)$$

Solution:

MATLAB program for this problem is given below:

```
%%%%%%%%%%%%%%%%%%%%%%%%
E=70e9;I=10e-6;w=1000;L=10;
x=0:1:10;
yc=-w*x.^2.*(6*L^2-4.*x*L+x.^2)/(24*E*I);% DEFLECTION OF CANTILEVER
ys=w.*x.*(2*L.*x.^2-x.^3-L^3)/(24*E*I);% DEFLECTION OF SIMPLY SUPPORTED
plot(x,yc,'-p',x,ys,'-*');
xlabel('x(m)');
ylabel('Deflection (m)');
legend('Cantilever','Simply-supported',3);
grid on;
%%%%%%%%%%%%%%%%%%%%%%%
```

Output is shown in Fig. ES5.24 (*a*).

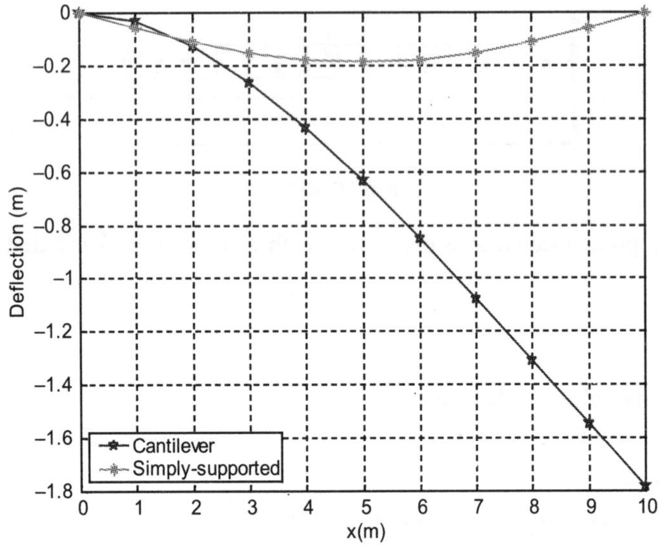

Fig. ES5.24 (a) MATLAB output

ES5.25: Figure ES5.25 shows a simply supported beam with a constant distributed load *w* over half of its length.

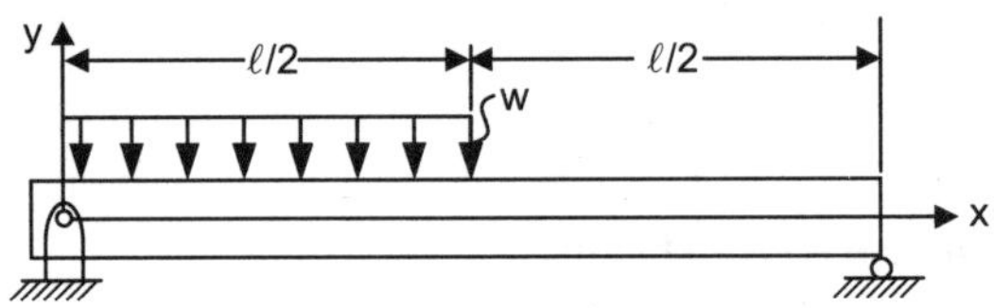

Fig. ES5.25

The deflection *y* as a function of *x* is given by

$$y = \frac{-wx}{384EI}(16x^3 - 24\ell x^2 + 9\ell^3); \qquad \text{for } 0 \le x \le \frac{\ell}{2}$$

$$y = \frac{-wx}{384EI}(8x^3 - 24\ell x^2 + 17\ell^2 x - \ell^3); \text{ for } \frac{\ell}{2} \le x \le \ell$$

where E = Young's modulus of elasticity,

I = moment of inertia and

ℓ = length of the beam.

Write a MATLAB program to plot the deflection of the beam *y* as a function of *x*.

Given $\ell = 25$ m, $E = 200 \times 10^9$ Pa, $I = 350 \times 10^{-6}$ m^4, and $w = 6 \times 10^3$ N/m.

Solution:

MATLAB listing is shown below:

```
%%%%%%%%%%%%%%%%%%%%
L=25;E=200e9;I=350e-6;w=6e3;
i=0;
for x=0:L
    i=i+1;
    if x<=L/2
     y(i)=-w*x*(16*x^3-24*L*x^2+9*L^3)/(384*E*I);
     else
      y(i)=-w*x*(8*x^3-24*L*x^2+17*L^2*x-L^3)/(384*E*I);
     end
    X(i)=x;
     end
plot(X,y,'-p');
xlabel('x(m)');
ylabel('y(m)');
grid on;
%%%%%%%%%%%%%%%%%%%%
```

Output is shown in Fig. ES5.25 (*a*) below:

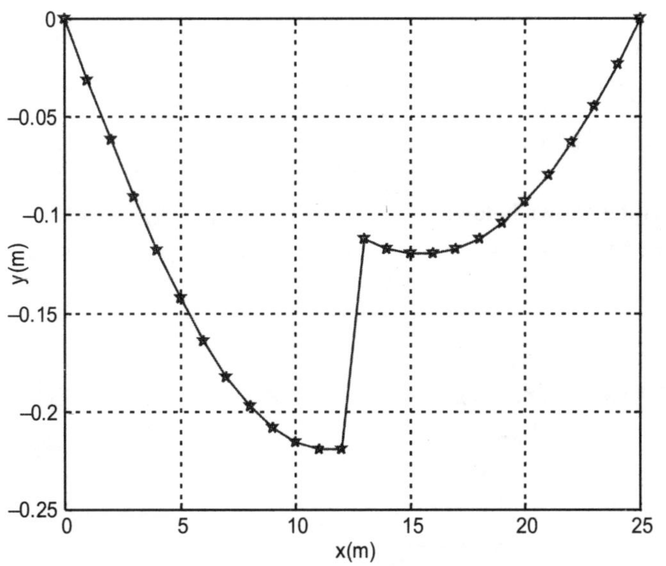

Fig. ES5.25 (a) MATLAB output

ES5.26: Derive the equations for the shear and bending moment curves for the beam shown in Fig.ES5.26. Write a MATLAB program to plot the shear and bending moment diagrams for the beam with w_0 = 20 kN/m and ℓ = 4 m.

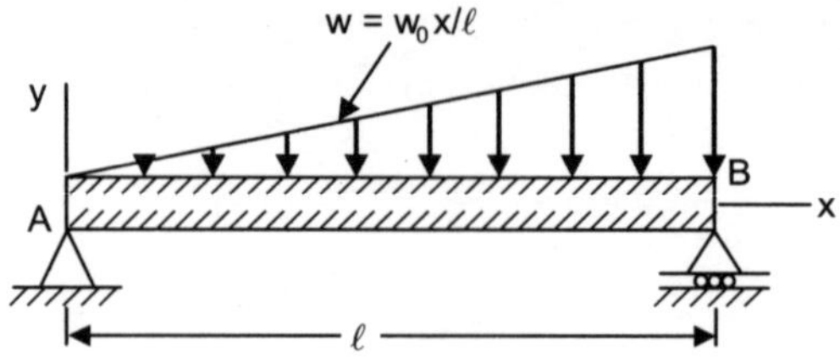

Fig. ES5.26

Solution: Shear force is given by $V = -\int w\,dx = -\int w_0(x/\ell)\,dx = -\frac{1}{2} w_0 \frac{x^2}{\ell} + c_1$

Bending moment $M = \int V dx = \int\left(-\frac{1}{2} w_0 \frac{x^2}{\ell} + c_1\right) dx = -\frac{1}{6} w_0 \frac{x^3}{\ell} + c_1 x + c_2$

The boundary conditions are:

$$\text{At } x = 0, M = 0 \quad \Rightarrow \quad -\frac{1}{6} w_0 \frac{0^3}{\ell} + c_1 0 + c_2 \quad \Rightarrow \quad c_2 = 0$$

$$\text{At } x = L, M = 0 \quad \Rightarrow \quad -\frac{1}{6} w_0 \frac{\ell^3}{\ell} + c_1 \ell = 0, \text{ which gives } c_1 = \frac{1}{6} w_0 \ell$$

Therefore

$$V = -\frac{1}{2} w_0 \frac{x^2}{\ell} + \frac{1}{6} w_0 \ell$$

and

$$M = -\frac{1}{6} w_0 \frac{x^3}{\ell} + \frac{1}{6} w_0 \ell x$$

Use $w_0 = 20$ kN/m, $\ell = 4$ m and plot.

Complete MATLAB script is shown below:

```
%%%%%%%%%%%%%%%%%%%%%%%%%%
w0=20; L=4;
x=[0:0.2:4];
V=-0.5*w0.*x.^2/L+w0*L/6;
M=-w0.*x.^3/(6*L)+w0*L.*x/6;
subplot(2,1,1);
plot(x,V,'-*');
xlabel('x(m)');
ylabel('Shear force (N)');grid on;
subplot(2,1,2);
plot(x,M,'-p');
xlabel('x(m)');
ylabel('Bending moment (N·m)');
grid on;
%%%%%%%%%%%%%%%%%%%%%%%%
```

Figure ES5.26 (*a*) shows the output of the script.

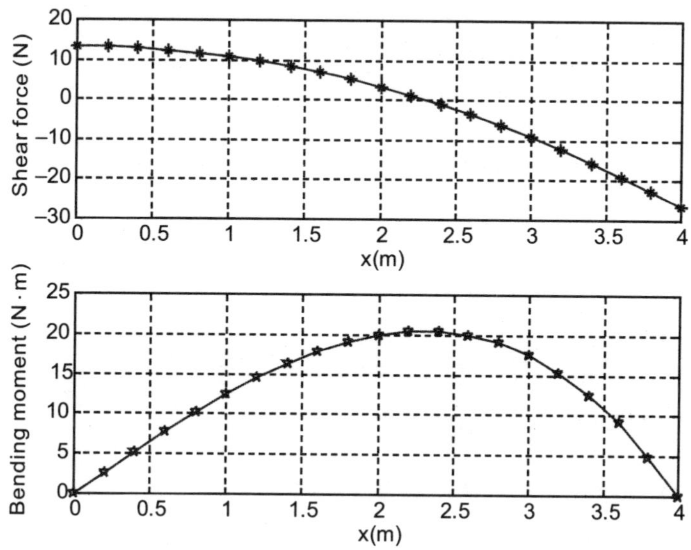

Fig. ES5.26 (a) MATLAB output

ES5.27: Figure ES5.27 shows a typical transmission-time installation suspended between two points lying at the same elevation. The length of the cable is ℓ_{AB} and mass per unit length is m'. Write a MATLAB program to plot the dimensionless quantities: H/ℓ, ℓ_{AB}/ℓ, $T_0/m'g\ell$ and $T_{max}/m'g\ell$ for values of c/ℓ from 0.1 to 0.6 using 0.01 increments and from 0.6 to 5 using 0.2 increments. Here c is deflection of cable.

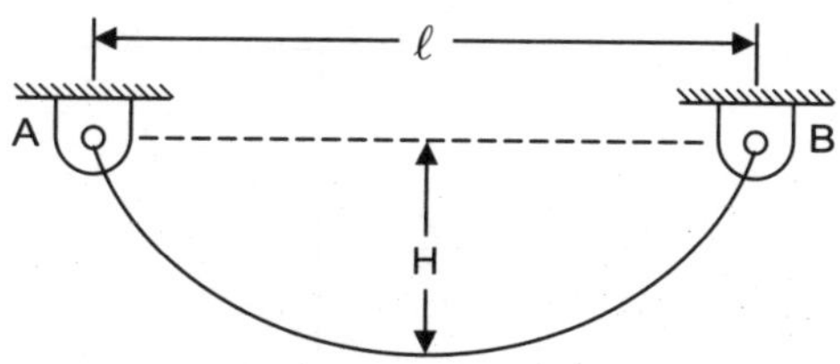

Fig. ES5.27

Solution: Let X and Y be the coordinate directions as shown in Fig. ES5.27 (*a*):

It can be seen that $x_B = L/2$ and $s_B = c \sinh(x_B/c) \Rightarrow 2\, s_{AB} = c \sinh(x_B/c)$

Also the expression for $y_B = c \cosh(x_B/c) = H + c$ or $H = y_B - c$

Tensions in the string: $T_0 = m'gc$ and

maximum tension is $T_{\max} = m'gy_B$

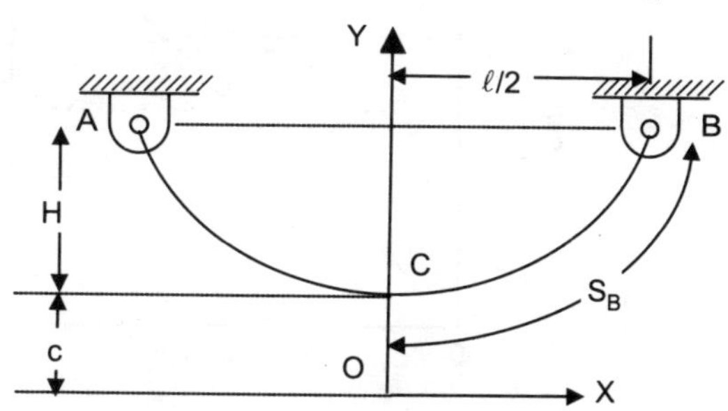

Fig. ES5.27 (a) Geometric deformation

Simple MATLAB script is given below:

```
%%%%%%%%%%%%%%%%%%%%%%%%%
L=1;mg=1;
xB=0.5*L;
c=[0.1*L:0.01*L:0.6*L];
```

```
sAB=2.*c.*sinh(b);
yB=c.*cosh(b);
h=yB-c;
T0=mg.*c;
Tmax=mg.*yB;
c=[0.6*L:0.2*L:5*L];
b=xB./c;
sAB=2.*c.*sinh(b);
yB=c.*cosh(b);
h=yB-c;
T0=mg.*c;
Tmax=mg.*yB;
subplot(2,2,1);
plot(c/L,h/L);ylabel('h/L');grid on;
subplot(2,2,2);
plot(c/L,sAB/L);ylabel('s-AB/L');grid on;
subplot(2,2,3);
plot(c/L,T0./(mg*L));ylabel('T_o/mgL');grid on;
subplot(2,2,4);
plot(c/L,Tmax/(mg*L));ylabel('T-max/(mgL)');grid on;
xlabel('c/L');
%%%%%%%%%%%%%%%%%%%%%%%%%%%
```

Output is shown in Fig. ES5.27(*b*).

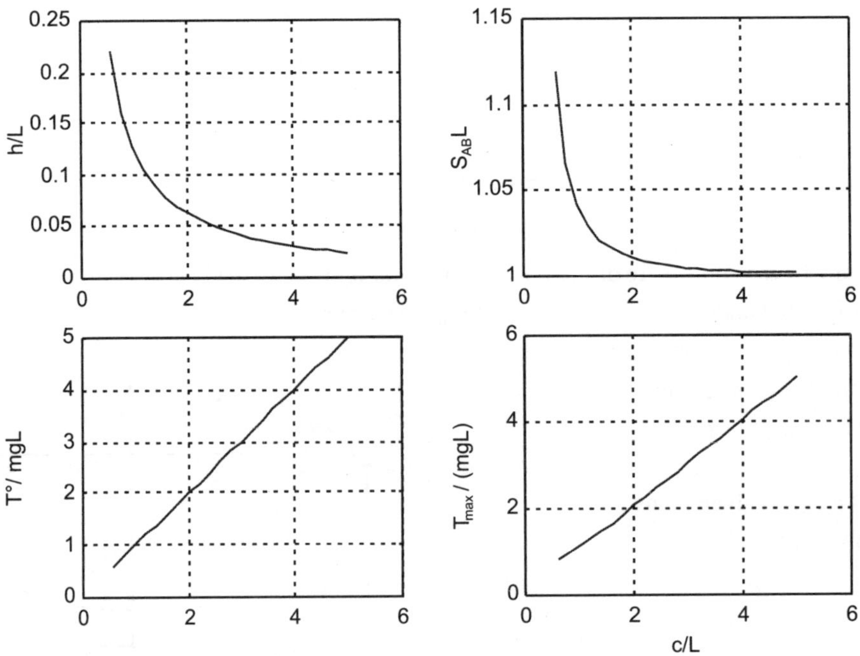

Fig. ES5.27(b) MATLAB output

ES5.28: Figure ES5.28 shows a planar mechanism with two uniform links *OQ* and *OP* each of mass *m* and are connected and constrained as shown. When a horizontal force *F* is applied, the angle α between the links increases, the light rod connected at *M* and passing through a pivoted collar at *N* compresses the spring. The stiffness of the spring is *k* N /m. Determine

(*a*) the force *F* which will produce equilibrium at the angle α if the spring is uncompressed in the position when α = 0

(*b*) Write a MATLAB program for determine the equilibrium value of α corresponding to a given force *F*.

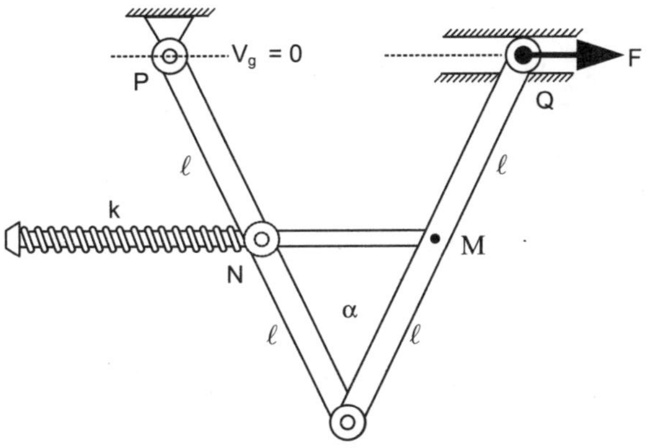

Fig. ES5.28

Solution: The body or system will be at rest (in the inertial rest frame of its centre of mass), meaning that it occupies one point in configuration space for all time. The problem is to find that point. One criterion for finding that point, which makes use of the calculus of variations, is called the principle of virtual work.

According to the principle of virtual work, any infinitesimal http://www.britannica.com/memberlogin virtual displacement in configuration space, consistent with the constraints, requires no work. A virtual displacement means an instantaneous change in coordinates (a real displacement would require finite time during which the body might move and forces might change). To express the principle, labelling the generalized coordinates as $r_1, r_2, \ldots, r_i, \ldots$ and if F_i is the net component of generalized force acting along the coordinate r_i, we can write:

$$\Sigma F_i \, dr_i = 0$$

Here, $F_i \, dr_i$ is the work done when the generalized coordinate is changed by the infinitesimal amount dr_i. If r_i is a real coordinate (say, the *x* coordinate of a particle), then F_i is a real force. If r_i is a generalized coordinate (say, an angular displacement of a rigid-body), then F_i is the generalized force such that $F_i \, dr_i$ is the work done (for an angular displacement, F_i is a component of torque).

The necessary and sufficient condition for the equilibrium is zero virtual work done by all working forces acting on the body during any virtual displacement δ*S* consistent with the constraints imposed on the particle. When using the principle of virtual work for a system of connected rigid bodies (mechanism) we must keep in mind that no virtual work is done by internal forces, by reactions in smooth constraints, or by forces normal to the direction of motion. The virtual work is done by reactions when friction is present.

In Fig. ES5.28, the compression x of the spring is the distance which M has moved away from N, which is

$$x = 2b \sin\frac{\alpha}{2}.$$

Thus, the elastic potential energy of the spring is

$$V_e = \frac{1}{2}k(2\ell \sin\frac{\alpha}{2})^2 = 2k\ell^2 \sin^2\frac{\alpha}{2}$$

With the datum for zero gravitational potential energy taken through the support at P for convenience the expression for V_g becomes:

$$V_g = 2mg(-\ell \cos\frac{\alpha}{2})$$

The distance between points P and Q is $4b \sin\frac{\alpha}{2}$, so that the virtual work done by external force F is

$$\delta U = F\delta(4\ell \sin\frac{\alpha}{2}) = 2F\ell \cos\frac{\alpha}{2}(\delta\alpha)$$

The virtual work equation $\delta V_g + \delta V_e - \delta U = 0$, becomes

$$\delta(-2mg\ell \cos\frac{\alpha}{2}) + \delta(2k\ell^2 \sin^2\frac{\alpha}{2}) - 2F\ell \cos\frac{\alpha}{2}(\delta\alpha) = 0$$

$$\Rightarrow \quad (\text{mg}\,\ell \sin\frac{\alpha}{2})\,\delta\alpha + 2k\ell^2 \sin\frac{\alpha}{2}\cos\frac{\alpha}{2}\delta\alpha - (2F\ell \cos\frac{\alpha}{2})\,\delta\alpha = 0$$

Simplifying this, finally gives

$$F = k\,\ell \sin\frac{\alpha}{2} + \frac{1}{2}mg \tan\frac{\alpha}{2}$$

For a given force F, finding the angle θ explicitly is not possible. But by computer solution, we can plot θ using above equation.

MATLAB script is given below:

```
%%%%%%%%%%%%%%%%%%%%%%%%%%%
L=0.3;
k=25;
mg=60;
alfa=[0:0.5:80];
al=alfa*pi/180;
F=k*L.*sin(al/2)+0.5*mg.*tan(al/2);
plot(F,alfa);
xlabel('Force applied F(N)');
ylabel('Equilibrium angle(degree)');
grid on;
%%%%%%%%%%%%%%%%%%%%%%%%%%%
```

Output is shown in Fig. ES5.28 (*a*).

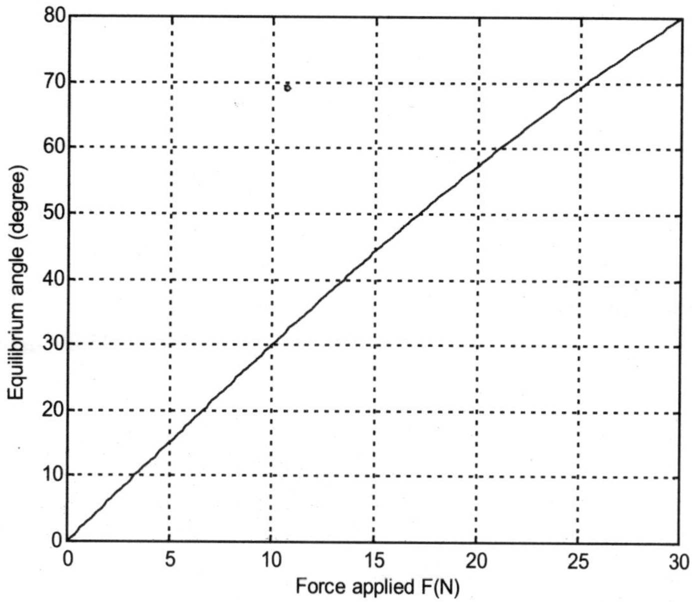

Fig. ES5.28 (a) MATLAB output

ES5.29: A mechanism is shown in Fig. ES5.29. Use MATLAB to plot the force in member *BD* as a function of θ for values of θ ranging from 20° to 120°.

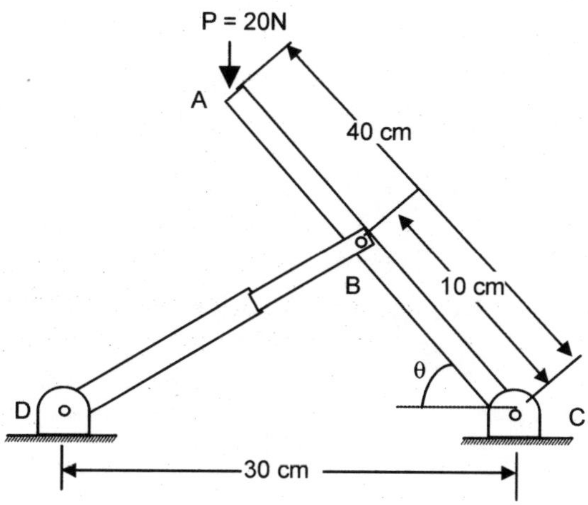

Fig. ES5.29

Solution: First of all the free-body diagram is drawn as shown in Fig. ES5.29(*a*).

First let us compute virtual displacements:

$$y_A = 40 \sin \theta \Rightarrow \delta y_A = 40 \cos \theta \, (\delta\theta) \quad ...(1)$$

Applying the principle of virtual work:

$$\delta U = 0 \Rightarrow -P\,\delta y_A + F_{BD}\,\delta BD = 0$$

$$\Rightarrow \quad -P(40\cos\theta\,\delta\theta) + F_{BD}\,\delta BD = 0$$

$$\Rightarrow \quad F_{BD} = P(40\cos\theta\,\delta\theta)/\delta BD \qquad ...(2)$$

The term $\delta(BD)$ is virtual deformation in member BD computed from geometry as follows:

$$BD^2 = BC^2 + CD^2 - 2(BC)(CD)\cos\theta$$

$$= 10^2 + 30^2 + -2(10)(30)\cos\theta$$

$$= 1000 - 600\cos\theta \qquad ...(3)$$

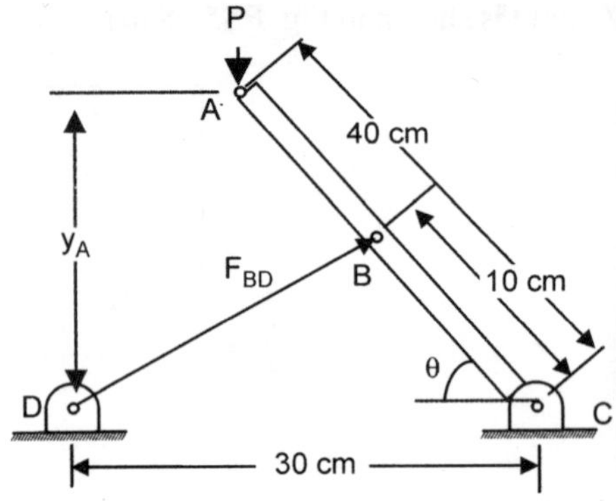

Fig. ES5.29 (a) Free-body diagram

Differentiating:

$$2\,BD(\delta BD) = 600\sin\theta\,(\delta\theta)$$

$$\Rightarrow \quad \delta BD = \frac{300\sin\theta}{BD}(\delta\theta) \qquad ...(4)$$

Equations (1) to (4) are used to plot the variation of force in member BD.

MATLAB script is given below:

```
%%%%%%%%%%%%%%%%%%%%%%%%%%%%%%%
P=20;
theta=20:5:120;
th=theta*pi/180;
BD=(1000-600.*cos(th)).^0.5;
dBD=300.*sin(th)/BD;
FBD=P*40.*cos(th)./dBD;
plot(theta,FBD,'-p');
xlabel('\theta(degrees)');
ylabel('Force in BD (N)');
grid on;
%%%%%%%%%%%%%%%%%%%%%
```

Output is shown in Fig. ES5.29 (*b*).

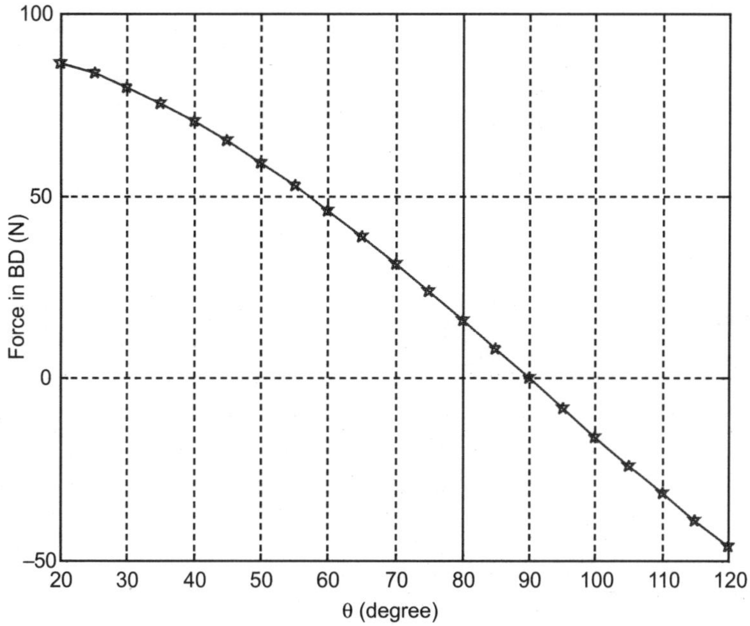

Fig. ES5.29 (b) MATLAB output

ES5.30: An 10 kg block is hung from the midpoint *C* of the cable *AB* which is attached to two springs as shown in Fig. ES5.30. Knowing that the springs are unstretched when $y = 0$, plot the distance y corresponding to equilibrium, as a function of the spring constant k_1 ranging from 500 N/m to 800 N/m. Take $k_2 = 1000$ N/m.

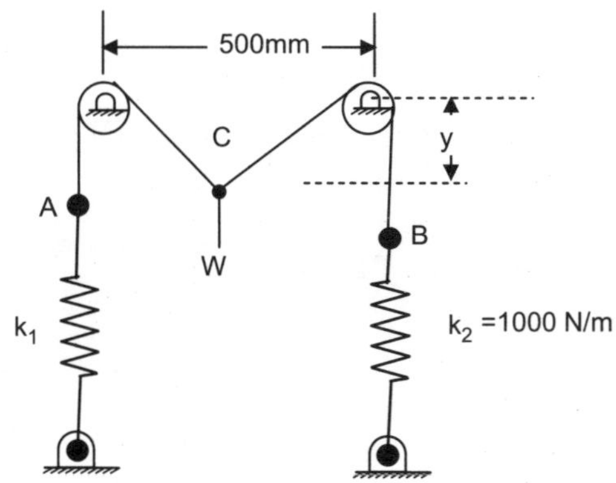

Fig. ES5.30

Solution: There are two degrees of freedom: u_A and u_B as shown in Fig. ES5.30(*a*). The total potential energy of the system is:

$$V = \frac{1}{2}k_1 u_A^2 + \frac{1}{2}k_2 u_B^2 - W_y$$

where $x^2 + y^2 = L_A^2$ ...(1)

$(0.5 - x)^2 + y^2 = L_B^2$ ...(2)

Eliminating x from above two equations:

$$y = \sqrt{L_A^2 - \left(L_A^2 - L_B^2 + 0.25\right)^2} \quad ...(3)$$

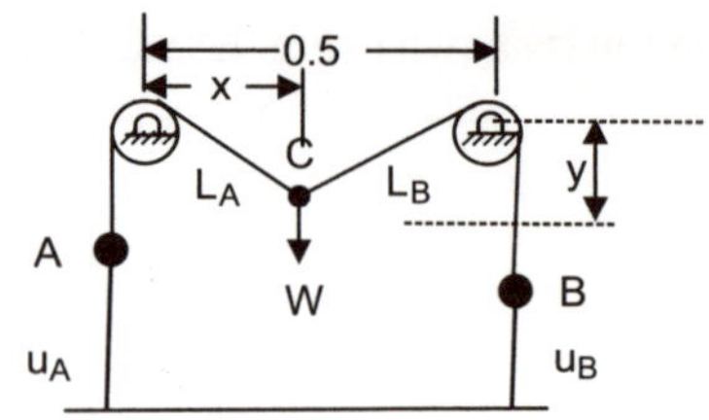

Fig. ES5.30 (a)

At equilibrium, derivatives of potential energy

$$\frac{\partial V}{\partial u_A} = 0 \quad \Rightarrow \quad k_1 u_A = W \frac{\partial y}{\partial u_A} \quad ...(4)$$

and

$$\frac{\partial V}{\partial u_B} = 0 \quad \Rightarrow \quad k_2 u_B = W \frac{\partial y}{\partial u_B} \quad ...(5)$$

Equations (4) and (5) are solved to obtain u_A and u_B respectively.

The displacements of the springs are related to L_A and L_B according to:

$$L_A = u_A + 0.25 \text{ and } L_B = u_B + 0.25.$$

These are substituted again in eq.(3) to get equilibrium value y_{eq}.

Complete MATLAB program for obtaining y_{eq} is given below:

```
%%%%%%%%%%%%%%%%%%%%%%%%%%%%%%%%%%%%%%
syms uA uB y
m=10;g=9.81;
k2=1000;
W=m*g;
i=0;
for k1=500:50:800
i=i+1;
LA=0.25+uA;
LB=0.25+uB;
y=sqrt(LA.^2+(LA.^2-LB.^2+0.25).^2);
dyA=diff(y,uA);
dyB=diff(y,uB);
eqn1=k1*uA-W*dyA;
eqn2=k2*uB-W*dyB;
g=solve(eqn1,eqn2);
z1=double(abs(g.uA)); z2=double(abs(g.uB));
x1=min(z1);x2=min(z2);
% FINDING EQUILIBRIUM VALUES
L1=0.25+x1;L2=0.25+x2;
s=sqrt(L1^2+(L1^2-L2^2+0.25)^2);
```

```
d(i)=s; k(i)=k1;
end
plot(k,d,'-p');
xlabel('Stiffness k_1(N/m)');
ylabel('Distance y for equilibrium');
grid on;
%%%%%%%%%%%%%%%%%%%%%%%%%%%%%
```

Output of this *m*-file is shown in Fig. ES5.30 (*b*). As the program requires many computations, compared to other routines it took 15 second time to display the plot on Pentium-4 processor with 3 GHz machine.

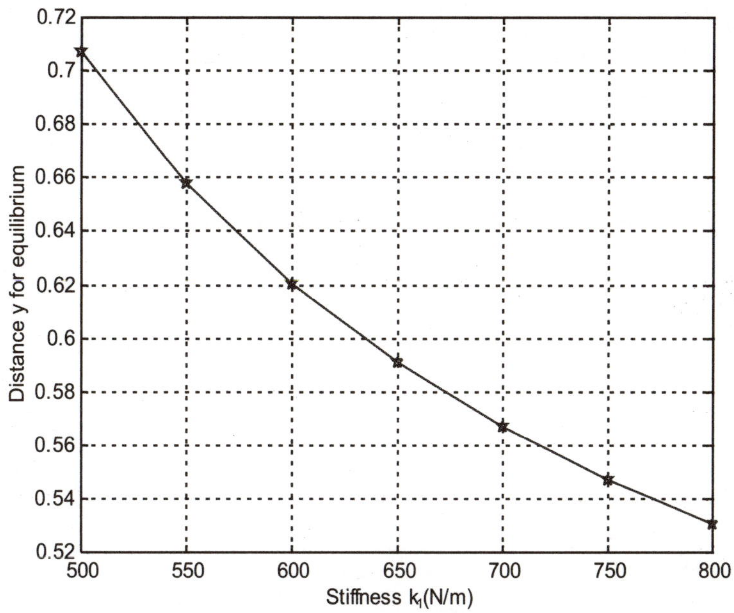

Fig. ES5.30 (b) MATLAB output

ES5.31: Determine the distance H for which the centroid of the shaded area in Fig.ES5.31 is as high above BB' as possible. Write a MATLAB program to plot the ratio H/B as a function of k for $0.1 \leq k \leq 0.9$.

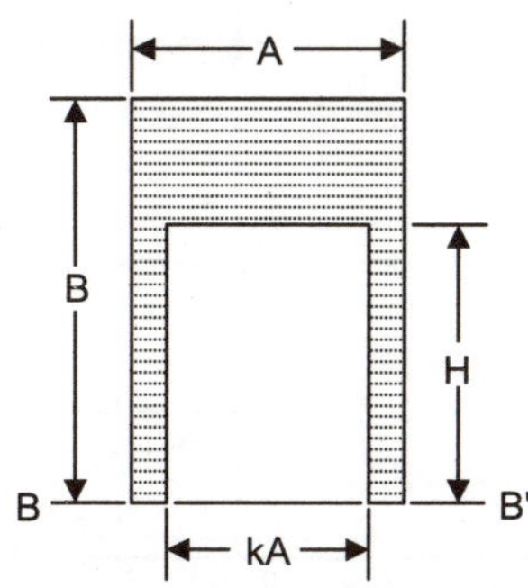

Fig. ES5.31

Solution: Following Table can be readily constructed from the dimensions.

Region	Area (a)	$\bar{y}$	$a\bar{y}$
Entire rectangle	AB	$B/2$	$AB^2/2$
Inner unshaded rectangle	kAH	$H/2$	$-kAH^2/2$
Shaded region $\sum a$	$AB - kAH$	?	$(AB^2 - kAH^2)/2$

By definition the centroid is $\bar{y} = \dfrac{a\bar{y}}{\sum a} = \dfrac{B^2 - kH^2}{2(B - kH)} = \dfrac{B}{2}\dfrac{(1 - k(H/B)^2)}{(1 - k(H/B))} = \dfrac{B}{2}\dfrac{(1 - k\alpha^2)}{1 - k\alpha}$

For maximum $\bar{y}$, $\dfrac{d\bar{y}}{d\alpha} = 0 \;\Rightarrow\; (1 - k\alpha)(-2k\alpha) + (1 - k\alpha^2)k = 0$

$$\Rightarrow \quad k\alpha^2 - 2\alpha + 1 = 0$$

Solving this for α, $\alpha = \dfrac{2 \pm \sqrt{4 - 4k}}{2k} = \dfrac{1 \pm \sqrt{1 - k}}{k}$

A complete MATLAB script to plot the variation of a as a function of k is given below:

```
%%%%%%%%%%%%%%%%%%%
k=0.1:0.05:0.9;
al_1=(1+sqrt(1-k))./k;
al_2=(1-sqrt(1-k))./k;
plot(k,al_1,'-p',k,al_2,'-*');
xlabel('k');
ylabel('H/B');
legend('alpha_1','alpha_2');
grid on;
%%%%%%%%%%%%%%%%%%%
```

Figure ES5.31(a) shows the output of the script.

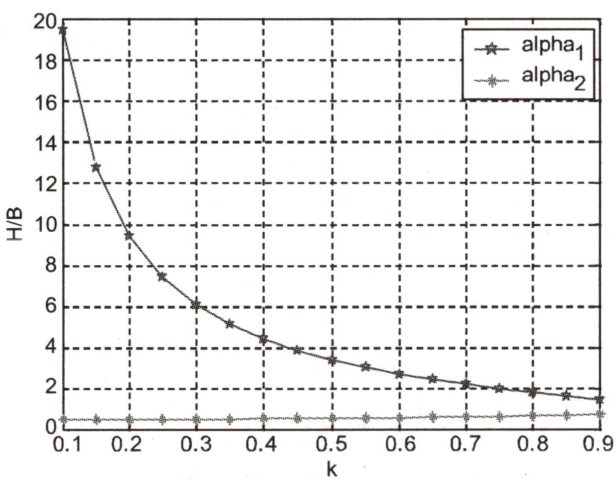

Fig. ES5.31 (a) MATLAB output

ES5.32: Figure ES5.32 shows an approximating the general spandrel using a series of n rectangles, each of width ΔA and of the form bcc′b'. Write MATLAB program to calculate the coordinates of the centroid of the area. Determine the centroid when A = 5cm, H = 5cm for m = 2 to 4.

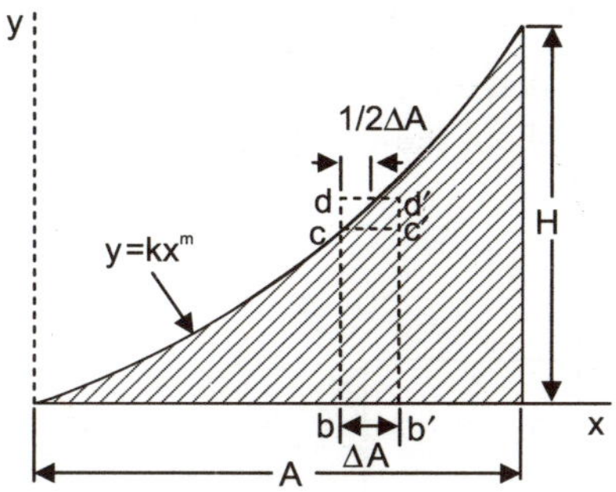

Fig. ES5.32

Solution: At $x = A, y = H,$ hence the equation of curve becomes

$$H = kA^m \text{ or } k = \frac{H}{A^m}$$

For rectangle p see Fig. ES5.32(a):

$$\bar{x}_p = (p-0.5)\Delta A \text{ and } \bar{y}_p = \frac{1}{2}k(x_{p-1})^m$$

$$= \frac{1}{2}\frac{h}{a^m}[(p-1)\Delta A]^m$$

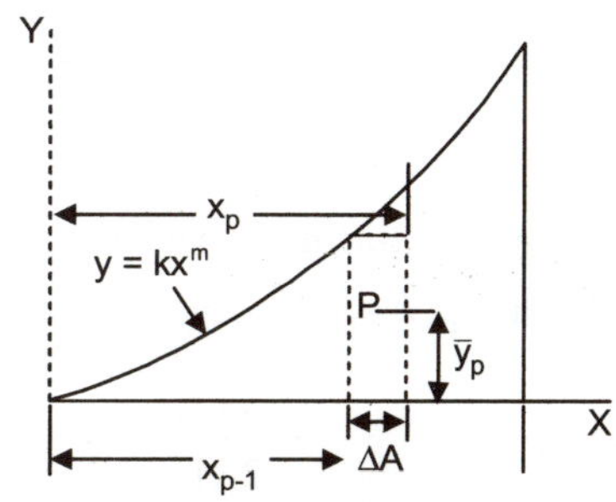

Fig. ES5.32 (a)

Complete MATLAB program for studying the variation of centroid with A and H is presented below:

```
%%%%%%%%%%%%%%%%%%%%%%%%%%%%%%
A=5;H=5;
m=[2;3;4];
N=20;% total sub-rectangular divisions
dA=A/N;
for i=1:length(m)
a=0;ax=0;ay=0;
for p=1:N
x_p=(p-0.5)*dA;
y_p=(H./(2*A^m(i)))*((p-1)*dA)^m(i);
a_p=dA*2*y_p;
a=a+a_p;%total Area
ax=ax+xp.*a_p;
ay=ay+yp.*a_p;
end
x=ax/a;y=ay/a;
fprintf('For m=%f: centroid is x=%f and y=%f\n,m(i),x,y);
end
%%%%%%%%%%%%%%%%%%%%%%%%%%%%%%
```

»

For $m = 2.000000$: centroid is $x = 3.778846$ and $y = 1.423750$
For $m = 3.000000$: centroid is $x = 4.021579$ and $y = 1.319735$
For $m = 4.000000$: centroid is $x = 4.183047$ and $y = 1.247985$.

ES5.33: Write a MATLAB program to determine the volume and the surface area of the solid obtained by rotating the area shown in Fig.ES5.33 about the y-axis when $A = 100$mm and $n = 1$.

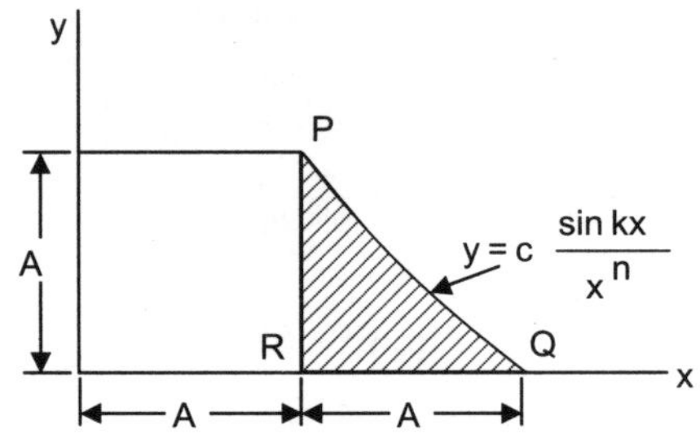

Fig. ES5.33

Solution: Using initial and boundary conditions:

At point Q, $y = 0$

Hence $$c\frac{\sin k(2A)}{(2A)^n} = 0 \quad \Rightarrow \quad 2kA = \pi, \quad \text{or} \quad k = \frac{\pi}{2A}$$

At point A: $y = A$.

Hence $$c\frac{\sin k(A)}{(A)^n} = A \quad \Rightarrow \quad \frac{c}{(A)^n} = A \quad \Rightarrow \quad c = (A)^{n+1}$$

Volume of solid obtained by rotating the area is

$$V = 2\pi a\,\overline{x} = 2\pi\int_0^{2A} xy\,dx = 2\pi\int_0^{2A} xA^{n+1}\frac{\sin kx}{x^n}dx \qquad ...(1)$$

$$\text{Surface area } a = \sum_{i=1}^{3} 2\pi\overline{x}L = \text{area on } PR + \text{area on } QR + \text{area on } PQ$$

$$= 2\pi A^2 + 2\pi A(3A/2) + 2\pi\int_0^{2A} x\,ds \qquad ...(2)$$

where ds = length of arc $PQ = \sqrt{(dx)^2 + (dy)^2} = dx\sqrt{1 + (dy/dx)^2}$ where $y = c\dfrac{\sin kx}{x^n}$

Hence $$dy/dx = kc\frac{\cos kx}{x^n} - nc\frac{\sin kx}{x^{n+1}}$$

Complete MATLAB program for evaluating equations (1) and (2) is presented below:

```
%%%%%%%%%%%%%%%%%%%%%%
syms x real
A=0.1;
A2=2*A;
k=pi/(2*A);
n=1;
c=A^(n+1);
VI=int(x*(c*sin(k*x)/x^n),x,0,A2);% integral term in equation(1)
V=2*pi*VI;
V=double(V);
% CALCULATION OF SURFACE AREA
f2= x*sqrt(1+(k*c*cos(k*x)/x^n-n*c*sin(k*x)/x^(n+1))^2);
Xb=int(f2,x,0,A2);
Xbb=double(Xb);
a=2*pi*A^2+3*pi*A^2+2*pi*Xbb;
fprintf('For order n=%g: The volume is %f m^3 and Surface area
is %f m^2\n',n,V,a);
%%%%%%%%%%%%%%%%%%%%%%
```

Output is shown below:

» For order $n=1$: The volume is 0.008000 m^3 and Surface area is 0.329227 m^2.

ES5.34: Calculate the centroid of the area found by deducting the quadrant of an ellipse from the rectangle of dimensions 10 cm × 5 cm as shown in Fig.ES5.34.

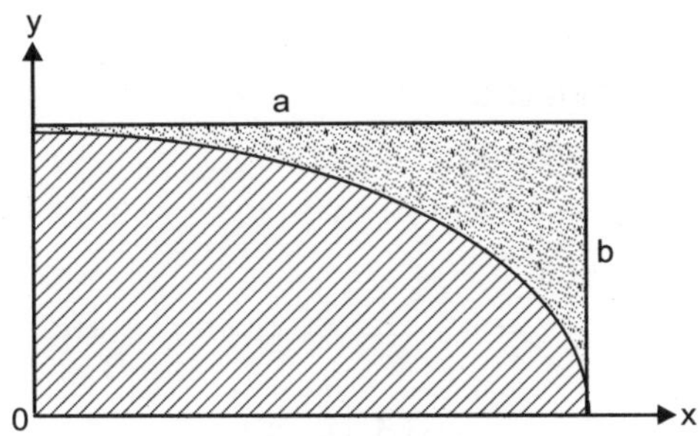

Fig. ES5.34

Solution: To calculate the centroid, use the following relations:

$$\bar{x} = \frac{A_1x_1 - A_2x_2}{A_1 - A_2} \text{ and } \bar{y} = \frac{A_1y_1 - A_2y_2}{A_1 - A_2}$$

where suffixes 1 and 2 respectively indicate the overall rectangle and the quadrant of ellipse. It is obvious that $A_1 = ab$ and $x_1 = a/2$ and $y_1 = b/2$, while for ellipse they should be computed as follows:

Consider a vertical strip of infinitesimally small thickness dx at any x having height of y. Let the area of this strip be $dA = ydx$. As this strip moves along the x-axis from 0 to a, it requires integration of above equation dA from 0 to a.

$$x_2 = \frac{\int_0^{\pi/2} xy\,dx}{\int_0^{\pi/2} y\,dx} \text{ and } y_2 = \frac{\int_0^{\pi/2} yx\,dy}{\int_0^{\pi/2} x\,dy},\ A_2 = \pi ab/4, \text{ where } x = a\cos\theta \text{ and } y = b\sin\theta$$

Hence $dx = -a\sin\theta\,d\theta$ and $dy = a\cos\theta\,d\theta$

Complete MATLAB program for computing the integrals and evaluating the centroid of overall area is given below:

```
%%%%%%%%%%%%%%%%%%%%%%%%%%%%%%
syms th
a=10;b=5;
x=a*cos(th);y=b*sin(th);
dx=diff(x,th);dy=diff(y,th);
dAx=x*dy; Ax=int(dAx,0,pi/2);
dAy=y*dx;Ay=int(dAy,0,pi/2);
x2n=x*dAy;y2n=y*dAx;
x2=int(x2n,0,pi/2)/Ay;
y2=int(y2n,0,pi/2)/Ax;
A2=pi*a*b/4;
% CENTROID OF OVERALL RECTANGLE
x1=a/2;y1=b/2; A1=a*b;
% CENTROID OF INTERSECTION REGION
xbar=double((x1*A1-x2*A2)/(A1-A2));
ybar=double((y1*A1-y2*A2)/(A1-A2));
% DISPLAY THE RESULT
fprintf('\nThe area formed between the rectangle and ellipse is
%g cm^2\n',A1-A2);
fprintf('The x coordinate of centroid is: %f cm\n',xbar);
fprintf('The y coordinate of centroid is: %f cm',ybar);
%%%%%%%%%%%%%%%%%%%%%%%%%%%%%%
```

Output is as follows:

The area formed between the rectangle and ellipse is 10.7301 cm^2.

The x coordinate of centroid is 7.766321 cm.

The y coordinate of centroid is 3.883160 cm.

ES5.35: A 3-dimensional structure is fabricated during a project work from four steel-rods of equal diameter as shown in Fig. ES5.35. Using MATLAB, study the variation of center of gravity of the structure as a function of h and R.

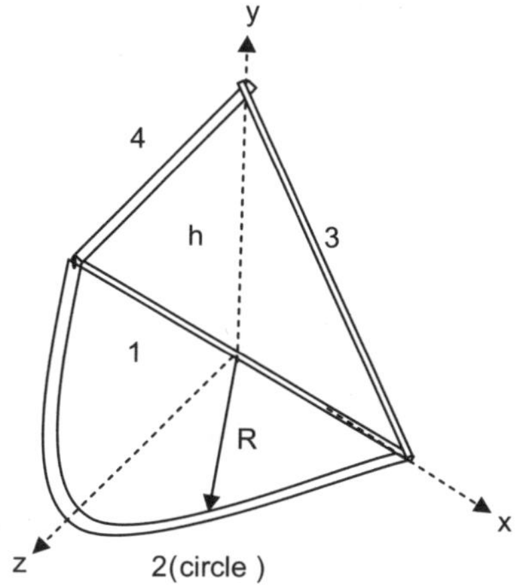

Fig. ES5.35 Three-dimensional structure

Solution: Assume that the rods are homogeneous so that the centers of gravity of the structure will coincide with the centroid of the corresponding line. Following table of centroids for individual rods can be prepared:

Rod#	L	$\bar{x}$	$\bar{y}$	$\bar{z}$
1	$2R$	0	0	0
2	πR	0	0	$2R/\pi$
3	$\sqrt{R^2+h^2}$	$R/2$	$h/2$	0
4	$\sqrt{R^2+h^2}$	$-R/2$	$h/2$	0

Now by definition: $\bar{x}_c = \dfrac{\Sigma L \times \bar{x}}{\Sigma L} = 0$

$$\bar{y}_c = \frac{\Sigma L \times \bar{y}}{\Sigma L} = \frac{h\sqrt{R^2+h^2}}{2R+\pi R+2\sqrt{R^2+h^2}}$$

$$\bar{z}_c = \frac{\Sigma L \times \bar{z}}{\Sigma L} = \frac{2R^2}{2R+\pi R+2\sqrt{R^2+h^2}}$$

MATLAB program for studying the variation of center of gravity of the structure is presented below:

```
%%%%%%%%%%%%%%%%%%%%%%%%
for h=10:20:100
for R=10:30:120
```

```
L=2*R(i)+pi*R(i)+2*sqrt(R(i)^2+h(i)^2);
xc=0;
yc=h(i)*sqrt(R(i)^2+h(i)^2)/L;
zc=2*R(i)^2/L;
fprintf('R\th\txc\tyc\tzc\n');
fprintf('%f\t%f\t%f\t%f\n%f\n',R(i),h(i),xc,yc,zc);
end
end
%%%%%%%%%%%%%%%%%%%%%%%
```

Output of the program is shown below:

R	*h*	*xc*	*yc*	*zc*
10	10	0	1.77442	2.5094
40	10	0	1.43101	11.1063
70	10	0	1.41045	19.5479
100	10	0	1.40527	27.9659
10	30	0	8.27378	1.74426
40	30	0	4.90735	10.469
70	30	0	4.46039	19.1321
100	30	0	4.33228	27.6638
10	50	0	16.6204	1.30381
40	50	0	9.59338	9.5887
70	50	0	8.08553	18.4225
100	50	0	7.57716	27.1089
10	70	0	25.668	1.03714
40	70	0	15.3814	8.72151
70	70	0	12.4209	17.5658
100	70	0	11.2682	26.3751
10	90	0	35.0495	0.860128
40	90	0	22.0146	7.94753
70	90	0	17.4533	16.6682
100	90	0	15.4594	25.5352

Following code is used to plot the y_c and z_c coordinates.

```
%%%%%%%%%%%%%%%%%%%%%%%%%%%%%%%
h=0:20:100;
R=0:30:120;
[h,R]=meshgrid(h,R);
L=(2+pi).*R+2.*sqrt(R.^2+h.^2);
xc=0;
yc=h.*sqrt(R.^2+h.^2)./L;
zc=2.*R.^2./L;
figure(1)
mesh(h,R,zc);
surf(h,R,yc);
xlabel('h(cm)');ylabel('R(cm)');zlabel('y_c(cm)');
```

```
figure(2)
mesh(h,R,zc);
surf(h,R,zc);
xlabel('h(cm)');ylabel('R(cm)');zlabel('z_c(cm)');
%%%%%%%%%%%%%%%%%%%%%%%%%%%%%
```

Outputs are shown in Figures ES5.35 (*a*) and (*b*).

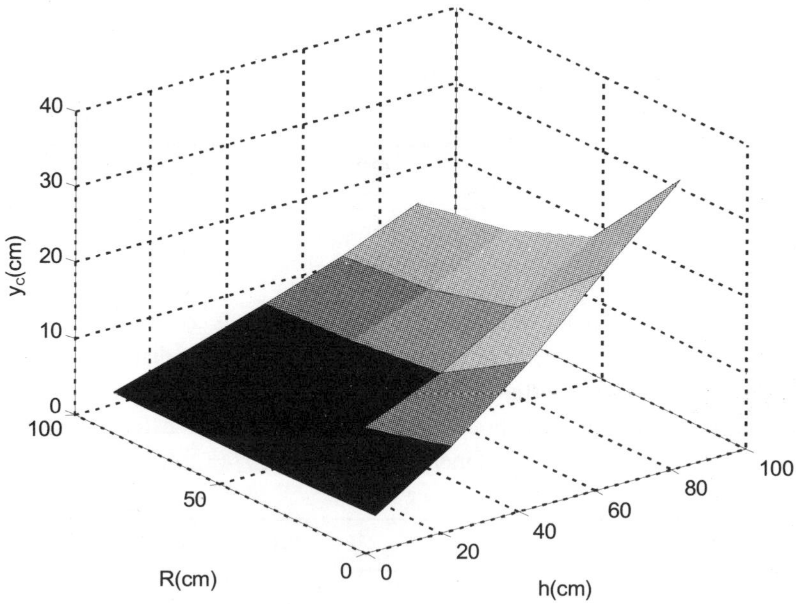

Fig. ES5.35 (a) Variation of y_C

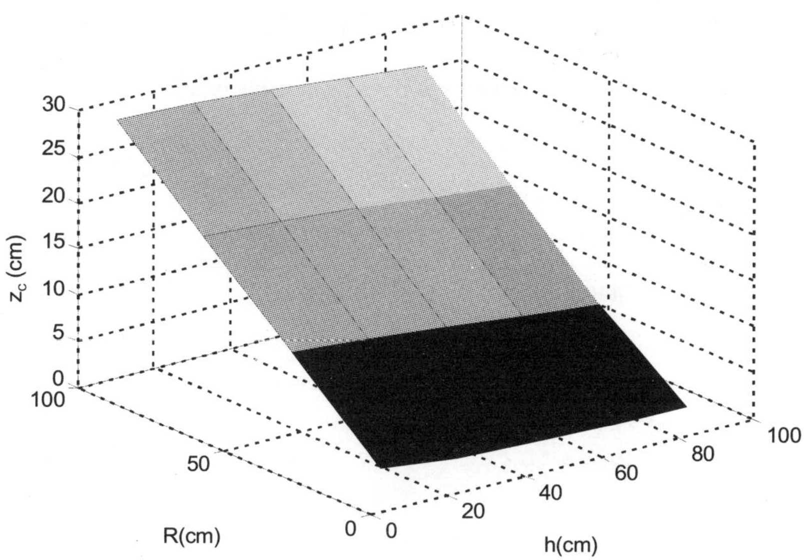

Fig. ES5.35 (b) Variation of z_C

ES5.36: Many cross-sections can be approximated by a series of rectangles as shown in Fig. ES5.36. Using MATLAB, calculate the moment of inertia and the radii of gyration of the cross-section with respect to horizontal and vertical centroidal axes. Take $b_1 = 3$ cm, $h_1 = 1$ cm, $b_2 = 1$ cm, $h_2 = 4$ cm, $b_3 = 5$ cm and $h_3 = 1$ cm. The centroids are: (0, –2.5), (0,0), (2.5,0)

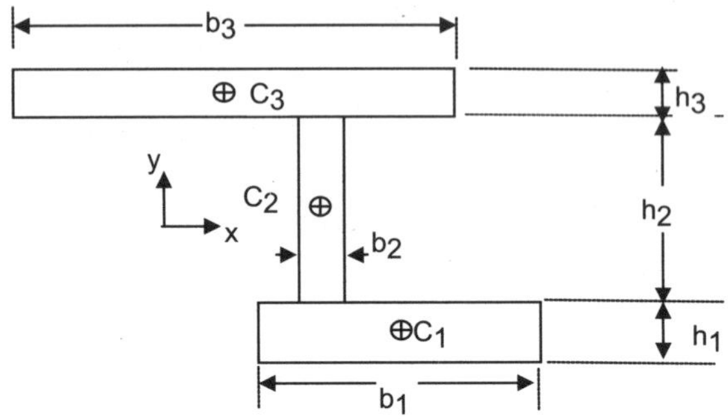

Fig. ES5.36

Solution: When a composite area is made of a large number of parts, it is convenient to tabulate the results for each of the parts in terms of its area A, its centroidal moments of inertia $\bar{I}$, the distance d from its centroidal axis to the axis about which the moment of inertia of the entire section is being computed and the product Ad^2. For any one of the parts, the moment of inertia about the desired axis by the transfer-of-axis theorem is $\bar{I} + Ad^2$. Thus for the entire section the desired moment of inertia becomes $I = \Sigma\bar{I} + \Sigma Ad^2$. Area in x-y plane for example the procedure can be tabulated as follows:

Part	A	d_x	d_y	Ad_x^2	Ad_y^2	$\bar{I}_x$	$\bar{I}_y$
...	...	...	...	...	...	...	...
...	...	...	...	...	...	...	..
Sums	ΣA			ΣAd_x^2	ΣAd_y^2	$\Sigma\bar{I}_x$	$\Sigma\bar{I}_x$

Moment of inertia for the composite area about x and y axis now becomes

$$I_x = \Sigma\bar{I}_x + \Sigma Ad_x^2$$

$$I_y = \Sigma\bar{I}_y + \Sigma Ad_y^2$$

Compute radius of gyrations k_x and k_y according to the relations: $k_x = \sqrt{\dfrac{I_x}{A_{\text{total}}}}$ and $k_y = \sqrt{\dfrac{I_y}{A_{\text{total}}}}$.

Complete MATLAB program for evaluating these parameters is presented below:

```
%%%%%%%%%%%%%%%%%%%%%%
N=3; % Number of rectangles in the figure
A total=0;
b=[3;1;3];
h=[1;4;1];
```

```
xb=[0;0;0];
yb=[-2.5;0;2.5];% SYMMETRICAL I-SECTION
 % X-Y COORDINATE FRAME COINCIDES WITH WEB
for i=1:N
Atotal=Atotal+b(i)*h(i);
end
xbA=0;ybA=0;
for i=1:N
xbA=xbA+xb(i)*b(i)*h(i);
ybA=ybA+yb(i)*b(i)*h(i);
end
Xbar=xbA/Atotal
Ybar=ybA/Atotal
Ix=0;Iy=0;
for i=1:N
Ix=Ix+(b(i)*h(i)^3/12+b(i)*h(i)*(yb(i)-Ybar)^2);
Iy=Iy+(h(i)*b(i)^3/12+b(i)*h(i)*(xb(i)-Xbar)^2);
end
kx=sqrt(Ix/Atotal);
ky=sqrt(Iy/Atotal);
fprintf('Ix=%f  cm^4 and Iy=%f  cm^4\n',Ix,Iy);
fprintf('kx=%f cm and ky=%f cm',kx,ky);
%%%%%%%%%%%%%%%%%%%%%%%%
```

Output is as follows:

Xbar = 0

Ybar = 0

Ix = 43.333333 cm^4 and Iy = 4.833333 cm^4

kx = 2.081666 cm and ky=0.695222 cm.

ES5.37: Figure ES5.37 shows an area with known moments of inertia I_x, I_y and product of inertia I_{xy} .

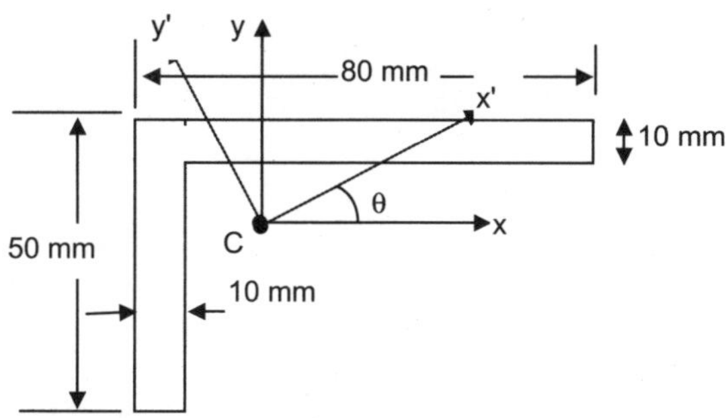

Fig. ES5.37

Calculate the moment and product of inertia of the area with respect to axes x' and y' obtained by rotating the original axes counterclockwise through an angle θ. Use MATLAB to plot $I_{x'}$, $I_{y'}$ and $I_{x'y'}$ as a function of θ for $0 \le \theta \le 90°$.

Solution: Inputs for the program should be moments of inertia and product of inertia. For each value of θ compute $I_{x'}$ and $I_{y'}$. That is

$$I_{x'} = \frac{1}{2}(I_x + I_y) + \frac{1}{2}(I_x - I_y)\cos 2\theta - I_{xy}\sin 2\theta$$

$$I_{y'} = (I_x + I_y) - I_{x'}$$

$$I_{x'y'} = \frac{1}{2}(I_x - I_y)\sin 2\theta + I_{xy}\cos 2\theta$$

Product of inertia of each rectangle about its centroidal axes parallel to *x-y* axes is zero by symmetry. Thus the I_{xy} for part-1 is $d_{x1} \times d_{y1}$ and for part-2 is $d_{x2} \times d_{y2}$.

Entire MATLAB program for computing I_x, I_y and I_{xy} as well as $I_{x'}$, $I_{y'}$ and $I_{x'y'}$ is presented below:

```
%%%%%%%%%%%%%%%%%%%%%%%%%%%%%%%%%%%%%%%%%%%%%%
A=0;
% CHOOSE x AND y axes COINCIDING WITH CENTROID OF VERTICAL RECT.
b=[10;70];
h=[50;10];
xb=[0;40];
yb=[0;20];
for i=1:2
   A=A+b(i)*h(i);
 end
 xbA=0;ybA=0;
for i=1:2
  xbA=xbA+xb(i)*b(i)*h(i);
  ybA=ybA+yb(i)*b(i)*h(i);
end
Xbar=xbA/A
Ybar=ybA/A
Ix=0;Iy=0;Ixy=0;
fprintf('Original moment of Inertia and product of inertia are\n');
for i=1:2
 dx=(xb(i)-Xbar);dy=(yb(i)-Ybar);
 Ix=Ix+(b(i)*h(i)^3/12+b(i)*h(i)*dy^2);
 Iy=Iy+(h(i)*b(i)^3/12+b(i)*h(i)*dx^2);
 Ixy=Ixy+b(i)*h(i)*dx*dy;% IXY=Σ(Ai)(dxi)(dyi)
```

```
end
fprintf('%e cm^4\t%e cm^4\t%e cm^4\n',Ix,Iy,Ixy);
theta=[0:10:90];
th=theta*pi/180;
Ixb=0.5*(Ix+Iy)+0.5*(Ix-Iy).*cos(2*th)-Ixy.*sin(2*th);
Iyb=(Ix+Iy)-Ixb;
Ixyb=0.5*(Ix-Iy).*sin(2*th)+Ixy.*cos(2*th);
subplot(3,1,1);
plot(theta,Ixb,'-p');ylabel('I-xb(cm^4)');grid on;
subplot(3,1,2);
plot(theta,Iyb,'-*');ylabel('I-yb(cm^4)');grid on;
subplot(3,1,3);
plot(theta,Ixyb,'-p');ylabel('I-xyb(cm^4)');grid on;
xlabel('\theta(degree)');
%%%%%%%%%%%%%%%%%%%%%%%%%%%%%%%%%%
```

Output is shown in Fig. ES5.37 (*a*) along with following command prompt result:

» Original moment of Inertia and product of inertia are

2.266667e + 005 cm^4   7.566667e + 005 cm^4   2.333333e + 005cm^4

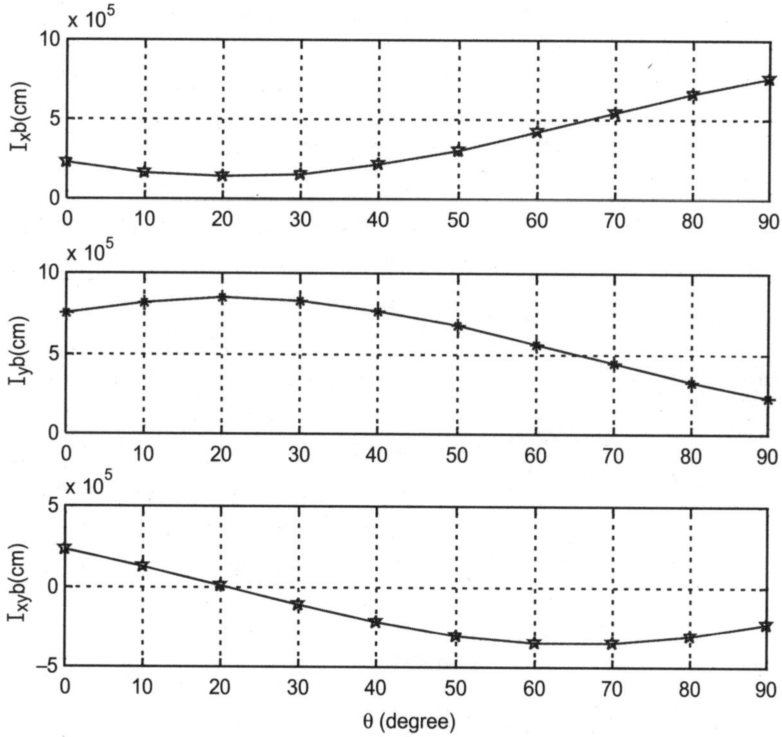

Fig. ES5.37 (a) MATLAB output

ES5.38: Obtain the mass moment of inertia with respect to x-axis of the homogeneous wire with mass per unit length as 0.1 kg/m by approximating using 12 straight line segments as shown in Fig.ES5.38. Write a MATLAB program to determine I_x of the wire with respect to the x-axis when p = 25 mm, ℓ=250 mm, and H = 100 mm.

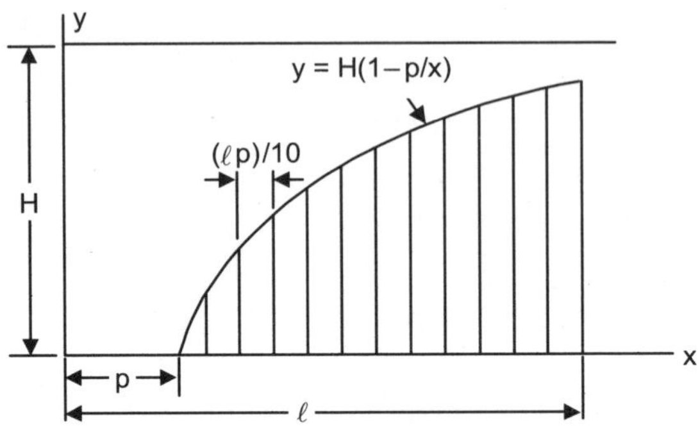

Fig. ES5.38

Solution: Consider any one of the segment S, the nomenclature of which is shown in Fig. ES5.38(a).

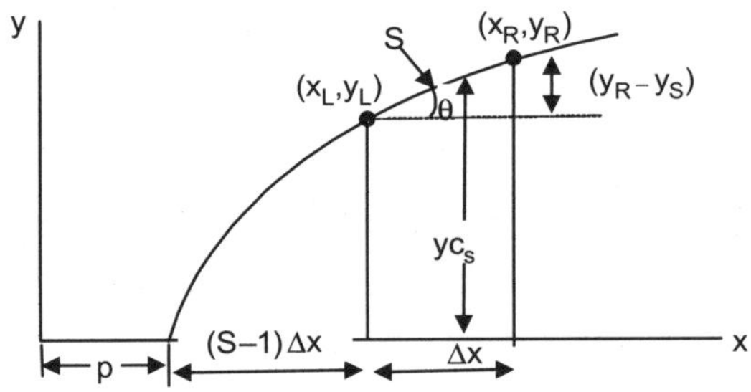

Fig. ES5.38 (a)

The mass moment of inertia of the segment is given by:

$$I_{xS} = \frac{1}{12} m_S (L_S \sin\theta)^2 + m_S\, yc_S^{\,2} = \frac{1}{12} m_S (y_R - y_L)^2 + m_S \bar{y}_S^{\,2}$$

where L_S is length of segment and θ is angle of segment with x-axis. $\bar{y}_S$ is the distance of center of segment from x-axis.

We have $\quad y_R - Y_L = H\left(I - \dfrac{p}{X_R}\right) - H\left(I - \dfrac{p}{X_L}\right) = pH\left(\dfrac{1}{X_L} - \dfrac{1}{X_R}\right)$

Complete MATLAB program for finding the overall moment of inertia is presented below:

```
%%%%%%%%%%%%%%%%%%%%%%%%%%%%%%%%%%%%
m=0.1;% MASS PER UNIT LENGTH OF SEGMENT
p=25;
p=p/1000;
```

```
L=250;
L=L/1000;
H=100;
H=H/1000;
dx=(L-p)/12;
Ix=0;
for i=1:12
xL=p+(i-1)*dx;
xR=xL+dx;
% COORDINATES OF CENTROID OF SEGMENT
xbarS=xL+0.5*dx;
ybarS=H*(1-p/xbarS);
dy=p*H*(1/xL-1/xR);
Ls=sqrt(dx^2+dy^2);
Ix=Ix+m*Ls*1/12*dy^2+m*Ls*ybarS^2;
end
fprintf('Ix=%7.3e kg·m^2\n,Ix);
%%%%%%%%%%%%%%%%%%%%%%%%%%%%%%
```

Output is as follows:

» Ix=1.392e–004 kg · m^2

Here the number of straight line segments are varied and the changes in moment of inertia can be studied.

5.2 DYNAMICS

This section provides further use of MATLAB in kinematics and kinetics of particles and rigid bodies.

5.2.1 Kinematics of Particles

ED5.1: The total acceleration y of the nose cone of a small experimental rocket as it moves up and down is given by

$$\begin{aligned} \ddot{y} &= -g - 0.00016v^2 \text{ up} \\ &= -g + 0.00016v^2 \text{ down} \end{aligned} \quad \text{in m/s}^2$$

where v = velocity in m/s.

(*a*) Derive an expression for the speed of the cone as a function of height as the cone moves up to its maximum height and as it returns to the ground. The nose cone is projected vertically from the ground with an initial velocity of 400 m/s.

(*b*) Write a MATLAB program to plot the speed of the cone as a function of the height for the upward motion and for the downward motion.

Solution: (*a*) This is an example of acceleration as a function of velocity. Thus we can use the relation

$$a\,dy = v\,dv$$

That gives $[-g - 0.00016v^2]dy = vdv$ for upward motion

$$\Rightarrow \int_0^y dy = -\int_{400}^{v} \frac{v}{g + 0.00016v^2}\,dv$$

$$\Rightarrow y = -\frac{1}{2 * 0.00016}\ln(g + 0.00016v^2)\Big|_{400}^{v} \qquad \text{(see Table 3.1 for integration)}$$

$$= -3125\ \ln(6250g + v^2)\Big|_{400}^{v} = 38460 - 3125\ \ln(6250g + v^2)$$

$$\Rightarrow v = 0.5 \times [-245250 + 4\ e^{(12.3 - 0.00032y)}]^{0.5}$$

For $y = y_{max}$, $v = 0$, thus: $y_{max} = 3988.31$ m

For downward motion, we have:

$$v\frac{dv}{dy} = (-g + 0.00016v^2)$$

Taking integrals both sides and separating

$$\Rightarrow \int_{y_{max}}^{y} dy = \int_0^v \frac{v}{-g + 0.00016v^2}dv = 3125\ \ln(6250\text{g} - v^2)\Big|_0^v = 3125\ \ln(6250\text{g} - v^2) - 34449$$

$$y - y_{max} = 3125\ \ln(6250g - v^2) - 34449$$

To get the v in terms of $y - y_{max}$, use the following MATLAB commands:

```
»eqn = 'y - ymax = 3125*ln(6250*g - v^2) - 34449';
» solve(eqn, 'v')
```

This gives two solutions in which positive solution is chosen as follows:

$$v = (-e^{0.00032*(y - y_{max}) + 34449/3125} + 61313)^{0.5}$$

$$\Rightarrow \qquad v = \left[-e^{-0.00032(y_{max} - y) + 11.02} + 61313\right]^{1/2}$$

For $y = 0$, we find $v = 210.24$ m/s. This can be obtained from following simple symbolic commands:

```
» v = '(- exp(1/3125*y - 1/3125*4010.8 + 34449/3125) + 6250*9.81)^(1/2)';
» subs(v, 'y',0)
  ans =
      210.2438 m/s
```

(*b*) **Complete MATLAB program for plotting is shown below:**

```
%%%%%%%%%%%%%%%%%%%%%%%%
g=9.81;
ymax=3988.3;
i=0;y=[0:100:4000];
i=i+1;
v_up=0.5*sqrt(-245250+4*exp(-1/3125*y+12.3));
v_down=sqrt(-exp(0.00032*y-0.00032*ymax+11.02)+6250*g);
if(ymax-y(i))<0 v_down=0;
end
figure(1)
plot(y,v_up,'-*');
xlabel('y(m)');ylabel('v(m/s)');
legend('Velocity up'); grid on;
figure(2)
plot(y,v_down,'-p');
xlabel('y(m)');ylabel('v(m/s)');
legend('Velocity down'); grid on;
%%%%%%%%%%%%%%%%%%%%%%%%%%%
```

Output is shown in Fig. ED5.1 for upward and downward motions.

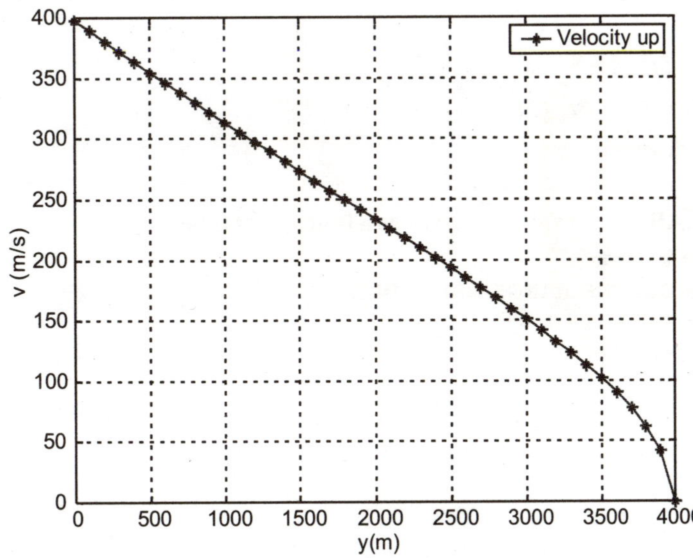

(a) Upward velocity

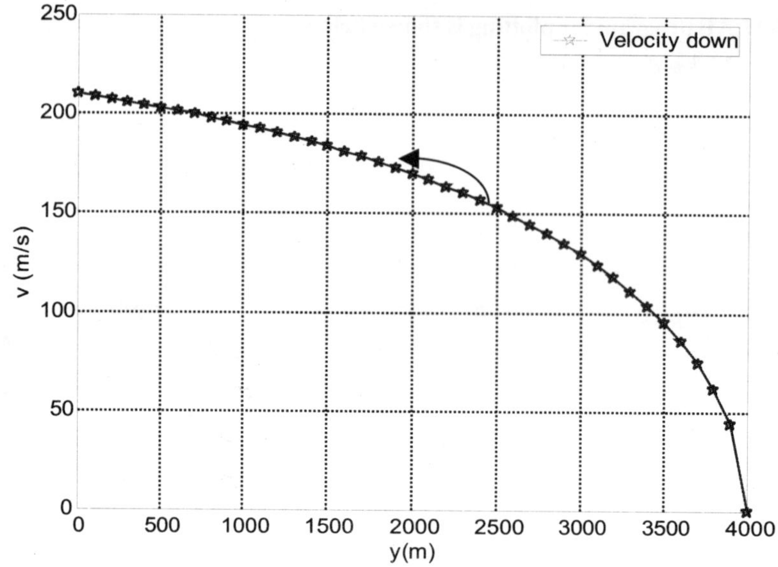

(b) Downward velocity

Fig. ED5.1 MATLAB output

ED5.2: The position x as a function of time of a particle that moves along a straight line is given by

$$x(t) = 0.5t^3 - 2.5t^2 - 6t + 15$$

(*a*) Derive expressions for the velocity and acceleration of the particle.

(*b*) Write a MATLAB program to plot the position, velocity and acceleration as a function of time for $0 \leq t \leq 10$ seconds.

Solution: (*a*) Given:

Displacement $x = 0.5t^3 - 2.5t^2 - 6t + 15$

Velocity $v = dx/dt = 1.5t^2 - 5t - 6$

Acceleration $a = dv/dt = 3t - 5$

(*b*) **Following MATLAB script evaluates x, v, a as a function of time t.**

```
%%%%%%%%%%%%%%%%%
t=0:0.5:10; % DEFINE RANGE VARIABLE
x=0.5*t.^3-2.5*t.^2-6.*t+15;
v=1.5*t.^2-5.*t-6;
a=3.*t-5;
subplot(3,1,1); plot(t,x,'-*');grid on;ylabel('x(m)');
title('\bfKINEMATIC ANALYSIS OF PARTICLE');
subplot(3,1,2);
plot(t,v,'-p');grid on;ylabel('v(m)');
subplot(3,1,3);
plot(t,a,'-*');grid on;ylabel('a(m)');
xlabel('Time(s)');
%%%%%%%%%%%%%%%%%
```

Output is shown in Fig. ED5.2.

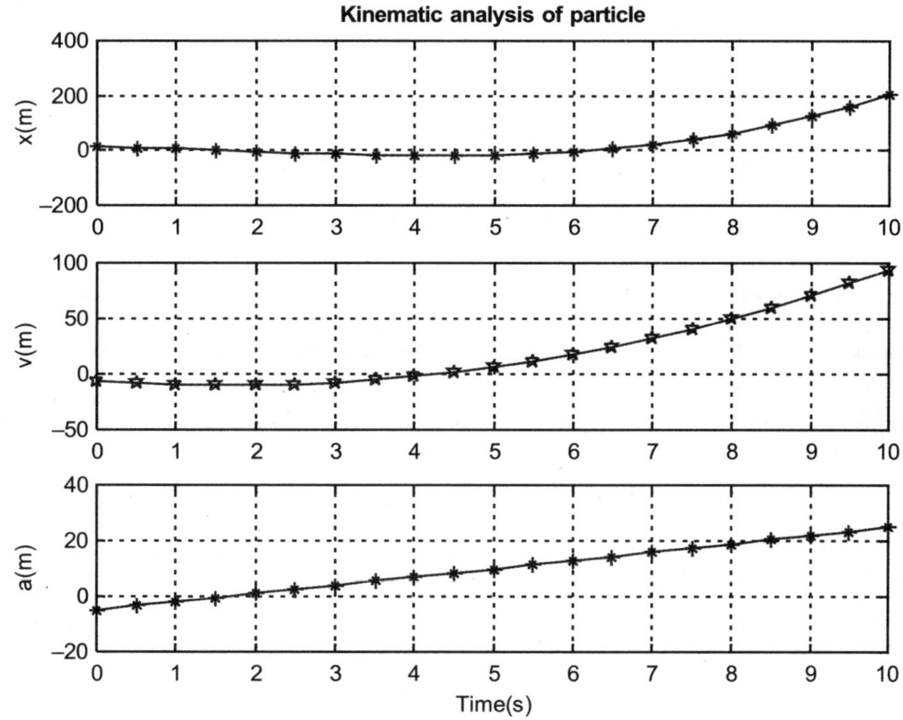

Fig. ED5.2 MATLAB output

ED5.3: The height of the ball at any time t after it is thrown is given by

$$y(t) = y_0 + v_{y0}t - \frac{1}{2}gt^2$$

where y_0 = the initial height of the object above the ground

v_{y_0} = the initial vertical velocity of the object

g = acceleration due to gravity

The horizontal distance traveled by the ball after it is thrown is given by

$$x(t) = x_0 + v_{x_0}t$$

where x_0 = the initial horizontal position of the ball on the ground and v_{x_0} = the initial horizontal velocity of the ball.

In the above, air friction and earth's curvature are neglected. Assume that the ball is initially thrown from $(x_0, y_0) = (0,0)$ with 25 m/s at an initial angle of θ degree.

(*a*) Write a MATLAB program to plot the trajectory of the ball and obtain the horizontal distance traveled before it touches the ground, by taking the range of θ as: $0 \leq \theta \leq 90°$.

(*b*) Determine the angle θ that maximizes the range of the ball and plot the trajectory.

Solution: A projectile in motion is an example of curvilinear motion of particle. Figure ED5.3 shows the motion of the projectile and its starting velocities.

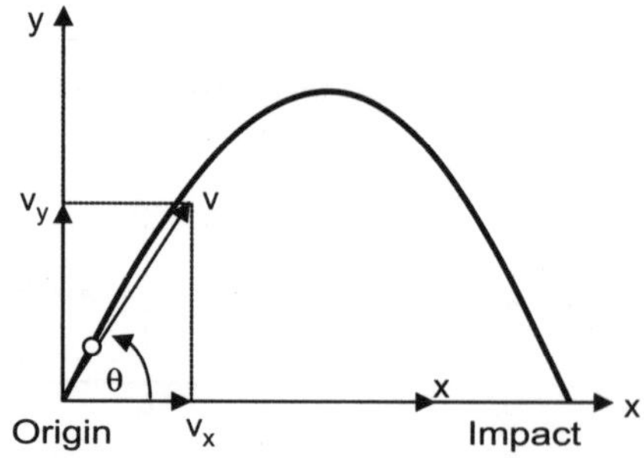

Fig. ED5.3 (a) Parabolic trajectory and components of vector v

In motion of projectile: $ax = 0$ and $ay = -g$ (for upward motion) and $= g$(downward motion). Horizontal distance traveled is range and vertical distance traveled is called height attained by it.

All the equations are given. Hence directly it could be possible to plot the trajectory and find the angle θ that maximizes the range of ball.

(a) **MATLAB Program:**

```
%%%%%%%%%%%%%%%%%%%%%%%%%%%%%
% Initialization of constant data
v0=25;g=9.81;
i=1;
fprintf('\nTheta\tRange\n');
for theta=0:10:90
    th=theta*pi/180;
    vx0=v0*cos(th);vy0=v0*sin(th);
tflight=2*vy0/g;
range(i)=vx0*tflight;
    fprintf('%f\t%f\n',theta,range(i));
    j=1;
    for t=0:tflight/10:tflight
    x(j)=vx0*t;
    y(j)=vy0*t-0.5*g*t^2;
    j=j+1;
end
plot(x,y);
hold on;
i=i+1;
end
[maxrange index]=max(range);
TETA=[0:10:90];
fprintf('\nThe maximum range %f occurs when theta=%f', maxrange,
 TETA(index));
```

```
title('\bfTrajectory of ball versus Initial angle \theta');
xlabel('Horizontal distance(m)');
ylabel('vertical distance(m)');
axis([0 65 0 35]);
grid on;
%%%%%%%%%%%%%%%%%%%%%%
```

Output is shown in Fig. ED5.3(*b*).

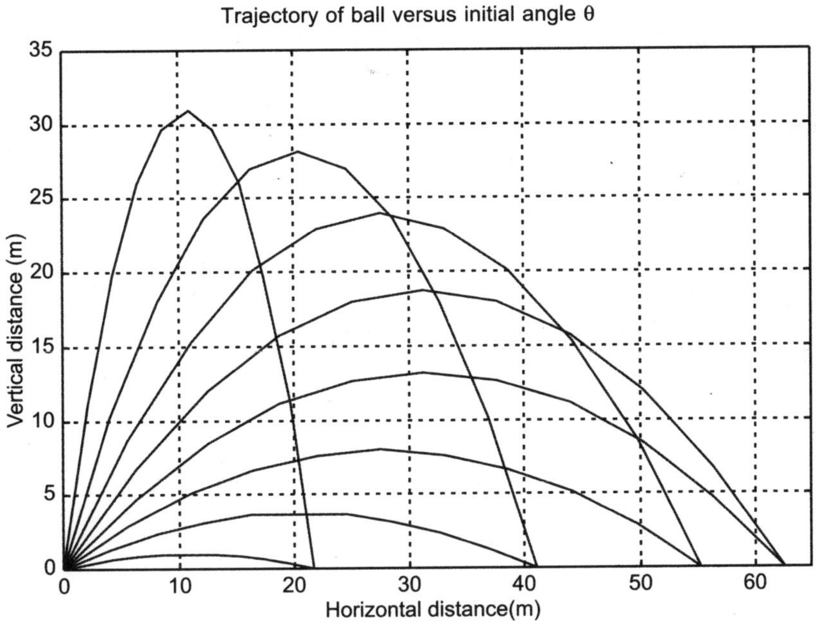

Fig. ED5.3 (b)

(*b*) Following data is also obtained at the command prompt.

Theta	Range
0.000000	0.000000
10.000000	21.790274
20.000000	40.952320
30.000000	55.174911
40.000000	62.742594
50.000000	62.742594
60.000000	55.174911
70.000000	40.952320
80.000000	21.790274
90.000000	0.000000

The maximum range 62.742594 occurs when theta=50.000000.

ED5.4: A motorcycle is moving along a circular path having a radius of 6 m such that its position as a function of time is given by $\theta = \sin 3t$, find the acceleration of particle as a function of time t and plot the components of acceleration.

Solution: This is an example of curvilinear motion with polar (r and θ) coordinates.

Acceleration has two components:

Radial acceleration $$a_r = \ddot{r} - r\dot{\theta}^2 = -r\frac{d}{dt}(\sin 3t) = -18 \cos 3t$$

Transverse acceleration $$a_\theta = r\ddot{\theta} + 2\dot{r}\dot{\theta} = r\frac{d^2}{dt^2}(\sin 3t) = -54 \sin 3t$$

MATLAB program is shown below:

```
%%%%%%%%%%%%%%%%%
ar=inline('-3*r*cos(3*t)');
at=inline('-9*r*sin(3*t)');
vectorize(ar);      % Makes the function as a vector compatible for
 multiplication
vectorize(at);
t=0:0.1:1;
plot(t,ar(6,t),'-o',t,at(6,t),'-p');
xlabel('Time(s)');
ylabel('Acceleration in mpers^2');
legend('Radial acceleration', 'transverse acceleration',2);
grid on;
%%%%%%%%%%%%%%%%%%%%%%%%%%%%
```

The output is shown in Fig. ED5.4.

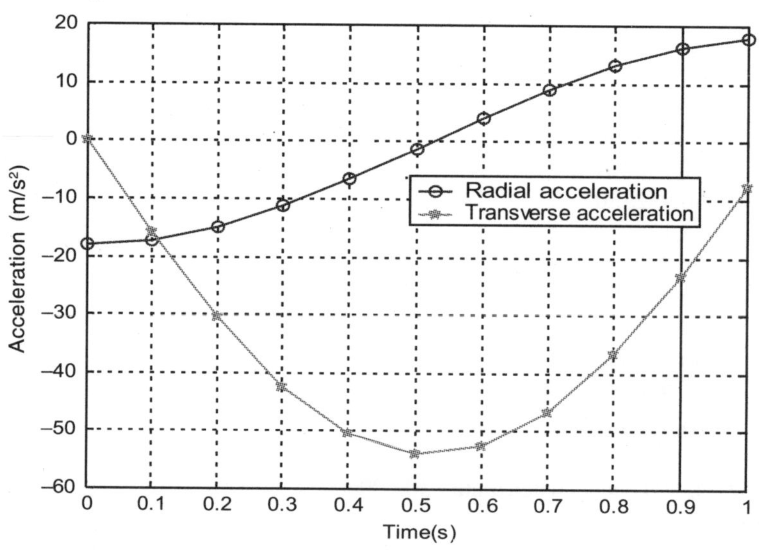

Fig. ED5.4 MATLAB response

ED 5.5: The height, h, horizontal distance, x and speed of a projectile, v, launched with a speed, v at an angle θ to the horizontal axis are given by

$$h(t) = vt\sin\theta - \frac{1}{2}gt^2$$

$$x(t) = vt\cos\theta$$

$$v(t) = \sqrt{v^2 - 2vgt\sin\theta + g^2t^2}$$

The projectile will strike the ground when $h(t) = 0$, and the time of the hit is given by

$$t_{\text{hit}} = 2\frac{v}{g}\sin\theta$$

Assuming $\theta = 35°$, $v = 50$ m/s, and $g = 9.81$ m/s^2, write a MATLAB program using logical operators to find (*a*) the height is no less than 16 m, (*b*) the height is no less than 16 m and the speed is no greater than 45 m/sec.

Solution: Following script is developed to solve the problem:

```
%%%%%%%%%%%%%
A=35;v0=50;g=9.81;
% to find the times at which the height is no less than 16 m
% Increment time by 0.01
time=[0:0.01:5];
% find the height h for different values of time
h=v0*time.*sin(A*pi/180)-0.5*g*time.^2;
% determine times when height is greater than 16 m
times=find(h>=16);
%find the maximum and minimum value for times
max_time=max(times)/100; min_time=min(times)/100;
fprintf('\nThe height is at least 16 m when the time is between\n')
fprintf('%f seconds and %f seconds \n', min_time, max_time);
%%%%%%%%%%%%%%%%%%%%%%
Output is as follows:
»The height is at least 16 m when the time is between
0.640000 seconds and 5.010000 seconds
% DETERMINING T AT WHICH h is no less than 16 m and v is no more than 45 m/s
A=35;v0=50;g=9.81;
time=[0:0.01:5];
h= v0*time.*sin(A*pi/180)-0.5*g*time.^2;
v=sqrt((v0^2)-(2*v0*g*time.*sin(A*pi/180))+(g^2*time.^2));
vector=find(h>=16&v<=45);
```

```
% find the maximum and minimum value for times
max_time=max(vector)/100;
min_time=min(vector)/100;
fprintf('\n\nThe height is at least 16 m and the velocity is no less
than 45 m/s\n');
fprintf('when the time is between %f seconds and %f seconds\n\n',
min_time,max_time)
%%%%%%%%%%%%%%%%%%%%%
```

Output is as follows:

The height is at least 16 m and the velocity is no less than 45 m/s when the time is between 1.040000 and 4.830000 seconds.

ED 5.6: The gravitational force F between two bodies of masses m_1 and m_2 is given by

$$F = \frac{G\,m_1\,m_2}{r^2}$$

where,

G = gravitational constant (6.67×10^{-11} Nm^2/kg^2)

m_1, m_2 = masses of the bodies (kg)

r = distance between the two bodies (m).

Write a MATLAB program to calculate the gravitational force between two bodies given their masses and distance between them. Determine the force on an 1,000 kg satellite in orbit 40,000 km above the earth's surface. The mass of the earth is 5.98×10^{24} kg.

Solution: Following lines are typed in a file gr.m and executed.

```
%%%%%%%%%%%%%%%%%%%%%%%%%%%
m1=1000; % Mass of the object-1 in kg
m2=5.98e24; % Mass of object-2 in kg
r=40000e3; % Enter distance between the two bodies in m
G=6.67e-11;% Gravitational constant
force=G*m1*m2/r^2;
fprintf('The force due to gravity is %0.4f N.\n',force);
%%%%%%%%%%%%%%%%%%%%%%%%%%%%
```

Output is as follows:

The force due to gravity is 249.2912 N.

ED 5.7: A cameraman standing at A (see Fig. ED5.7) is following the movement of a race car B which is traveling along a straight track at a constant speed v. Plot the variation of angular velocity (with which he must turn in order to keep the camera directed on the car) as a function of angle θ shown in the figure. Given: v = 24 m/s and a = 30 m.

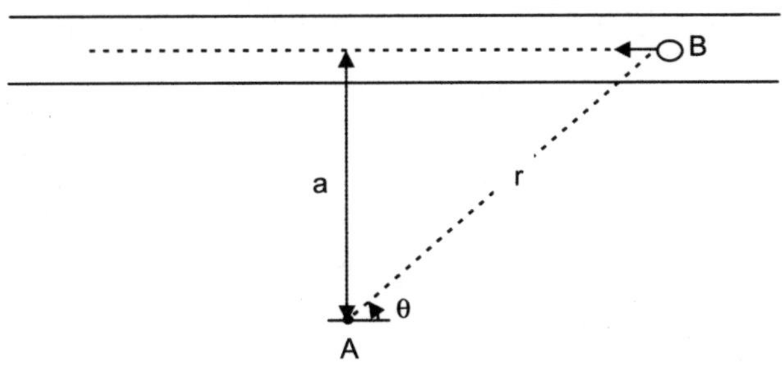

Fig. ED5.7

Solution: Here r and θ are variables. Thus their first time derivatives exist. Let us frame from geometry, the relations.

$$r \sin\theta = a \quad \text{...(1)}$$

and $$r\cos\theta = b \text{ (say)}$$

a is constant, but b varies. Differentiating the equations

$$r(\cos\theta)\,\dot{\theta} + \dot{r}\sin\theta = 0 \quad \text{...(2)}$$

$$r(-\sin\theta)\,\dot{\theta} + \dot{r}\cos\theta = \dot{b} = -v \quad \text{...(3)}$$

Solving (1)–(3) for r, $\dot{r}$ and $\dot{\theta}$ in terms of θ,

$$r = a/\sin\theta,$$

$$\dot{r} = -v\cos\theta$$

and $$r\dot{\theta} = v\sin\theta$$

Complete MATLAB program is given below:

```
%%%%%%%%%%%%%%%%%%%%%%%%%%%%%
a=30;v=24;
theta=1:5:179;
thrad=(pi/180).*theta;
r=a./sin(thrad);
rd=-v.*cos(thrad);
thd=v*sin(thrad)./r;
figure(1);
subplot(2,1,1); %SUB-PLOT-1
plot(theta,rd,'-p');
title('\bfvelocity in radial direction in m/s');
grid on;
subplot(2,1,2); % SUB-PLOT-2
plot(theta,thd,'-*');
xlabel('Angle revolved by cameraman theta (degrees)');
title('\bfangular velocity (rad/s)');
grid on;
%%%%%%%%%%%%%%%%%%%%%%%%%%%%%%%
```

Output is shown in Fig. ED 5.7(*a*).

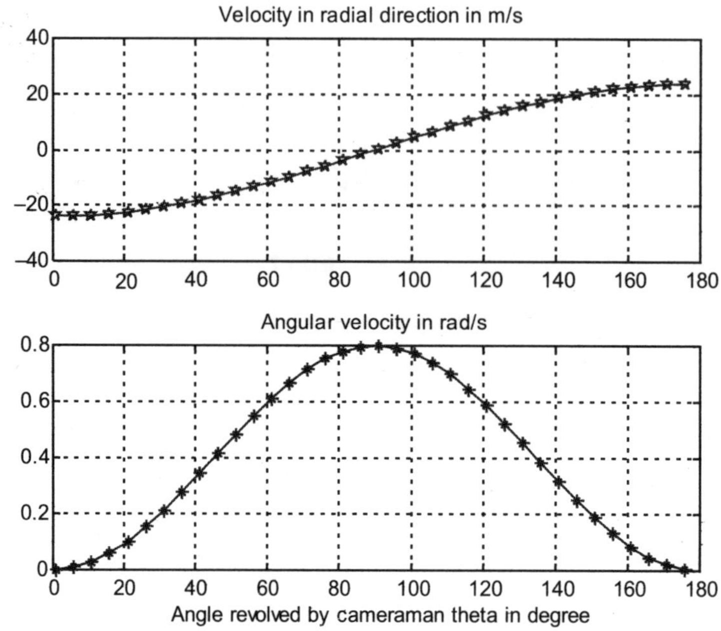

Fig.ED 5.7(a)

ED 5.8: The satellite's orbit in polar coordinates is given by

$$r = \frac{p}{1 - \varepsilon \cos\theta}$$

where r = distance of the satellite from the center of the earth

θ = angle of the satellite from the center of the earth

p = a parameter specifying the size of the orbit

ε = a parameter specifying the eccentricity of the orbit

Note that when a satellite orbits the earth, the satellite's orbit will form an ellipse with the earth located at one of the focal points of the ellipse. Write a MATLAB program to plot the orbit of a satellite for (*a*) $\varepsilon = 0$, (b) $\varepsilon = 0.25$ and (*c*) $\varepsilon = 0.5$. The size parameter, p = 1100 km.

Solution: When a single force acts upon a particle, such as during the free-flight trajectory of a satellite in a gravitational field, then the motion is referred to as central-force motion. The orbit depends upon the eccentricity *e* and as a result the trajectory can be circular, elliptical or hyperbolic (Fig. ED5.8 (*a*)).

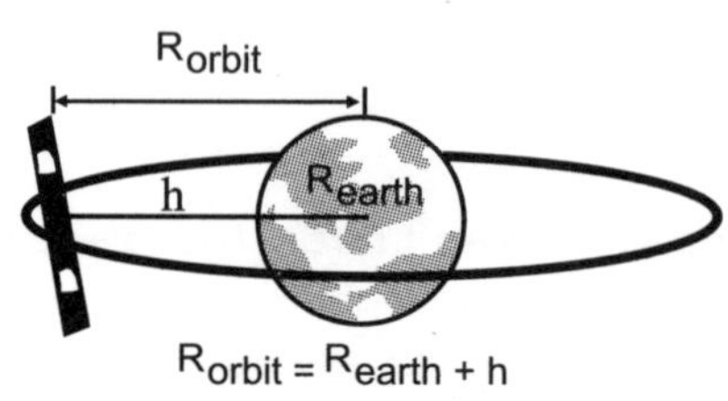

Fig. ED5.8 (a) Satellite in motion

MATLAB program for this formulation is as follows:

```
%%%%%%%%%%%%%%%%%%%%%%%%%%
% Initialize values
theta=0:pi/20:2*pi;
p=1100;
% Calculate orbits for e=0
e=0.0;
r=p./(1-e.*cos(theta));
% Plot orbit
figure(1)
polar (theta,r,'b-');
title ('\bfOrbit for eccentricity=0');
% Calculate orbits for e=0.25
e=0.25;
r=p./(1-e.*cos (theta));
%plot orbit
figure(2)
polar (theta,r,'b-');
title ('\bfOrbit for eccentricity =0.25');
%Calculate orbits for e=0.5
e=0.5;
r=p./(1-e.*cos(theta));
%Plot orbit
figure(3)
polar (theta,r,'b-');
title('\bfOrbit for eccentricity=0.5');
%%%%%%%%%%%%%%%%%%%%%%%%
```

Output is shown in Fig. ED5.8 (*b*) to (*d*).

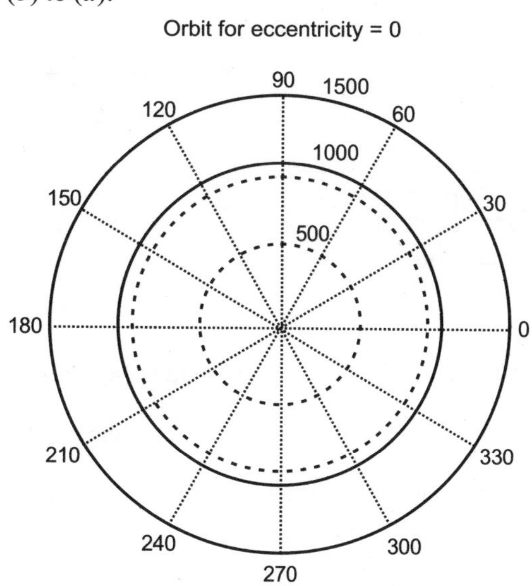

Fig. ED5.8 (b)

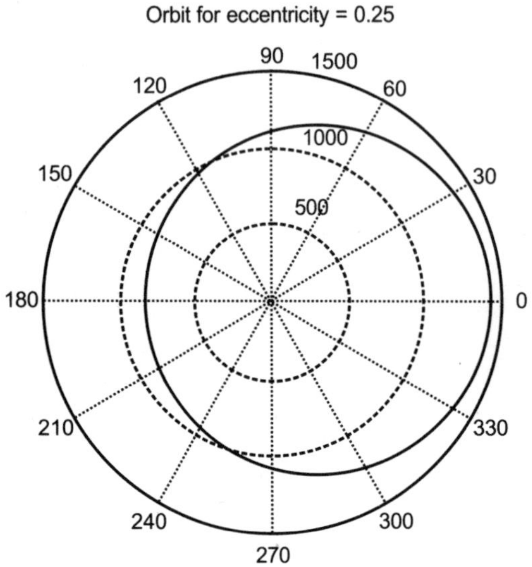

Fig. ED5.8 (c)

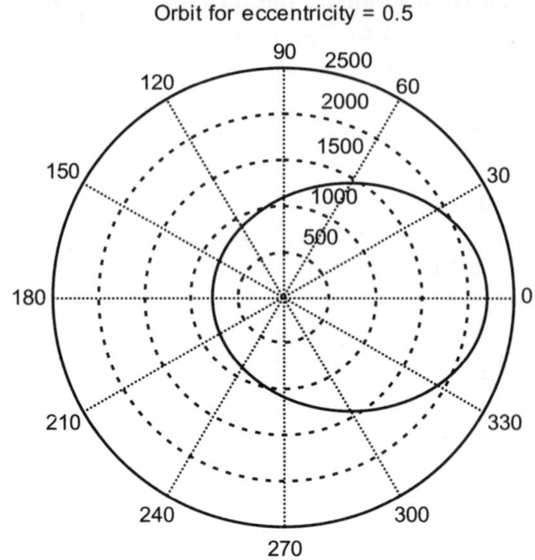

Fig. ED5.8 (d)

5.2.2 Particle Kinetics

Particle Kinetics can be attempted with Newton's law, energy principle or momentum principle depending on the data.

ED5.9: Figure ED5.9 shows a block of mass attached to a spring of spring constant k. The block is released from rest when the spring is in a horizontal and undeformed position. Write (a) MATLAB program to determine and plot the length of the spring, the magnitude and direction of velocity of the block as the block passes directly under the point of suspension of the spring for $k/m = 14s^{-2}$, $22s^{-2}$, and $25s^{-2}$, (b) the value of k/m for which that velocity is horizontal. Assume $r_0 = 2$m.

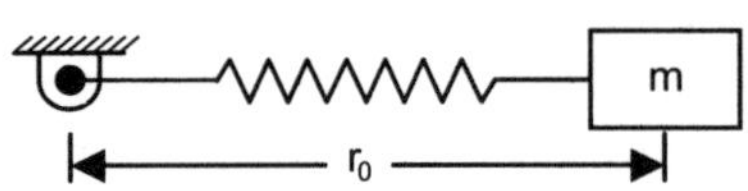

Fig. ED5.9

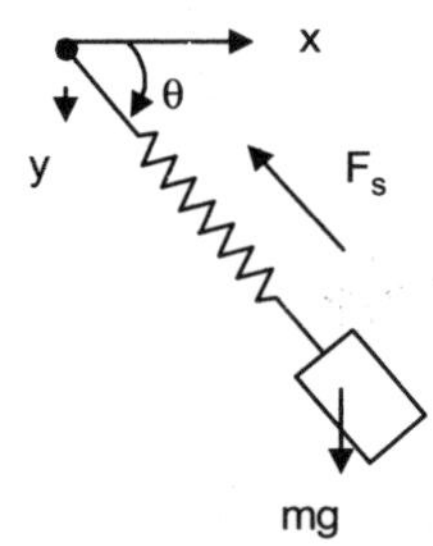

Fig. ED5.9 (a) Free-body diagram at an instant

Solution: Figure ED5.9 (a) shows the free-body diagram of the system after it has rotated at an angle θ.

Applying the equations of motion in x and y direction, we get:

$$\Sigma F_x = m\bar{a}_x \Rightarrow -F_s \cos\theta = ma_x \quad ...(1)$$

$$\Sigma F_y = m\bar{a}_y \Rightarrow mg - F_s \sin\theta = ma_y \quad ...(2)$$

Here F_s is spring force $F_s = k(r - r_0)$, where r_0 is undeformed length of the spring and r is deformation given by $r = \sqrt{x^2 + y^2}$ and the angle $\cos\theta = \dfrac{x}{\sqrt{x^2+y^2}}$ and $\sin\theta = \dfrac{y}{\sqrt{x^2+y^2}}$.

From eq.(1), $$a_x = \frac{dv_x}{dt} = -\frac{k(r-2)}{m}\cos\theta = -\frac{k\left(\sqrt{x^2+y^2}-2\right)}{m}\frac{x}{\sqrt{x^2+y^2}} \quad ...(3)$$

From eq.(2), $$a_y = \frac{dv_y}{dt} = g - \frac{k(r-2)}{m}\sin\theta = g - \frac{k\left(\sqrt{x^2+y^2}-2\right)}{m}\frac{y}{\sqrt{x^2+y^2}} \quad ...(4)$$

From eqs.(3) and (4) we define x and y as

$$\frac{dx}{dt} = v_x \text{ and } \frac{dy}{dt} = v_y \quad ...(5)$$

These equations define the motion of the mass. The differential equations (3)–(5) are solved using MATLAB built-in function.

Complete MATLAB script for solving the differential equations and plotting velocity vector at various values of k/m is given below:

```
%%%%%%%%%%%%%%%%%%
global r0 g kdm
r0=2;g=9.81;
 for j=1:3
   switch(j)
   case 1
     kdm=14;
   case 2
     kdm=22;
   case 3
     kdm=25;
   end
   z0=[2;0;0;0]; % INITIAL CONDITIONS OF X,Y,VX,VY
tspan=[0 2];
[t,z]=ode45('f',tspan,z0);
[x,k]=min(abs(z(:,1)));
r_max=z(k,2); % LENGTH OF SPRING IN VERTICAL DIRECTION
v=sqrt(z(k,3)^2+z(k,4)^2); angle=atan(z(k,4)/z(k,3))*180/pi;
fprintf('r=%fm\n',r_max);
fprintf('v=%fm\n',v);
fprintf('Angle v forms with the horizontal=%f degrees\n',angle);
    switch(j)
       case 1
             plot(t,z(:,3),'-');
```

```
        case 2
                plot(t,z(:,3),'-p');
        case 3
                plot(t,z(:,3),'o-');
        end
        legend('k/m=14','k/m=22','k/m=25');
        hold on;
xlabel('Time t');
ylabel('Solution v_x');
  grid on;
end
%%%%%%%%%%%%%%%%%
```

This runs with function 'f.m' defined as follows based on eqs.(3)–(5).

```
function dydt=f(t,z)
global r0 g kdm
dydt=[z(3);z(4);-kdm*sqrt(z(1)^2+z(2)^2-r0)*(z(1)/sqrt(z(1)^2+z(2)^2));
g-kdm*sqrt(z(1)^2+z(2)^2-r0)*(z(2)/sqrt(z(1)^2+z(2)^2))];
%%%%%%%%%%%%%%%%%%%
```

Output is shown below:

» r = 0.172034m

v = 5.485309m

Angle v forms with the horizontal = –43.304357 degrees

Figures ED5.9 (b) and (c) shows the variation of v_x and v_y as a function of time for different values of k/m ratios.

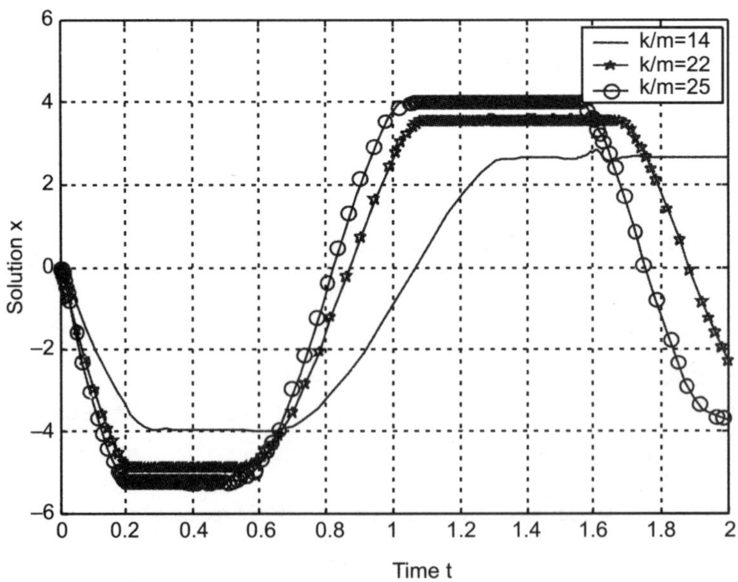

Fig. ED5.9 (b) Variation of velocity v_x

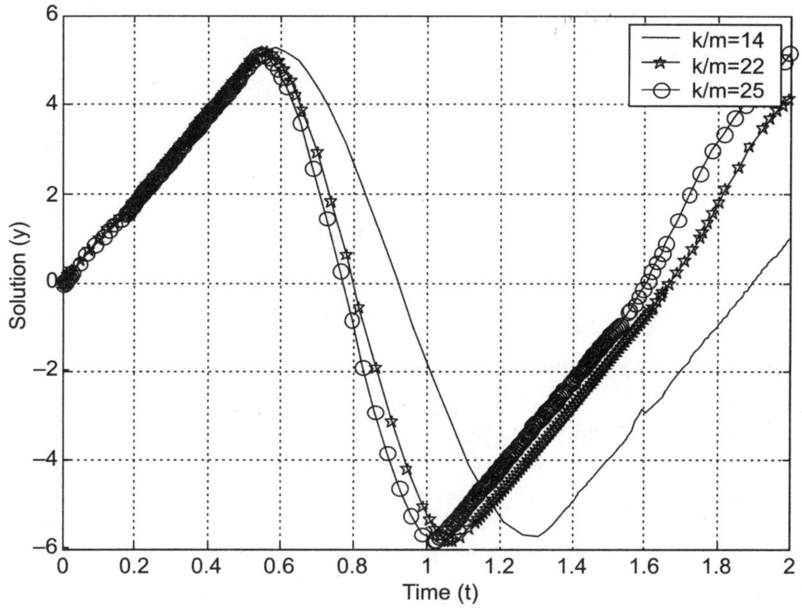

Fig. ED5.9 (c) Variation of velocity v_y

ED5.10: Figure ED5.10 shows a block of mass *m* initially at rest acted upon by a force *P* which varies in magnitude with time as shown in Fig.ED5.10(*a*). Write a MATLAB program to determine and plot the speed of the block as a function of time *t* for $0 \le t \le 20$ s. Determine the maximum velocity of the block and the corresponding value of *t*. The coefficient of static and kinetic friction between the block and the horizontal surface are $\mu_s = 0.45$ and $\mu_k = 0.25$. The mass of the block $m = 30$ kg.

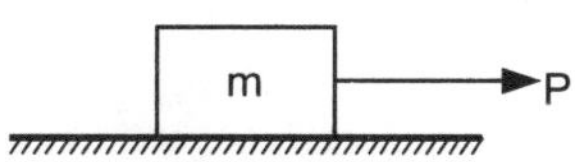

Fig. ED5.10

Solution: The static friction force is $F_s = \mu_s W = 0.45(30)(9.81) = 132.435$ N

From the *P-t* diagram, we have

$$P = 50t\,,\ \text{for}\ 0 \le t \le 10.$$

$$= -\frac{500}{(20-10)}(t-20) = -50(t-20)\,,\ \text{for}\ 10 < t \le 20$$

The value 132.432 N resists the applied force and block does not start until *F* exceeds this value. That is till $t = 2$ seconds block does not start. As at $t = 3$ seconds, $P = 150$ N, so motion should have started from $t = 2$ seconds to 3 seconds. The initial and final momentum of the block are shown in Fig.ED5.10(*b*).

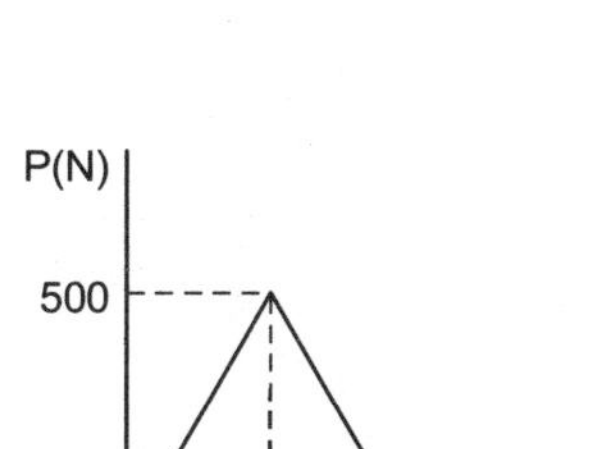

Fig. ED5.10 (a)

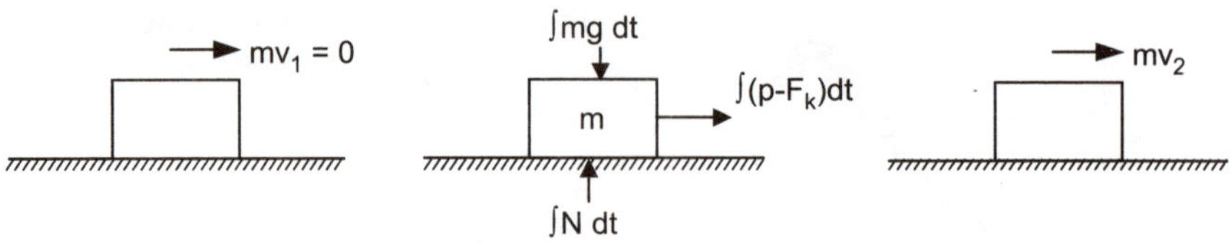

Fig. ED5.10 (b)

The dynamic friction force is

$$F_k = \mu_k W = 0.25(30)(9.81) = 73.575 \text{ N}$$

For $2 \le t \le 10$, applying impulse-momentum principle:

$$mv = \int(P - F_k)dt = \int_2^t (50t - 73.575) = 25t^2 - 73.575t + 47.15$$

when $t = 10$ second, $v = 60.38$ m/s

For $10 < t \le 20$, applying impulse-momentum principle with starting velocity as 60.38 m/s:

$$mv - m(60.38) = \int_{10}^t (1000 - 50t - 73.575)dt = -25t^2 + 926.425t - 6764.25$$

or $$mv = -25t^2 + 926.425t - 4953.1$$

Complete MATLAB program for plotting velocities and finding maximum values of velocities is given below:

```
%%%%%%%%%%%%%%%%%%%%
m=30;g=9.81;
i=1;
for t=0:0.1:20
if t<2 v(i)=0;
elseif ((t>2)&(t<=10)) v(i)= (25*t^2-73.575*t+47.15)/m;
elseif ((t>10)&(t<=20)) v(i)= (-25*t^2+ 926.425*t -4953.1)/m;
end
i=i+1;
end
[maxvel,index]=max(v)
T=[0:0.1:20];
fprintf('\nMaximum value of velocity is %f and corresponding time is %f
 sec\n',maxvel,T(index));
plot(T,v);
xlabel('Time(s)');ylabel('Velocity(m/s)');
grid on;
%%%%%%%%%%%%%%%%%%%%%%
```

Complete output is shown below:

» Maximum value of velocity is 120.983750 and corresponding time is 18.500000 s

Figure ED5.10(*c*) shows the MATLAB plot.

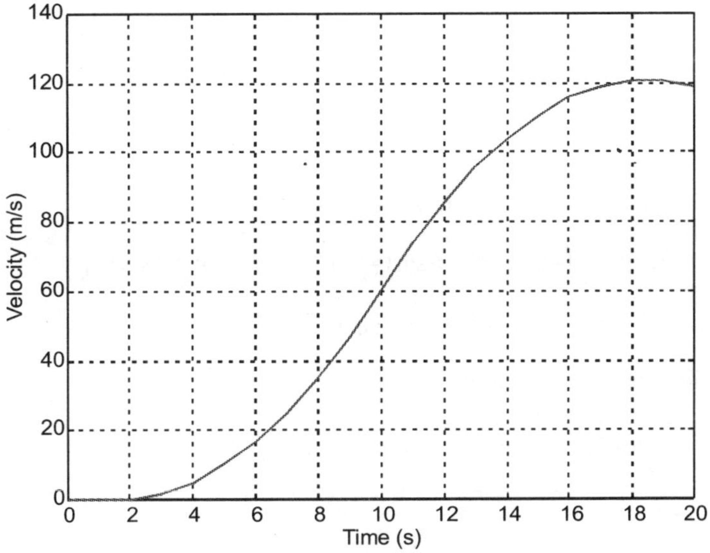

Fig. ED5.10 (c) MATLAB output

ED5.11: Figure ED5.11 shows two hemispheres held together by a cord only which maintain the spring under compression. Write a MATLAB program to calculate and plot the magnitude of the resulting velocity of each hemisphere as a function of θ for $10° \leq \theta \leq 150°$ knowing that the cord is severed causing the hemispheres to fly apart. The potential energy of the compressed spring is 120 J and the assembly has an initial velocity of magnitude v_0 = 8 m/s.

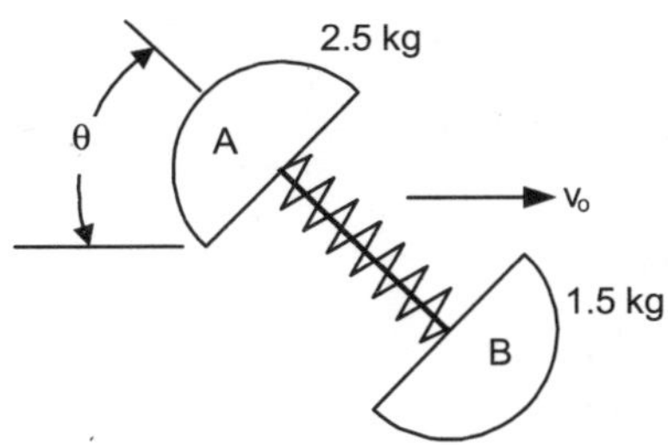

Fig. ED5.11

Solution: This problem gives an idea of solving problem where the existing mass explodes into two masses like in projectile blasting problem. Applying the principle of conservation of momentum with respect to *X*-*Y* coordinate frame as shown in Fig. ED5.11(*a*), we get:

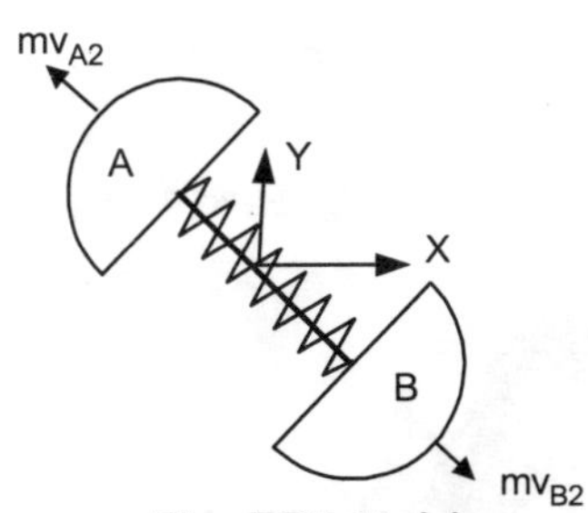

Fig. ED5.11 (a)

$$0 = m_A v_{A2} - m_B v_{B2}$$

or $$v_{B2} = m_A v_{A2} / m_B = 5/3\ v_{A2} \qquad ...(1)$$

Also using conservation of energy principle:

$$T_1 + V_1 = T_2 + V_2$$

$\Rightarrow$ $$V_1 - V_2 = 120 = T_2 - T_1 = T_2 = \frac{1}{2}m_A v_{A2}^2 + \frac{1}{2}m_B v_{B2}^2$$

or $$1.25v_{A2}^2 + 0.75v_{B2}^2 = 120 \quad ...(2)$$

Solving eqs.(1) and (2), we get:

$$v_{A2} = 6 \text{ m/s and } v_{B2} = 10 \text{ m/s}$$

Now the velocities of A and B with respect to fixed reference frame X-Y are:

$$v_A = v_O + v_A' = 8\hat{i} + \left(-6\cos\theta\hat{i} + 6\sin\theta\hat{j}\right) = \left((8 - 6\cos\theta)\hat{i} + 6\sin\theta\hat{j}\right)$$

$$v_B = v_O + v_B' = \left((8 + 10\cos\theta)\hat{i} - 10\sin\theta\hat{j}\right)$$

The magnitudes of velocities are given as:

$$|v_A| = \sqrt{(8 - 6\cos\theta)^2 + (6\sin\theta)^2}$$

$$|v_B| = \sqrt{(8 + 10\cos\theta)^2 + (10\sin\theta)^2}$$

The magnitudes of resulting velocities of each hemisphere as a function of angle at which the cord is severed is obtained from the following MATLAB script.

```
%%%%%%%%%%%%%%%%%%%%%%%%
g=9.81;v0=8;
theta=[10:10:150];
t=theta*pi/180;
vAx=8-6*cos(t); vAy=6*sin(t);
vBx=8+10*cos(t);vBy=10*sin(t);
vA=sqrt(vAx.^2+vAy.^2);
vB=sqrt(vBx.^2+vBy.^2);
plot(theta,vA,theta,vB);
xlabel('Theta(degrees)');
ylabel('Hemisphere velocity (m/s^2)');
legend('\bfVelocity of Hemisphere A','Velocity of Hemisphere B');
grid on;
%%%%%%%%%%%%%%%%%%%%%%%%
```

Output is presented in Fig. ED5.11(*b*).

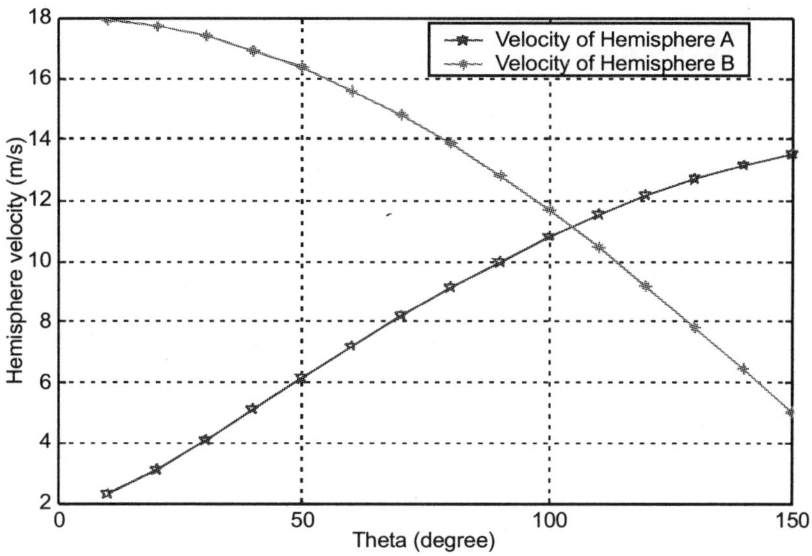

Fig. ED5.11 (b) MATLAB output

ED5.12: A freely rolling base (*B*) has 60 kg mass. A box (*A*) of 40 kg is sliding over it from rest, 5 m down the base as shown in Fig. ED5.12. Determine base speed, when the box reaches bottom of the crate assuming it as smooth surface.

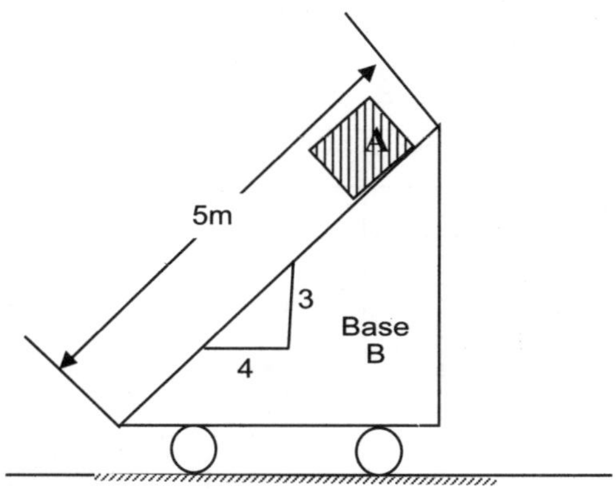

Fig. ED5.12

Solution: Here it requires applying principle of conservation momentum for this system of particles.

$$0 + 0 = m_A(v_B - v_A \cos\theta) + m_B v_B.$$

or $$v_A = (m_A v_B + m_B v_B)/\cos\theta \qquad ...(1)$$

Now as distance and velocity are known, conservation energy equation can be applied.

i.e., $$\frac{1}{2}m_B v_B^2 + \frac{1}{2}m_A\left((v_B - v_A\cos\theta)^2 + (v_A\sin\theta)^2\right) = m_A g d\sin\theta \qquad ...(2)$$

Hence velocity of base v_B varies as a function of distance travelled by the box A over it at a fixed angle of inclined plane.

MATLAB program for this given below:

```
%%%%%%%%%%%%%%%%%%%%%%%%%%%
ma=40;
mb=60;
g=9.81;
theta=atan(3/4);
mul=(ma+mb)/(ma*cos(theta));
d=0:0.2:5;
pe=ma*g*d*sin(theta);
den=0.5*mb+0.5*ma*((1-mul*cos(theta))^2+(mul*sin(theta))^2));
vb=sqrt(pe./den);
va=mul*vb;
plot(d,vb,'-',d,va,'-p');
xlabel('Distance moved by box (m)');
ylabel('Velocity (m/s)');
legend('Velocity of base-B','Velocity of box-A');
grid on;
%%%%%%%%%%%%%%%%%%%%%%%%%%%%%
```

The output is shown in Fig. ED5.12(a).

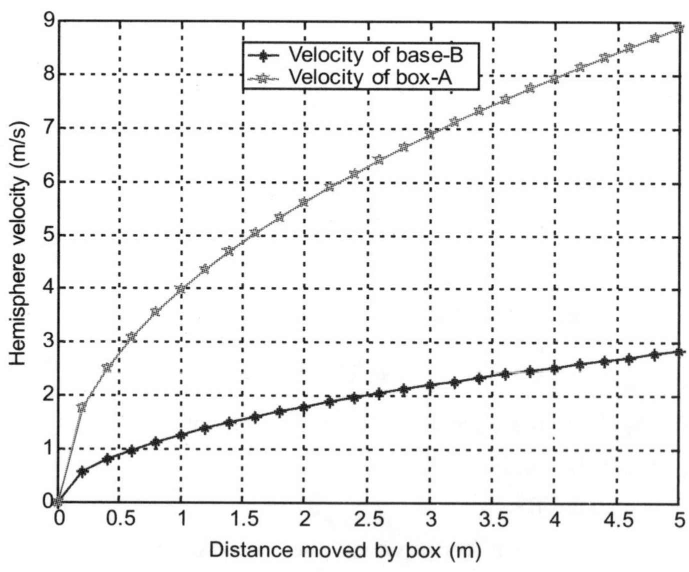

Fig. ED5.12 (a) MATLAB output

ED5.13: A chain has a mass per unit length 0.2 kg/m. If it is raised with a constant speed 5cm/s starting from rest, plot the force required to raise it as a function of height.

Solution: This is an example of systems gaining mass (see Fig. ED5.13(*a*)).

Net external force acting on chain is

$$\Sigma F_s = (F - Wy) = F - (0.2g)y,$$

where y is distance to be moved upwards.

Relative velocity = velocity of chain (v_c) = 5 cm/s = 0.05 m/s

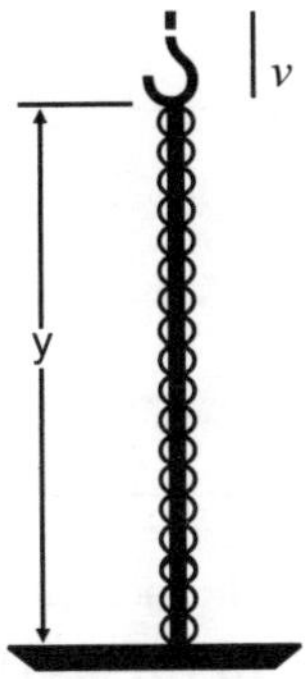

Fig. ED5.13 (a)

Rate of mass added to chain $= \dfrac{dm}{dt} = 0.2\, v_c$

Hence $\qquad \Sigma F_s = m\dfrac{dv}{dt} + u\dfrac{dm}{dt}$, where u is relative velocity of chain.

or $\qquad F - (0.2g)y = m\,(0) + v_c\,(0.2v_c) = 0.2\, v_c^2$

or $\qquad F = (0.2)(0.05)^2 + (0.2g)y$

```
%%%%%%%%%%%%%%
u=0.05;
m=0.2;
g=9.81;
y=5:-0.5:0.5;
F=m*u^2+m*g.*y;
plot(y,F,'-p');
xlabel('Length of chain hanging (m)');
ylabel('Force required (N)');
grid on;
%%%%%%%%%%%%%%%%%
```

Output is shown in Fig. ED 5.13 (*b*).

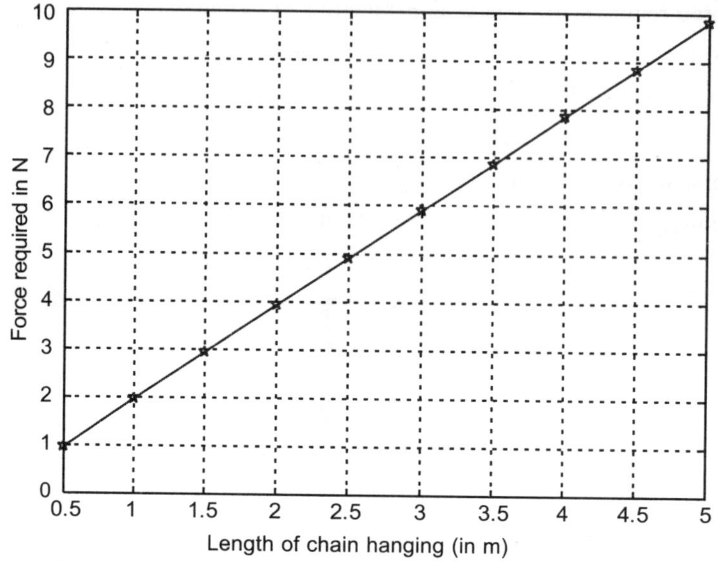

Fig. ED5.13 (b) MATLAB output

5.2.3 Planar Kinematics of Rigid Bodies

There are various methods of solving rigid-body kinematics. Some methods include, vector analysis, graphical approach (velocity and acceleration diagrams). In vector approach, a moving frame of reference is attached to one of the points on the body and motion of other point is described with reference to this point. The rotation is assumed to be along *z*-axis perpendicular to the plane containing the connected rigid bodies.

ED5.14: The angular velocities of links 3 and 4 of the four-bar mechanism at an instant shown in Fig.ED5.14 are given by

$$\begin{bmatrix} -r_3 \sin\theta_3 & r_4 \sin\theta_4 \\ r_3 \cos\theta_3 & -r_4 \cos\theta_4 \end{bmatrix} \begin{bmatrix} \omega_3 \\ \omega_4 \end{bmatrix} = \begin{bmatrix} \omega_2 r_2 \sin\theta_2 \\ -\omega_2 r_2 \cos\theta_2 \end{bmatrix}$$

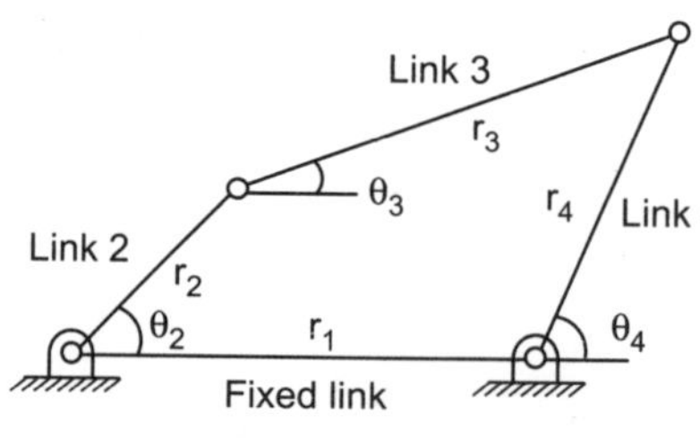

Fig. ED5.14 Four-bar mechanism

The link lengths and angles are given below:

Link	Length, r(cm)	Angle, θ (degree)
1	13.0	0
2	5.0	45
3	11.0	25
4	8.0	89

The angular velocity of the input link, ω_2 is given as + 110 rad/s. Write a MATLAB program to solve for the angular velocities of links 3 and 4.

Solution: The velocity and acceleration analysis are done based on the following equations:

$$v_B = v_A + \omega_3 \times AB,$$

where v_A and v_B are obtained from the two ends likewise.

$$a_B = a_A + \alpha_3 \times AB - \omega_3^2 AB.$$

Here it requires only velocities.

So the complete MATLAB script to solve the angular velocities ω_3 and ω_4 is presented below:

```
%%%%%%%%%%%%%%%%%
th2=45*pi/180;th3=25*pi/180;th4=89*pi/180;
r2=5;r3=11;r4=8;om2=110;
b=[om2*r2*sin(th2);-om2*r2*cos(th2)];
A=[-r3*sin(th3) r4*sin(th4);r3*cos(th3)-r4*cos(th4)];
om=inv(A)*b;
fprintf('Angular velocities of link 3 is %f rad/s and link 4 is %f rad
s\n',om(1),om(2));
%%%%%%%%%%%%%%%%%
```

Output will come like this:

» Angular velocities of link 3 is –38.643913 rad/s and link 4 is 26.161594 rad/s.
Minus sign represents opposite sense (clockwise). It should be noted that this value is at this angular instant. As input angle changes all other angles will change and angular speed also changes.

A plot of angular speeds of links-3 and 4 are shown as a function of angular speed of link-2 (input link) in Fig. ED5.14 (*a*).

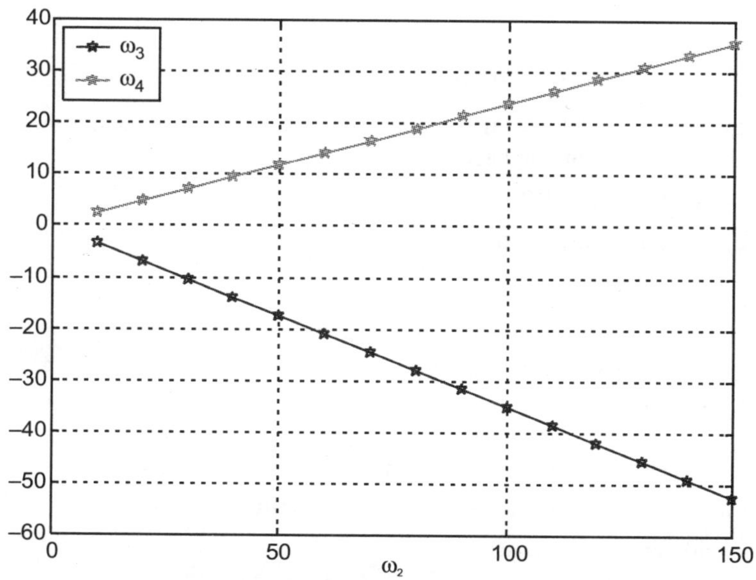

Fig. ED5.14 (a) Variation of angular speeds

ED5.15: Figure ED5.15 shows two rotating rods connected by a slider *P*. Write a MATLAB program to determine and plot for values of θ from 0 to 180°,

(*a*) the angular velocity and angular acceleration of rod *AE*

(*b*) determine the value of θ for which the angular acceleration α_{AE} of rod *AE* is maximum and the corresponding value of α_{AE}.

Assume that the rod *BP* rotates counterclockwise with a constant angular velocity of 10 rad/s.

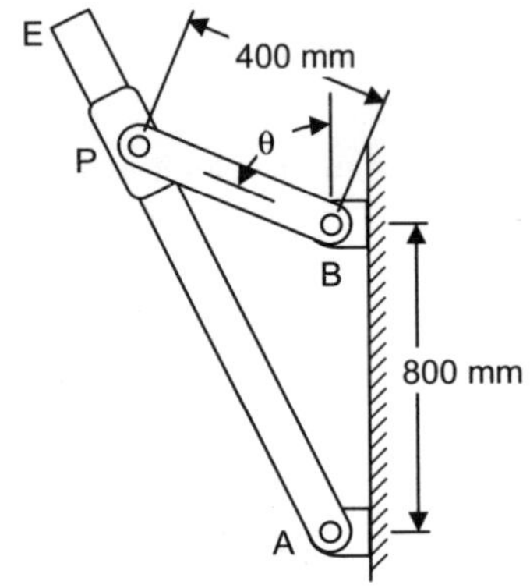

Fig. ED5.15

Solution: Formulation of the problem is explained first. Here the slider is translating on another rotating link. Thus moving coordinate frame is not sufficient. We require a rotating frame of reference.

First of all from geometry in triangle ΔABP

$$\frac{AP}{\sin(180° - \theta)} = \frac{BP}{\sin \angle BAP} \quad \text{or} \quad \ell = AP = \frac{BP \sin\theta}{\sin \angle BAP}$$

where the angle $\angle BAP = \beta = \tan^{-1}\left[\frac{BP\sin\theta}{AB + BP\cos\theta}\right]$

The slider *P* moves along *AE* so its velocity and acceleration should be chosen in this direction. First consider the point *P* to be on *BP*, so its velocity and acceleration are given by

$$v_P = v_B + \omega \times BP = \omega \times BP$$

$$a_P = a_B + \alpha \times BP - \omega^2 BP = \alpha \times BP - \omega^2 BP$$

Also if we consider point *P* as on *AE*, then

$$v_P = v_A + \omega_{AE} \times AP + v_{PP'} = \omega_{AE} \times AP + v_{PP'}$$

$$a_P = a_A + \alpha_{AE} \times AP - \omega^2{}_{AE} \times AP + a_{PP'} + (2\,\omega_{AE} \times v_{PP'}) = \alpha_{AE} \times AP - \omega^2{}_{AE} \times AP + a_{PP'} + (2\,\omega_{AE} \times v_{PP'})$$

Here $v_{PP'}$ and $a_{PP'}$ are the velocity and acceleration of the moving block along *AE*.

Choosing rightwards as +ve *x* and upwards as +ve *y*, we can write

$$BP = -400\sin\theta\hat{i} + 400\cos\theta\hat{j}$$

$$AP = -\ell \sin\beta \hat{i} + \ell \cos\beta \hat{j}$$
$$v_{PP'} = -v_{PP'} \sin\beta \hat{i} + v_{PP'} \cos\beta \hat{j}$$
$$a_{PP'} = -a_{PP'} \sin\beta \hat{i} + a_{PP'} \cos\beta \hat{j}$$

From velocity equivalence $\omega \times BP = \omega_{AE} \times AP + v_{PP'}$
Performing cross product of two vectors and equating *i* and *j* terms, we get:

$$\omega_{AE} \ell \cos\beta + v_{PP'} \sin\beta = 0.4\omega \cos\theta \quad ...(1)$$

and $$-\omega_{AE} \ell \sin\beta + v_{PP'} \cos\beta = -0.4\omega \sin\theta \quad ...(2)$$

These two equations are solved first:

$$\ell\, \omega_{AE} = 0.4\omega \cos\theta \cos\beta + 0.4\omega \sin\theta \sin\beta = 0.4\omega \cos(\beta - \theta)$$
$$v_{PP'} = 0.4\omega \cos\theta \sin\beta - 0.4\omega \sin\theta \cos\beta = 0.4\omega \sin(\beta - \theta)$$

Then from acceleration equivalence $\alpha \times BP - \omega^2 BP = \alpha_{AE} \times AP - \omega^2_{AE} \times AP + a_{PP'} + (2\, \omega_{AE} \times v_{PP'})$, with $\alpha = 0$, we get

$$\ell \alpha_{AE} \cos\beta + a_{PP'} \sin\beta = -0.4\omega^2 \sin\theta + \ell\omega^2_{AE} \sin\beta - 2\omega_{AE} v_{PP'} \cos\beta \quad ...(3)$$

$$\ell \alpha_{AE} \sin\beta - a_{PP'} \cos\beta = 0.4\omega^2 \cos\theta - \ell\omega^2_{AE} \cos\beta - 2\omega_{AE} v_{PP'} \sin\beta \quad ...(4)$$

Solving the equations (3) and (4), we get

$$\ell \alpha_{AE} = 0.4\omega^2 \cos(90 + \beta - \theta) - 2\omega_{AE} v_{PP'}$$

and $$a_{PP'} = -0.4\omega^2 \cos(\beta - \theta) + \ell\omega^2_{AE}$$

Complete MATLAB program is presented below:

```
%%%%%%%%%%%%%%%%%%%%%%%%%%%%%
om=10;BP=0.4;AB=0.8;
theta=[0:10:180];
th=theta*pi/180;
beta=atan(BP*sin(th)./(AB+BP*cos(th)) );
AP=BP*sin(th)./sin(beta);
omAE=BP*om*cos(beta-th)./AP;
vPP= BP*om*cos(pi/2+beta-th);
alAE=(BP.*om.^2.*cos(pi/2+beta-th)-2.*omAE.*vPP)./AP;
aPP=-BP.*om.^2.*cos(beta-th) + AP.*omAE.^2;
[maxacc,I]=max(alAE);
fprintf('Max.Angular Acceleration is%f rad/s^2 at theta=%f
deg\n',maxacc,theta(I));
subplot(4,1,1);
plot(theta,omAE,'-p');
ylabel('\omega-AE (rad/s)');grid on;
title('ANALYSIS OF QUICK-RETURN MECHANISM')
subplot(4,1,2);
plot(theta,alAE,'-*');
```

```
ylabel('\alpha-AE(rad/s^2)' );grid on;
subplot(4,1,3);
plot(theta,vPP,'-*');
ylabel('vPP(m/s)' );grid on;
subplot(4,1,4);
plot(theta,aPP,'-*');
ylabel('aPP(m/s^2)' );
xlabel('\Theta(degree)');grid on;
%%%%%%%%%%%%%%%%%%%%%%%%%%%%%%%
```

Output is given below:

» Max.Angular Acceleration is133.198320 rad/s^2 at theta=160.000000 deg

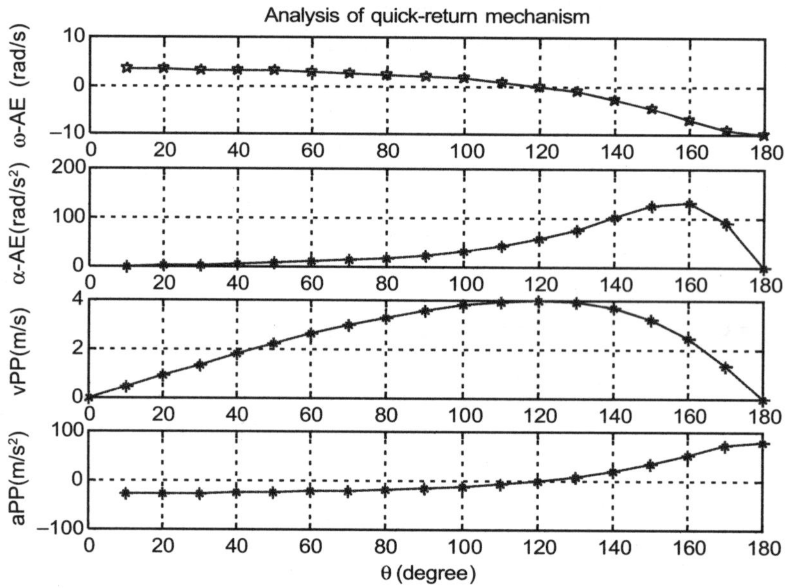

Fig. ED5.15 (a) MATLAB output

ED 5.16: A mechanism used to convert the constant circular motion ω of the rod *AB* into translating motion of the rod *CD* is shown in Fig. ED5.16. Plot the velocity and acceleration of *CD* as a function of angle θ of *AB*. Take *AB* =100 mm and ω = 10 rad/s.

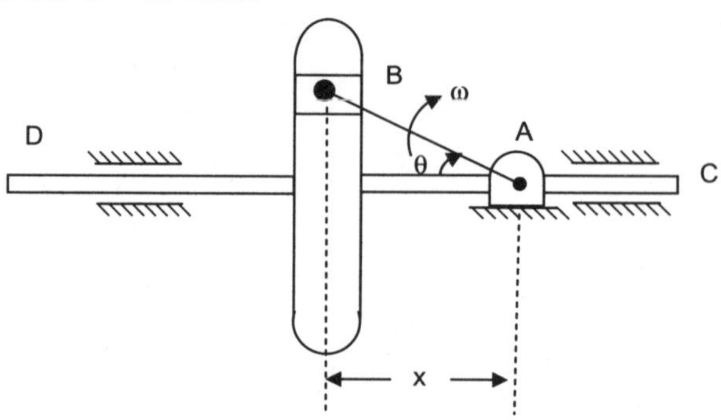

Fig. ED5.16

Solution: Analytically, the motion of rod *CD* can be written in terms of motion of *AB* as:

$$x = AB\cos(\theta) = \ell\cos(\theta)$$

Thus differentiating

$$\dot{x} = v = -\ell\dot{\theta}\sin\theta$$

$$\ddot{x} = a_x = -\ell\left(\ddot{\theta}\sin\theta + \dot{\theta}^2\cos\theta\right)$$

But $\quad v_{CD} = v_x = -\omega\ell\sin\theta$, and $a_{CD} = -\omega\ell^2\cos\theta$, since $\dot{\theta} = \omega$ and $\ddot{\theta} = \alpha = 0$

Now following MATLAB script is used to plot the velocity and acceleration of *CD*.

```
%%%%%%%%%%%%%%%%%%%%%%%
AB=0.1;
omega=10;
theta=[0:10:180];th=theta*pi/180;
vCD=-AB*omega.*sin(th);
aCD=-AB^2*omega.*cos(th);
figure(1)
plot(theta,vCD,'-p');
xlabel('Angle theta (degree)');
ylabel('v-CD in m/s');
title('VELOCITY PLOT');
figure(2)
plot(theta,aCD,'-p');
xlabel('Angle theta (degree)');
ylabel('a-CD in m/s^2');
title('ACCELERATION PLOT');
%%%%%%%%%%%%%%%%%%%%%%%
```

Output is given in Figs. ED5.16 (*a*) and (*b*).

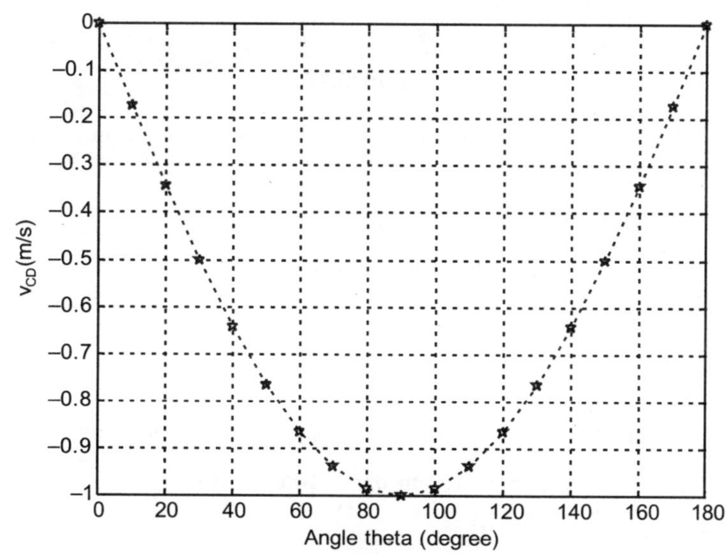

(a) Velocity plot

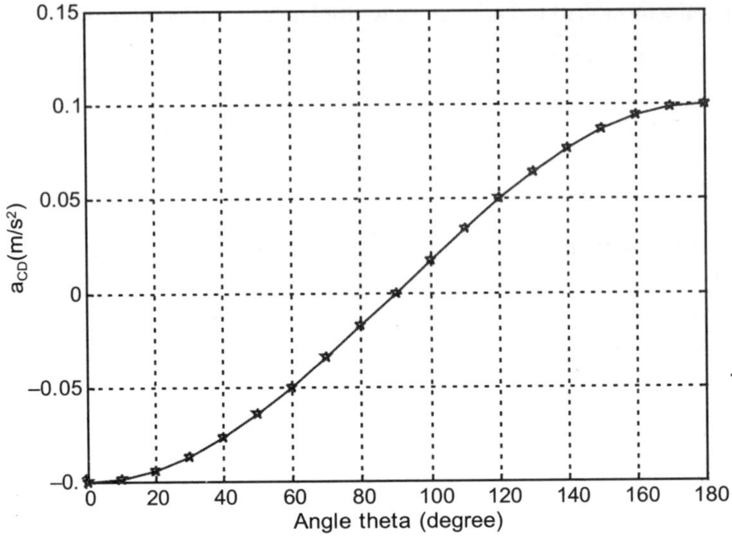

(b) Acceleration plot

Fig. ED5.16 (a) & (b) MATLAB output

This example illustrates that sometimes it is better to use the analytical method than relative motion analysis.

ED5.17: A rod *AB* shown in Fig. ED5.17 is confined to move along the inclined paths. If point *A* has an acceleration 3m/s² and velocity of 2m/s at any instant θ, study the variation of angular acceleration of rod as function of angle θ, using MATLAB. Take *AB* = 10 m.

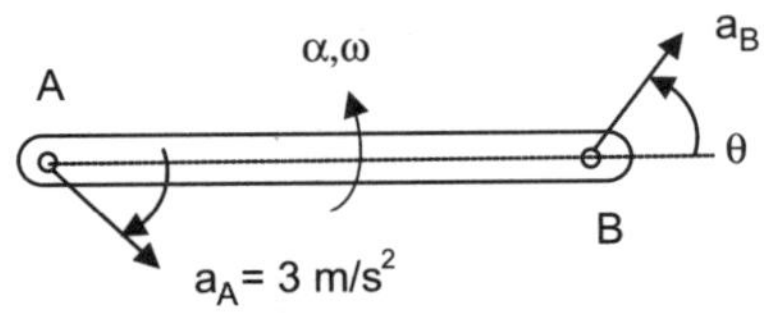

Fig. ED5.17

Solution: First of all it is required to find the angular velocity of the rod ω , from velocity equations:

$$v_B = v_A + v_{BA} = v_A + \omega \times AB$$

$$\Rightarrow \quad v_B \cos\theta\hat{i} + v_B \sin\theta\hat{j} = 2\cos\theta\hat{i} - 2\sin\theta\hat{j} + \omega\hat{k} \times 10\hat{i}$$

$$= 2\cos\theta\hat{i} + (10\omega - 2\sin\theta)\hat{j}$$

Hence $\quad v_B = 2$ and $\omega = 0.4 \sin\theta$

Now using acceleration equation:

$$a_B = a_A + a_{BA} = a_A + \alpha \times AB - \omega^2 AB$$

$$\Rightarrow \quad a_B \cos\theta\hat{i} + a_B \sin\theta\hat{j} = 3\cos\theta\hat{i} - 3\sin\theta\hat{j} + \alpha\hat{k} \times 10\hat{i} - (0.4 \sin\theta)^2\, 10\hat{i}$$

$$= (3\cos\theta - 1.6\sin\ \theta)\hat{i} + (10\alpha - 3\sin\theta)\hat{j}$$

So $\quad a_B \cos\theta = (3\cos\theta - 1.6\sin^2\theta) \qquad ...(1)$

$$a_B \sin\theta = (10\alpha - 3\sin\theta) \quad ...(2)$$

Solving the equations (1) and (2) for α, we get:

$$\alpha = \frac{(3\cos\theta - 1.6\sin^2\theta)\tan\theta + 3\sin\theta}{10}$$

```
%%%%%%%%%%%%%%%%%%%%%%
theta=[0:10:70];
th=theta*pi/180;
omega=0.4.*sin(th);
alfa=((3.*cos(th)-1.6.*sin(th).^2).*tan(th)+3.*sin(th))./10;
subplot(2,1,1)
plot(theta,omega,'-p');
ylabel('\omega(rad/s)');
grid on;
subplot(2,1,2)
plot(theta,alfa,'-*');
ylabel('\alpha(rad/s^2)');
grid on;
xlabel('\theta(degree)');
%%%%%%%%%%%%%%%%%%%%%%
```

Output is shown in Fig. ED5.17(*a*).

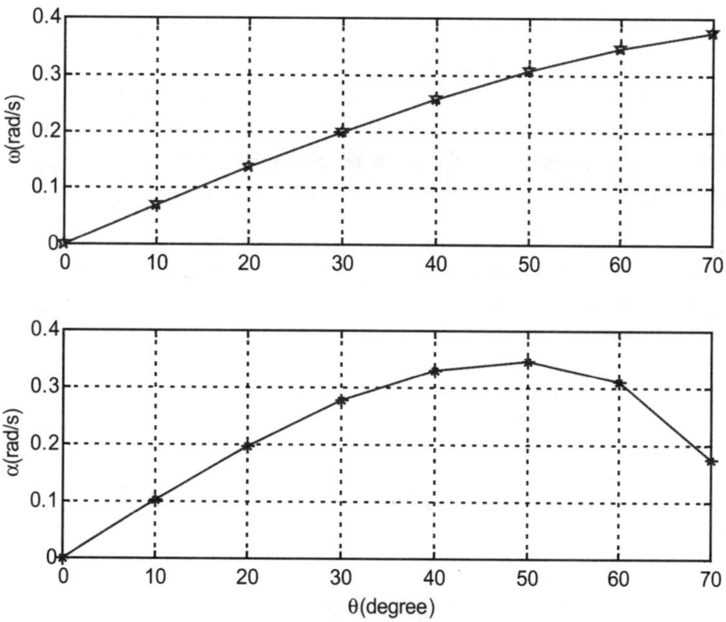

Fig. ED5.17 (a) MATLAB output

5.2.4 Rigid Body Kinetics in Plane Motion

Just like in particles, Kinetics problem is attempted either with method force and acceleration (when accelerations are given), energy method (when velocity and displacements are given) and momentum principle (when velocities and times are given). Following examples illustrate the methodology in all the cases.

ED5.18: Figure ED5.18 shows that the end A of the rod AB moved to the left with a constant speed v_A. Write a MATLAB program to calculate and plot the normal reactions at ends A and B of the rod for values of θ from 0° to 50°. Determine the value of θ at which end B of the rod loses contact with the wall.

The mass of the rod AB = 12 kg.

The constant speed of end A = 70 cm/s.

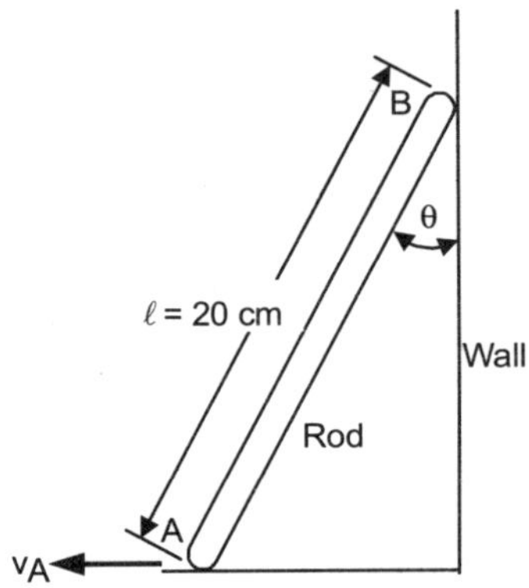

Fig. ED 5.18

Solution: Most engineering applications deal with rigid bodies which are moving under given constraints. Examples include: Cranks which should rotate about a fixed axis, wheels must roll without slipping and connecting rods must describe certain prescribed motions. In all such cases definite relations exist between the components of the acceleration a of mass center G of the body considered and its angular-acceleration α. Such a motion is called constrained motion. The solution of a problem involving a constrained plane motion requires first kinematic analysis problem.

First the angular velocity of rod has to be determined. This time let us find it by the instantaneous center method. Instantaneous center lies on the perpendiculars to the two velocities v_A and v_B as shown in Fig. ED5.18 (*a*).

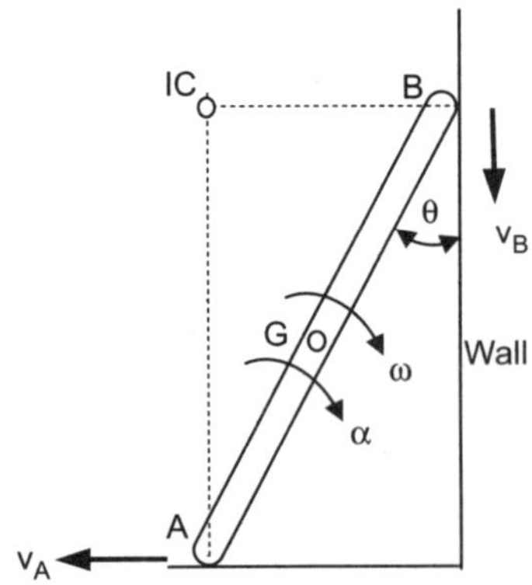

Fig. ED5.18
(a) Kinematic diagram

Thus we can write $\omega = \dfrac{v_A}{IC-A} = \dfrac{v_B}{IC-B} = \dfrac{0.7}{0.2\cos\theta} = \dfrac{3.5}{\cos\theta}$ ...(1)

Knowing that the acceleration of A is zero, we have

$$a_B = a_A + a_{BA}$$

$$\Rightarrow \quad -a_B\,\hat{j} = 0 + (a_{BA})_n + (a_{BA})_t$$

$$= -\omega^2 AB + (-\alpha)\hat{k} \times AB \text{ (because } \alpha \text{ is in anti-clockwise sense)}$$

$$= -\omega^2(0.2\sin\theta\hat{i} + 0.2\cos\theta\hat{j}) - \alpha\hat{k} \times (0.2\sin\theta\hat{i} + 0.2\cos\theta\hat{j})$$

Equating $\hat{i}$ and $\hat{j}$ terms both sides, we get

$$-0.2\omega^2 \sin\theta + 0.2\,\alpha\cos\theta = 0$$
$$-0.2\omega^2 \cos\theta - 0.2\,\alpha\sin\theta = -a_B$$

They give $\alpha = \omega^2 \tan\theta$ ...(2)

Further acceleration of center G is given by

$$a_G = a_A + a_{GA}$$

$$\Rightarrow \quad a_{Gx}\hat{i} + a_{Gy}\hat{j} = (a_{BA})_n + (a_{BA})_t = -\omega^2 AG + (-\alpha)\hat{k}\times AG$$
$$= -\omega^2(0.1\sin\theta\hat{i} + 0.1\cos\theta\hat{j}) - \alpha\hat{k}\times(0.1\sin\theta\hat{i} + 0.1\cos\theta\hat{j})$$
$$= (0.1\alpha\cos\theta - 0.1\omega^2\sin\theta)\hat{i} - (0.1\omega^2\cos\theta + 0.1\alpha\sin\theta)\hat{j}$$

$$\Rightarrow a_{Gx} = (0.1\alpha\cos\theta - 0.1\omega^2\sin\theta) = 0 \text{ and } a_{G_y} = (0.1\omega^2\cos\theta + 0.1\alpha\sin\theta) = -\frac{0.1\omega^2}{\cos\theta} \quad ...(3)$$

From kinetics, we find by summing the forces in x, y directions and moments, (see Fig. ED5.18 (*b*))

we get:

$$\Sigma F_x = m\,a_{Gx} \Rightarrow -N_B = 0 \quad \text{(dummy equation-no use)} \quad ...(4)$$

$$\Sigma F_y = m\,a_{Gy} \Rightarrow N_A - mg = m\times -\frac{0.1\omega^2}{\cos\theta}$$

$$\Rightarrow N_A = mg - m\frac{0.1\omega^2}{\cos\theta} \quad ...(5)$$

$$\Sigma M_G = I_G\alpha \Rightarrow -N_B\frac{\ell}{2}\cos\theta + N_A\frac{\ell}{2}\sin\theta = \frac{1}{12}m\ell^2\alpha$$

$$\Rightarrow N_B = \frac{N_A\frac{\ell}{2}\sin\theta - \frac{1}{12}m\ell^2\alpha}{\frac{\ell}{2}\cos\theta} \quad ...(6)$$

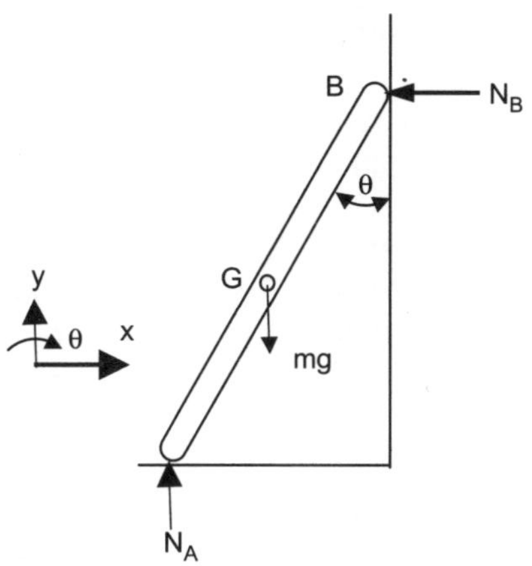

Fig. ED5.18 (b) Free-body diagram

Equations (1), (5) and (6) are enough to find the reactions.

Complete MATLAB program for plotting the reactions as a function of time is given below:

```
%%%%%%%%%%%%%%%%%%%%%%%%%%%%%
L=0.2;m=12;g=9.81;IG=m*L^2/12;
theta=[0:5:60];
th=theta*pi/180;
om=3.5./cos(th);
al=om.^2.*tan(th);
agy=-(L/2)*om.^2./cos(th);
NA=m*g +m.*agy;
NB=(NA*(L/2).*sin(th)-IG*al)./((L/2).*cos(th));
z=[theta;NA;NB];
fprintf('Theta(deg) Normal Reaction A (N) Normal reaction B(N)\n');
fprintf('%5.3f %6.3f %6.3f\n',z);
fprintf('\n');
figure(1)
plot(theta,NA,'-p',theta,NB,'-*');
xlabel('\Theta(degree)');
ylabel('Normal reaction (N)');
legend('Reaction at A','Reaction at B');
grid on;
%%%%%%%%%%%%%%%%%%%%%%%%%%%%
```

Output of the program is given below:

Theta(deg)	Normal Reaction A (N)	Normal reaction B(N)
0.000	103.020	0.000
5.000	102.851	8.565
10.000	102.329	17.139
15.000	101.409	25.716
20.000	100.004	34.249
25.000	97.973	42.616
30.000	95.088	50.543
35.000	90.976	57.460
40.000	85.019	62.193
45.000	76.142	62.283
50.000	62.370	52.342
55.000	39.819	19.783
60.000	0.120	–67.689

It is seen that the end of rod *B* loses contact with the wall approximately at an angle 56.62° where sign change occurs.

Figure ED5.18 (c) shows the variation of reactions as a function of θ.

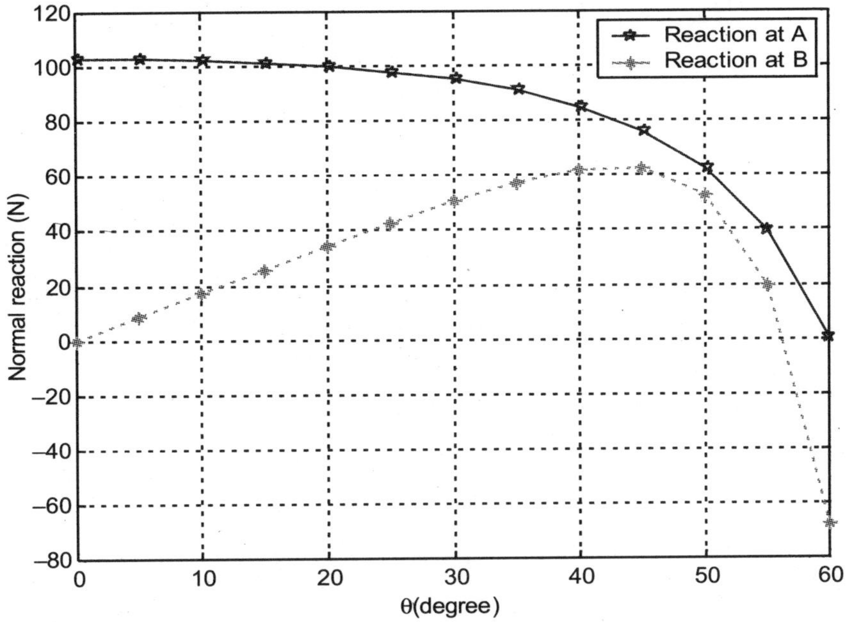

Fig. ED5.18(c) MATLAB output

ED5.19: Figure ED5.19 shows a uniform slender bar AB of weight W suspended from springs AC and BD. Write a MATLAB program to calculate and plot the accelerations of ends A and B immediately after spring AC has broken for values of θ from 0° to 90°.

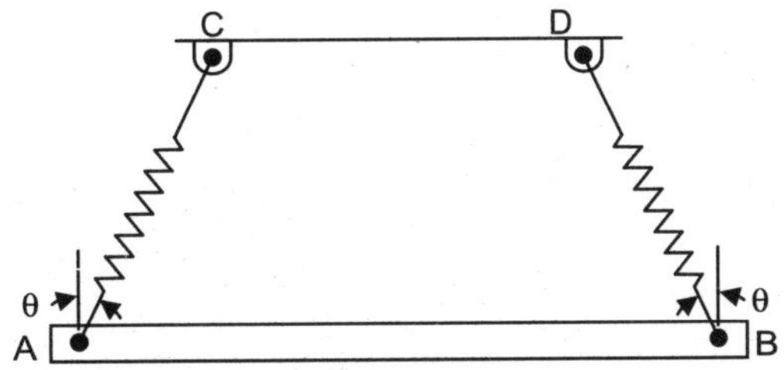

Fig. ED5.19

Solution: From statics, we find the initial spring tensions (see Fig. ED5.19 (a)).

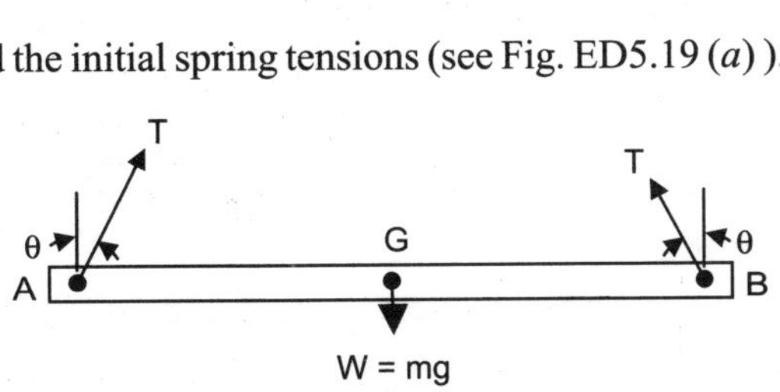

Fig. ED5.19 (a) Equilibrium diagram

Thus $\Sigma F_y = 0 \Rightarrow 2T\cos\theta - mg = 0$

$$\Rightarrow \qquad T = \frac{mg}{2\cos\theta}$$

Just after AC breaks we find from kinetics (see Fig. ED5.19(b))

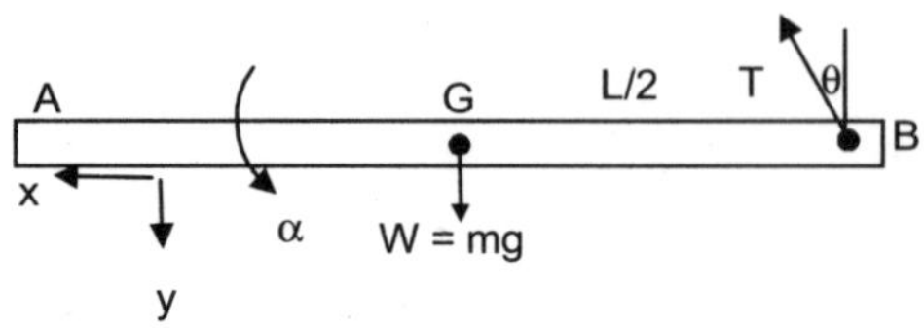

Fig. ED5.19 (b) Free-body diagram

$$\Sigma F_x = ma_{Gx} \quad \Rightarrow \quad T\sin\theta = ma_{Gx} \Rightarrow a_{Gx} = \frac{T\sin\theta}{m} = \frac{g\tan\theta}{2} \qquad \text{...(1)}$$

$$\Sigma F_y = ma_{Gy} \quad \Rightarrow \quad mg - T\cos\theta = ma_{Gy} \Rightarrow a_{Gy} = \frac{g}{2} \qquad \text{...(2)}$$

$$\Sigma M_G = I_G\alpha \quad \Rightarrow \quad (T\cos\theta)\frac{L}{2} = I_G\alpha \Rightarrow \frac{mg}{2\cos\theta} \times \cos\theta \times \frac{L}{2} = m\frac{L^2}{12}\alpha$$

or $$\alpha = \frac{3g}{L} \qquad \text{...(3)}$$

From kinematics, acceleration of points A and B along considered x and y directions:

$$a_A = a_G + a_{AG} = a_{Gx}\hat{i} + a_{Gy}\hat{j} + \alpha\,\hat{k} \times \frac{L}{2}\hat{i} = a_{Gx}\hat{i} + \left(a_{Gy} + \alpha\frac{L}{2}\right)\hat{j}$$

$$a_B = a_G + a_{BG} = a_{Gx}\hat{i} + a_{Gy}\hat{j} + \alpha\,\hat{k} \times -\frac{L}{2}\hat{i}' = a_{Gx}\hat{i} + \left(a_{Gy} - \alpha\frac{L}{2}\right)\hat{j}$$

Complete MATLAB script for plotting the accelerations is given below:

```
%%%%%%%%%%%%%%%%%%%%
theta=[0:5:85];g=9.81;L=1;
th=theta*pi/180;
aGx=0.5*g*tan(th);
aGy=0.5*g;
alfa=3*g/L;
aAx=aGx;aAy=aGy+0.5*alfa*L; aA=sqrt(aAx.^2+aAy.^2);
aBx=aGx;aBy=aGy-0.5*alfa*L; aB=sqrt(aBx.^2+aBy.^2);
plot(theta, aA,'-p', theta, aB, '-*');
xlabel('\theta(degree)');ylabel('Acceleration(m/s^2)');
legend('a_A','a_B');
grid on;
%%%%%%%%%%%%%%%%%%%%%%
```

Figure ED5.19 (*c*) shows the output of the script.

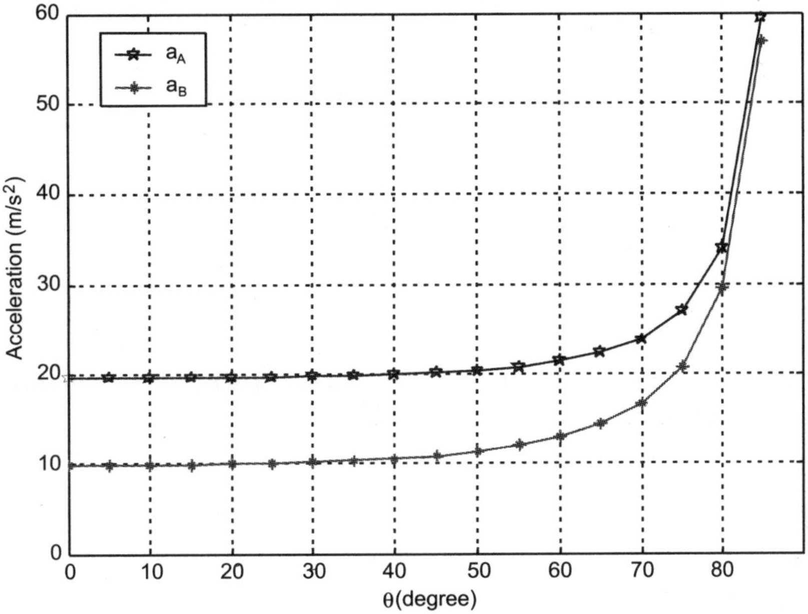

Fig. ED5.19 (c) MATLAB output

ED5.20: A bicycle and rider of total mass 80 kg with center of mass located *G*. If coefficient of kinetic friction at rear tire is 0.8, plot the deceleration of rider when brakes are applied as a function of height of *G* from ground. Also show the variation of normal reaction at the rear wheel. Assume *G* lies between two wheels separated by a distance 150 cm. Choose $0.9 \leq h \leq 1.3$ meters.

Solution: Assuming that the rear wheel locks for braking, the free-body diagram is shown in Fig.ED5.20(*a*) with all the forces acting on it.

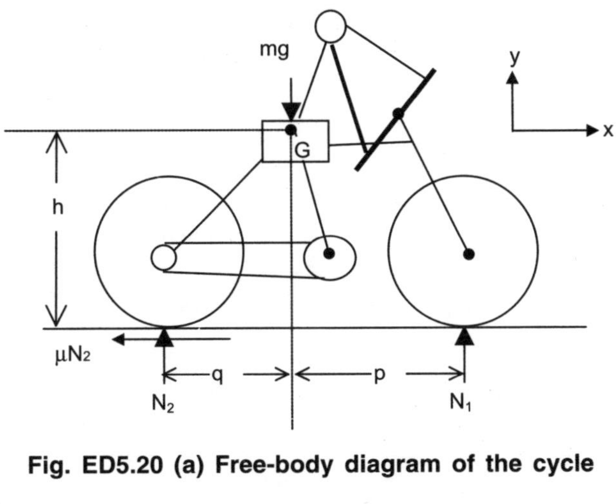

Fig. ED5.20 (a) Free-body diagram of the cycle

This is a problem of rigid-body translation. There is no rotation part.

Applying the D'Alembert's principle, we get the equations of motion as:

$$\Sigma F_x = ma_{Gx} \Rightarrow -\mu N_2 = ma_G \quad ...(1)$$

$$\Sigma F_y = ma_{Gy} \Rightarrow -mg + N_1 + N_2 = 0 \quad ...(2)$$

$$\Sigma M_G = 0 \quad \Rightarrow N_1 \times p - N_2 \times q - \mu N_2 \times h = 0 \quad ...(3)$$

The equations can be either solved manually or by using MATLAB. We get:

$$N_2 = \frac{mgp}{(p+q+\mu h)},\ a_G = \frac{-\mu gp}{(p+q+\mu h)} \quad \text{and} \quad N_1 = \frac{(q+\mu h)mg}{(p+q+\mu h)}$$

Following MATLAB script is employed to study the variation of a_G and N_2 as a function of h.

```
%%%%%%%%%%%%%%%%%%%%%%%%%%%%
m=80;
p=0.75;q=0.75;
mu=0.8;g=9.81;
h=[0.9:0.05:1.3];
aG=-mu*g*p./(p+q+mu.*h);
N2=m*g*p./(p+q+mu.*h);
N1=m*g*(q+mu.*h)./(p+q+mu.*h);
subplot(3,1,1)
plot(h,aG,'-p');
ylabel('a_G(m/s^2)');
grid on;
title('\bfVariation of parameters with height of G');
subplot(3,1,2)
plot(h,N1,'-*');
ylabel('N_1(N)');
grid on;
subplot(3,1,3)
plot(h,N2,'-+');
ylabel('N_2(N)');
grid on;
xlabel('height of G(m)');
%%%%%%%%%%%%%%%%%%%%%%%%%%%%%%
```

Figure ED5.20 (*b*) shows the output of the program.

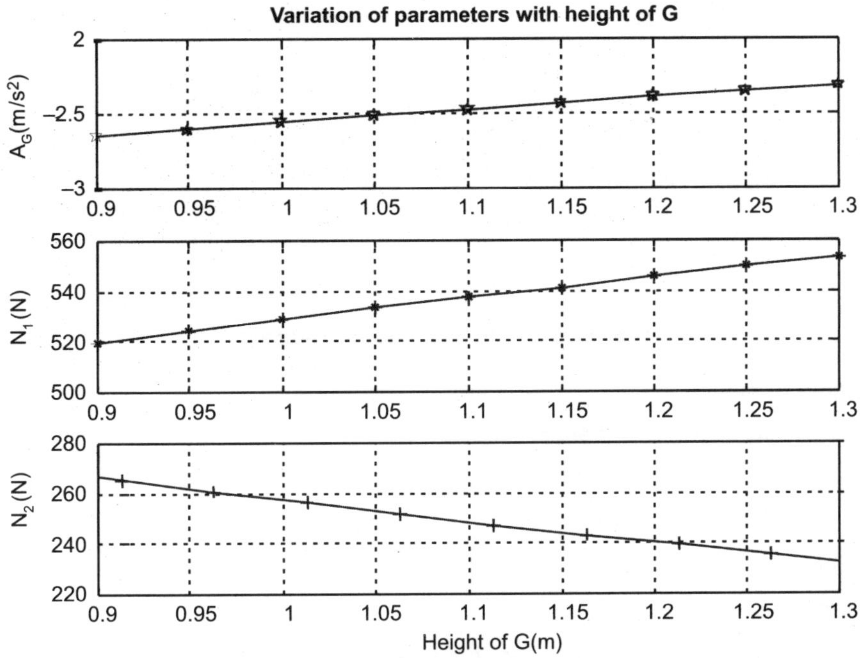

Fig. ED5.20 (b) MATLAB output

ED5.21: A spool has a mass of 75 kg and a radius of gyration $k_G = 0.38$ m. It rests on the inclined surface (Fig. ED5.21) for which coefficient of kinetic friction is $\mu_k = 0.15$. If the spool is released from rest and slips at *A*, plot the initial tension in the cord and angular acceleration of the spool as a function of angle of incline θ.

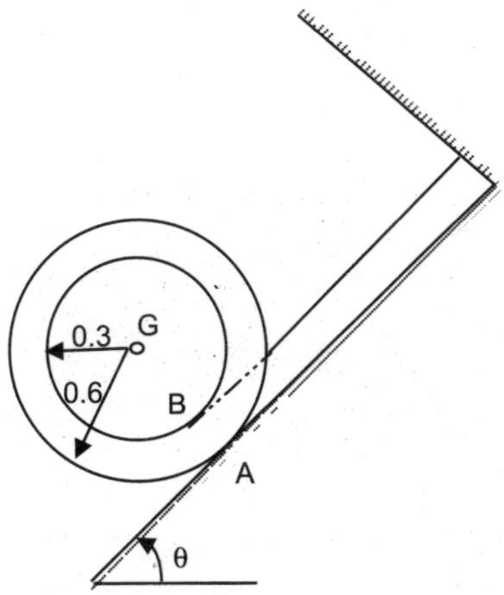

Fig. ED5.21

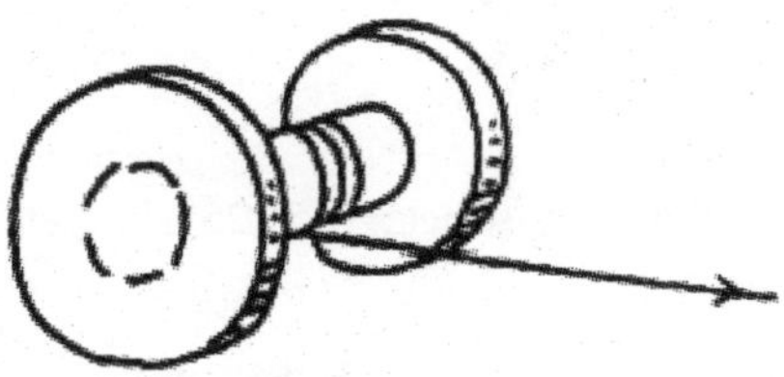

Fig. ED5.21 (a) Spool pulled towards right

Solution: It is well known from physics "when a spool is pulled to the right, it will roll toward the right only" as shown in Fig. ED5.21 (*a*). The spool accelerates both linearly and angularly as it descends.

The free-body diagram of the system is shown in Fig. ED5.21(*b*).

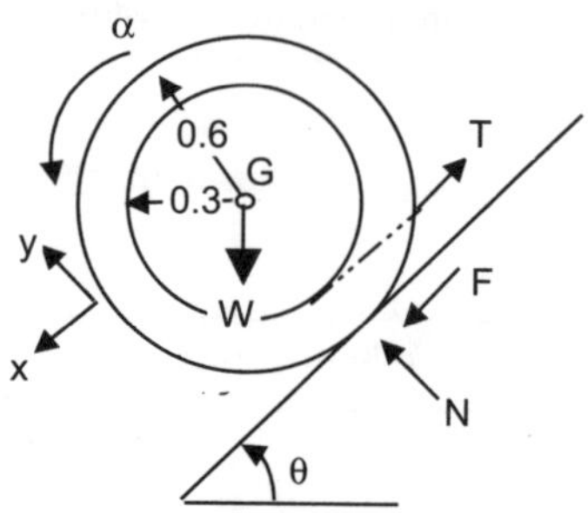

Fig. ED5.21 (b) Free-body diagram

Writing the equations of motion:

$$\Sigma F_x = ma_x \Rightarrow F - T + W\sin\theta = ma_G \quad ...(1)$$

$$\Sigma F_y = ma_y \Rightarrow N - W\cos\theta = 0 \quad ...(2)$$

$$\Sigma M_G = I_G\alpha \Rightarrow 0.3T - 0.6F = I_G\alpha \quad ...(3)$$

As spool is slipping at contact point A, $F = \mu_k N$ ...(4)

Also it is only rolling at point B, hence $a_G = 0.3\alpha$ ...(5)

Equations (1)–(5) are solved together to obtain F, N, T, a_G and α. Substituting $W = \text{mg} = 75\times9.81 = 735.75$ N, $I_G = mk^2{}_G = 75\times0.38^2 = 10.83$ kgm^2 and manipulating equations (4) and (5) in eqs.(1)–(3).

The following MATLAB script solves the equations in terms of μ_k and θ:

```
%%%% %%%%% Here x=T, y=aG and z=α %%%%%%
eqn1='-x + 735.75*muk*cos(theta) +735.75*sin(theta)=75*y';
eqn2='0.3*x -735.75*0.6*muk*cos(theta)=10.83*z';
eqn3='y-0.3*z';
[x,y,z]=solve(eqn1,eqn2,eqn3)
%%%%%%%%%%%%%%%%%%%%%%%%%%%%
```

Output is as follows:

```
» x =
    1018.24*muk*cos(theta)+453.25*sin(theta)
  y =
    –3.766*muk*cos(theta)+3.766*sin(theta)
  z =
    –12.555*muk*cos(theta)+12.555*sin(theta)
```

Now a complete MATLAB script for plotting the variation of tension in string and angular acceleration of the spool as a function of friction and angle of inclined plane is given:

```
%%%%%%%%%%%%%%%%%%%%%%%%%%%%%%
theta=[30*pi/180;45*pi/180;60*pi/180]; % at 3 different inclined surfaces
muk=[0:0.01:0.25];
for i=1:length(theta)
      Ten(i,:)= 1018.24*muk*cos(theta(i))+453.25*sin(theta(i));
      aG(i,:)= -3.766*muk*cos(theta(i))+3.766*sin(theta(i));
      alfa(i,:)=-12.555*muk*cos(theta(i))+12.555*sin(theta(i));
end
subplot(3,1,1);
plot(muk,Ten,'-p')
ylabel('T(N)');grid on;
subplot(3,1,2);
plot(muk,aG,'-p')
ylabel('a_G(m/s^2)'); grid on;
subplot(3,1,3);
plot(muk,alfa,'-p');
ylabel('\alpha(rad/s^2)');grid on;
xlabel('\mu_k');
legend('\theta=30','\theta=45','\theta=60');
%%%%%%%%%%%%%%%%%%%%%%%%%%%
```

Output is shown in Fig. ED5.21 (*c*).

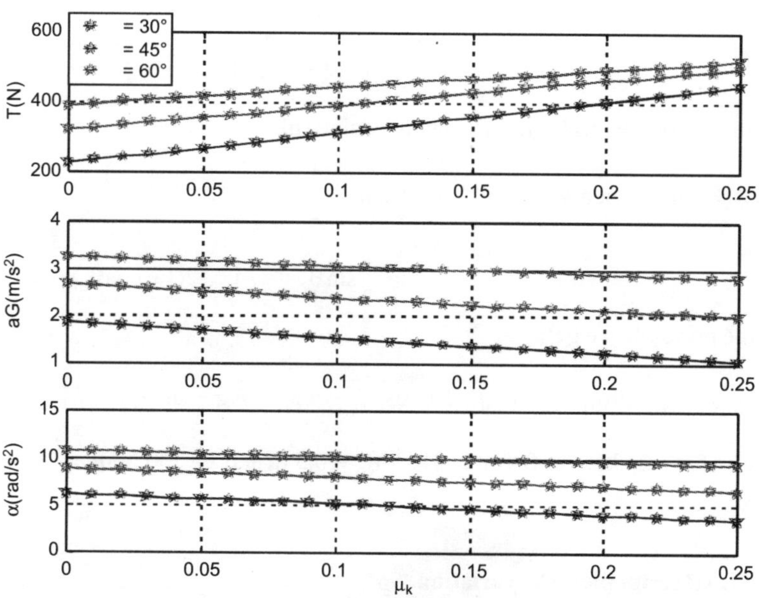

Fig. ED5.21 (c) MATLAB output

ED5.22: A 25 kg slender rod *AB* of 0.4 m long is attached to a spring *BC* (Fig. ED5.22) having unstretched length 0.05 m. If the rod is released from rest at $\theta = 40^o$, plot the variation of angular velocity of the rod as a function of angle θ. Consider *CA=AB* and spring stiffness as 500 N/m. Also find the angle at which the spring becomes unstretched.

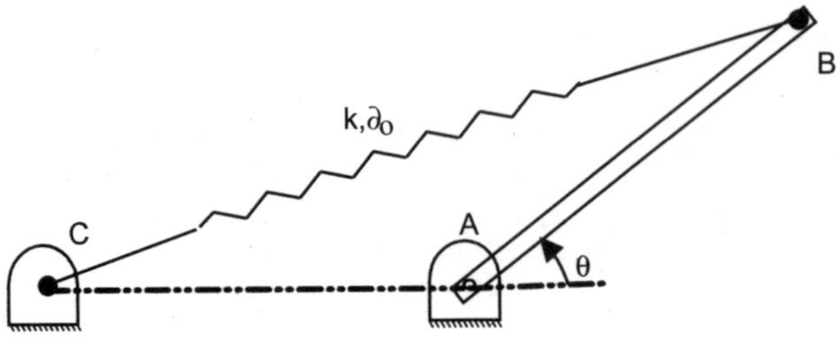

Fig. ED5.22

Solution: This problem can be solved by energy principle in a rigid-body as it requires to relate displacement, angular velocity and force. As there is no friction and drag, the principle of conservation of energy can be conveniently applied.

That is: $T_1 + V_1 = T_2 + V_2$

where T_1 = Initial kinetic energy $= 0$ (rod is released from rest)

T_2 = Final kinetic energy $= \frac{1}{2}I_A\omega^2 = \frac{1}{2}\left(\frac{1}{3}m\ell^2\right)\omega^2 = \frac{1}{6}m\ell^2\omega^2$

V_1 = Initial potential energy $= V_{g1} + V_{e1} = mg\,h_1 + \frac{1}{2}kx_1^2$

$$= mg\frac{\ell}{2}\sin\theta_1 + \frac{1}{2}k\left(\ell\sqrt{2(1+\cos\theta_1)} - \partial_o\right)^2 \text{ (since } x = CB - \partial_o\text{)}$$

Here *CB* is obtained from the triangle *ABC* using cosine rule.

V_2 = Final potential energy $= V_{g2} + V_{e2} = mg\,h_2 + \frac{1}{2}kx_1^2$

$$= mg\frac{\ell}{2}\sin\theta_2 + \frac{1}{2}k\left(\ell\sqrt{2(1+\cos\theta_1)} - \partial_o\right)^2$$

Now applying the principle we get:

$$\frac{1}{6}m\ell^2\omega^2 = mg\frac{\ell}{2}\sin\theta_1 + \frac{1}{2}k\left(\ell\sqrt{2(1+\cos\theta_1)} - \partial_O\right)^2 - mg\frac{\ell}{2}\sin\theta_2 + \frac{1}{2}k\left(\ell\sqrt{2(1+\cos\theta_2)} - \partial_O\right)^2$$

Taking the values of $m = 25$ kg, $\ell = 0.4$m, $\theta_1 = 10^o$, $\partial_o = 0.3$ m and $k = 500$ N/m, we can study the variation of ω with the angle θ_2.

Following MATLAB script plots the variation.

```
%%%%%%%%%%%%%%%%%%%
m=25;L=0.4;delta=0.05;k=500;g=9.81;
```

```
IA=m*L^2/6;
th1=40*pi/180;
theta2=[40:10:160];
th2=theta2*pi/180;
V1=0.5*m*g*L*sin(th1)+0.5*k*(L*sqrt(2*(1+cos(th1)))-delta)^2;
V2=0.5*m*g*L.*sin(th2)+0.5*k*(L*sqrt(2*(1+cos(th2)))-delta).^2;
omega=sqrt((V1-V2)/IA);
plot(theta2,omega,'-p');
xlabel('\theta(degree)')
ylabel('\omega(rad/s)');
grid on;
%%%%%%%%%%%%%%%%%%%%
```

Output is shown in Fig. ED5.22 (*a*).

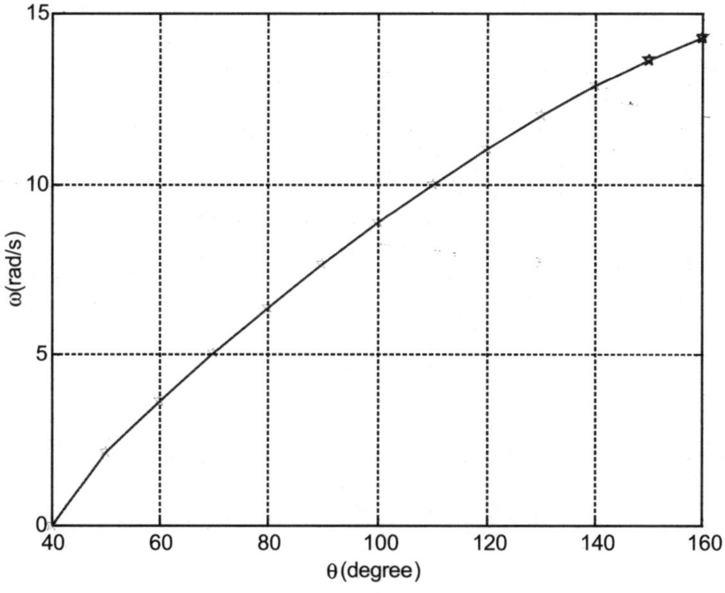

Fig. ED5.22 (a) MATLAB output

Here it can be observed that sometimes the term $V_1 - V_2$ becomes negative and it is not possible to compute the angular velocity of the rod. This situation represents the ideal situation.

Angle at which spring becomes unstretched is given by solving the equation:

$$\left(\ell\sqrt{2(1+\cos\theta_2)} - \partial_o\right) = 0$$

$$\Rightarrow \qquad \left(0.4\sqrt{2(1+\cos\theta_2)} - 0.05\right) = 0$$

This equation can be solved by MATLAB by just typing the following equations:

```
eqn = '0.4*sqrt(2*(1+x))-0.05';
x = solve(eqn);
theta = acos(x)*180/pi
```

Output is theta = 172.8334 degree

ED5.23: A 100 kg spool has a radius of gyration $k_G = 0.35$ m. A cable is wrapped around the hub of the spool (see Fig. ED5.23) and horizontal force $P = (2 + e^{-3t})$ is applied. Determine its angular velocity as a function of time $0 \le t \le 10$s, if spool starts from rest initially. Assume that there is no slipping at the floor during rolling.

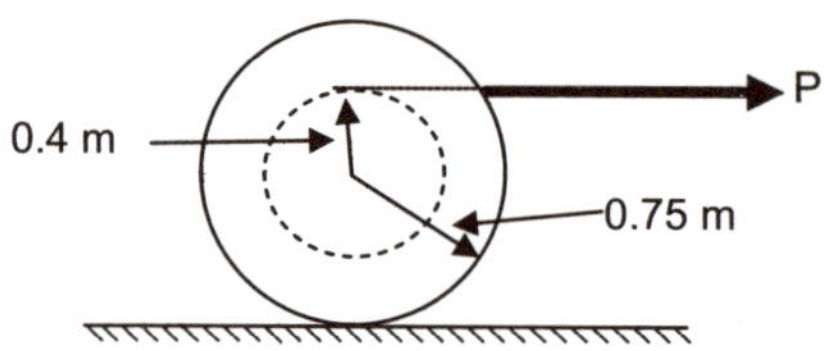

Fig. ED5.23

Solution: Here it requires to relate the velocity, time and force. Thus, impulse-momentum principle can be applied. For obtaining the impulse, free-body diagram is drawn as shown in Fig. ED5.23(*a*). The forces causing impulse are *P* and friction *f*. So the linear impulse of a force is given by:

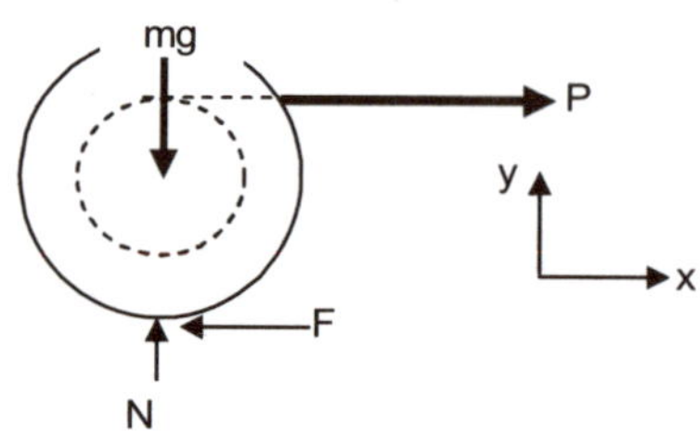

Fig. ED5.23 (a) Free-body diagram

$$\Sigma I = \int_0^t (P - F)dt = \int_0^t (2 + e^{-3t} - F)dt$$

Also angular impulse is given by:

$$\Sigma A = \int_0^t (P \times 0.4 + F \times 0.75)dt = \int_0^t \left(\{2 + e^{-3t}\} \times 0.4 + F \times 0.75\right)dt$$

The standard equations of impulse and momentum are written as follows:

$$m(v_G)_1 + \sum I = m(v_G)_2 \Rightarrow 0 + \int_0^t (2 + e^{-3t} - F)dt = m(v_G)_2 \Rightarrow 2t - \frac{1}{3}e^{-3t} - Ft = mv_G \quad ...(1)$$

$$I_G(\omega)_1 + \sum A = I_G(\omega)_2 \Rightarrow 0 + \int_0^t \left(\{2 + e^{-3t}\} \times 0.4 + F \times 0.75\right)dt = I_G\omega_2$$

$$\Rightarrow \left(0.8t - \frac{0.4}{3}e^{-3t} + 0.75Ft\right) = I_G\omega \quad ...(2)$$

Also as the disk is rolling without slipping $v_G = 0.75\omega$ ...(3)

Solving equations (1)–(3) in terms of time variable *t*, we get:

$$\omega = \frac{2.3t - 0.1167e^{-3t}}{I_G + 0.75^2 m}$$

Following MATLAB script is written based on this final relation.

```
%%%%%%%%%%%%%%%%%%%%%%%
m=100;g=9.81;kG=0.35;
IG=m*kG^2;
t=0:2:100;
omega=(2.3.*t-0.1167.*exp(-3.*t))/(IG+m*0.75^2);
plot(t,omega,'-p');
xlabel('time(s)');
ylabel('\omega(rad/s)');grid on;
%%%%%%%%%%%%%%%%%%%%%%%
```

Output is shown in Fig. ED5.23 (*b*).

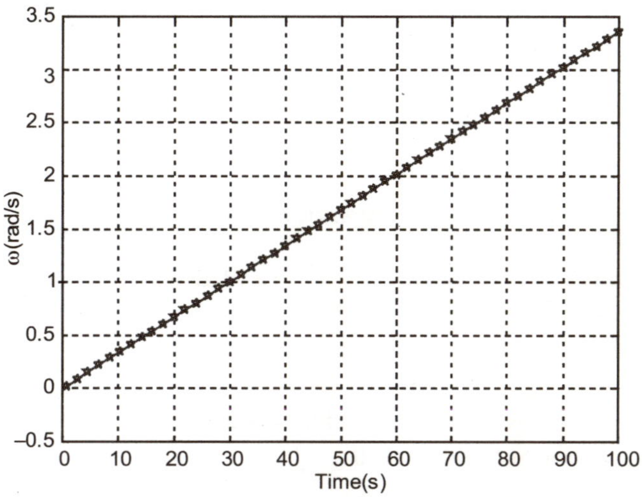

Fig. ED5.23 (b) MATLAB output

ED5.24: Figure ED5.24 shows a box of mass *m* pulled by a rope. The force required to move the box is given by

$$F = \frac{\mu mg}{\cos\theta + \mu \sin\theta}$$

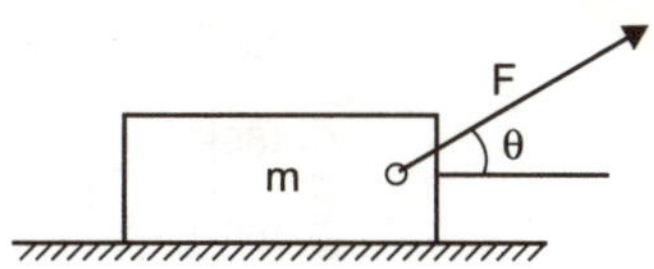

Fig. ED5.24

where μ = coefficient of friction
m = mass of the box
g = acceleration due to gravity (9.81 m/s^2)
θ = angle with horizontal.

If mass of box, $m = 25$ kg and coefficient of friction, $\mu = 0.40$, write a MATLAB program to determine the angle θ when the pulling force F is 100 N. Also write a MATLAB program to determine (*a*) the angle θ at which the force required to pull the box is the smallest, (*b*) the magnitude of the force in (*a*).

Solution: Entire data and equations are given. It requires to solve the equation for θ.

The following MATLAB program is self explanatory.

```
%%%%%%%%%%%%%%%%%%%%%%
eqn='100-0.4*25*9.81/(x+0.4*sqrt(1-x^2))';
x=solve(eqn);
theta=acos(x)*180/pi;
fprintf('Angle theta in degrees=%f\n',eval(theta))
%%%%%%%%%%%%%%%%%%%%%
```

Output is as follows:

» Angle theta in degrees=46.180323

Smallest force can be arrived by varying θ over a range.

For this MATLAB program is shown below:

```
%%%%%%%%%%%%%%%%%%%%%
m=25; g=9.81;mu=0.4;
theta=[0:0.1:90];
th=theta*pi/180;
F=mu*m*g./(cos(th)+mu*sin(th));
[F_min,i]=min(F);
th_min=th(i);
fprintf('Minimum force %f occurs at angle %f
degrees\n',F_min,th_min*180/pi);
plot(theta,F);xlabel('\theta(degrees)');ylabel('Force(N)');
grid on;
%%%%%%%%%%%%%%%%%%%%%
```

Output is as follows:

» Minimum force 91.083563 occurs at angle 21.800000 degrees.

Figure ED5.24(a) shows variation of F with angle θ.

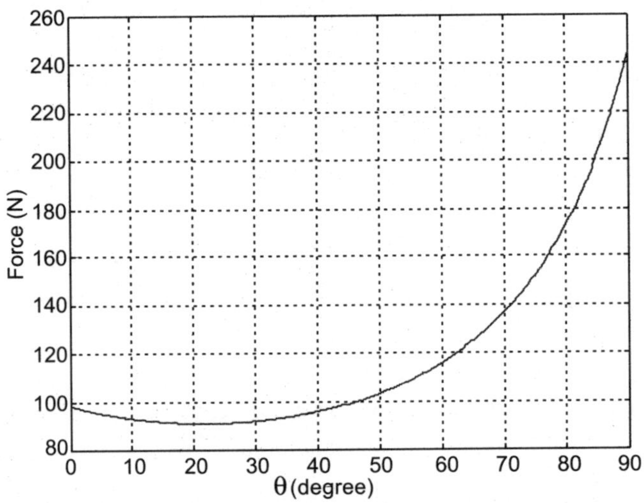

Fig. ED5.24 (a) MATLAB output

5.2.5 Three-dimensional Motion of Rigid Bodies

As presented in Chapter-1, a more general case of motion is in three dimensions. Kinematic analysis of robotic limb motion and kinetics of gyroscopes are some examples. Rotors revolving at high speed such as spindle of machine tools often require consideration of gyroscopic effect in dynamic analysis. The following simple examples illustrate the methodology.

ED5.25: In Fig. ED5.25 shown, the rod *BD* is rotating about vertical axis with an angular velocity of 7 rad/s and acceleration 4 rad/s^2. If the limb *AC* is rotating downwards, such that its angular speed $\dot{\theta} = 2$ rad/s and acceleration $\ddot{\theta} = 3$ rad/s^2 in clockwise sense. Plot the variation of velocity and acceleration of point *A* on the limb as a function of angle θ. Given length of $AC = 0.8$ m.

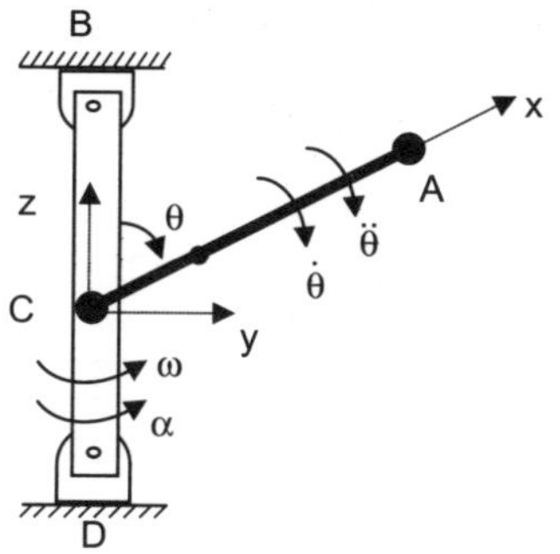

Fig. ED5.25

Solution: This is an example of general motion of rigid-body in three-dimensions for which the angular velocity is ω and angular acceleration is α. General equations for velocity and acceleration of point *A* are given as follows:

$$v_A = v_C + \omega \times CA \quad ...(1)$$

and
$$a_A = a_C + \alpha \times CA + \omega \times (\omega \times CA) \quad ...(2)$$

Here the point *C* is hinged so there is no motion. Thus $v_C = a_C = 0$.

By choosing *X, Y, Z* (fixed coordinate) frame coinciding at point *D* and *x, y, z* frame at *C* with *x*-axis parallel to *CA* and *z*-axis parallel to *BD*, the vectors α and ω are defined as follows:

$$\omega = \omega_x\hat{i} + \omega_y\hat{j} + \omega_z\hat{k} = -\dot{\theta}\hat{i} + 0\hat{j} + \omega\hat{k},$$

$$\alpha = -\ddot{\theta}\hat{i} - \omega\dot{\theta}\hat{j} + \alpha\hat{k}$$

With $CA = l\sin\theta\hat{j} + l\cos\theta\hat{k}$, equations (1) and (2) will give the velocity of *A* as a function of angle θ.

Complete MATLAB program is given below:

```
%%%%%%%%%%%%%%
%Initialization of variables
om=7;td=2;tdd=3;al=4;le=0.8;
i=1;
for theta=0:5:180
```

```
t=theta*pi/180;
speed=[-td 0 om]; % defining omega as vector
ace=[-tdd -om*td al]; % defining alpha as vector
CA=[0 le*sin(t) le*cos(t)]; % position vector CA is defined
vA(i)=norm(cross(speed,CA));
aA1=cross(ace,CA); aA2=cross(speed,cross(speed,CA));
aA(i)=norm(aA1+aA2);
i=i+1;
end
T=[0:5:180];
subplot(2,1,1);
plot(T,vA,'-p');
xlabel('angle theta');
ylabel('velocity of A in m/s');
grid on;
subplot(2,1,2);
plot(T,aA,'-*');
xlabel('Angle theta');
ylabel('Acceleration of A in m/s^2');
grid on;
%%%%%%%%%%%%%%%%%%%%%%%%%%%%%%%%
```

Output is shown in Fig. ED5.25(*a*).

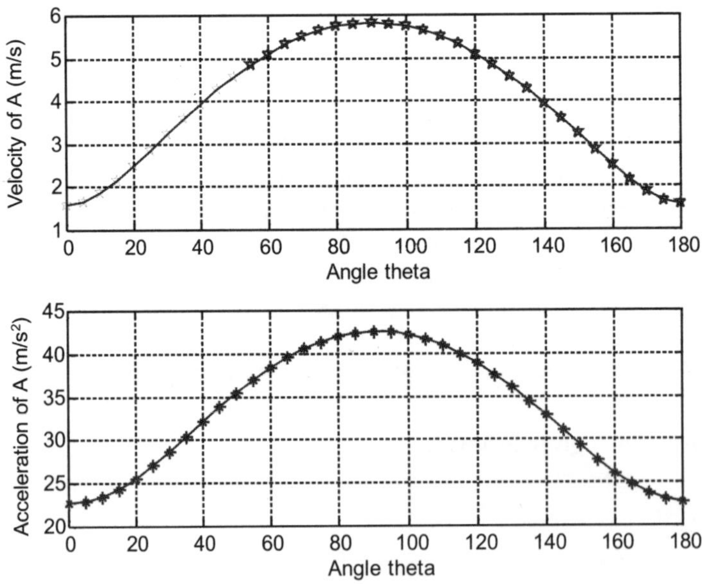

Fig. ED5.25(a) MATLAB output

ED5.26: Figure ED5.26 shows a thin homogeneous disk of mass 10 kg and radius 0.3 m mounted centrally on the horizontal axle AB. The constant angular velocity of the axle is 30 rad/s. The disk is laid at an angle θ with the plane normal to the rotating axle. If the axle has negligible mass, write a MATLAB program (for $0° \leq \theta \leq 90°$) to (***a***) determine and plot θ formed by the axle as a function of angular momentum of the disk about C, (***b***) the kinetic energy of the disk, (***c***) the magnitude of the rate of change of the angular momentum of the disk about C.

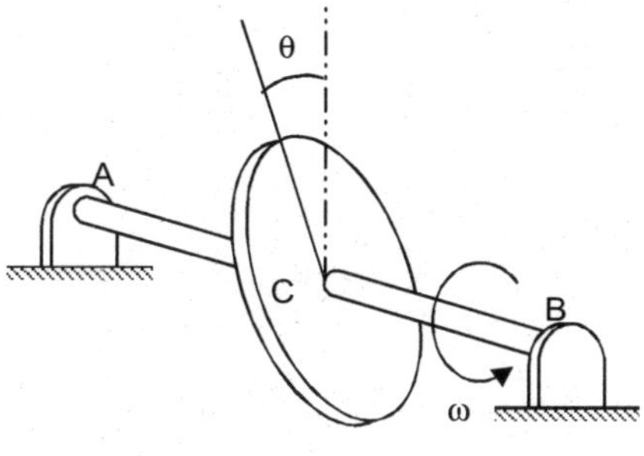

Fig. ED5.26

Solution: (*a*) The origin of the x, y, z coordinate system is located at the disk's center of mass C, which is also a fixed point. The axes are fixed in and rotate with the disk, since these axes will then always represent principal axes of inertia for the disk. Hence $\Omega = \omega$. Figure ED5.26 (*a*) shows the free-body diagram.

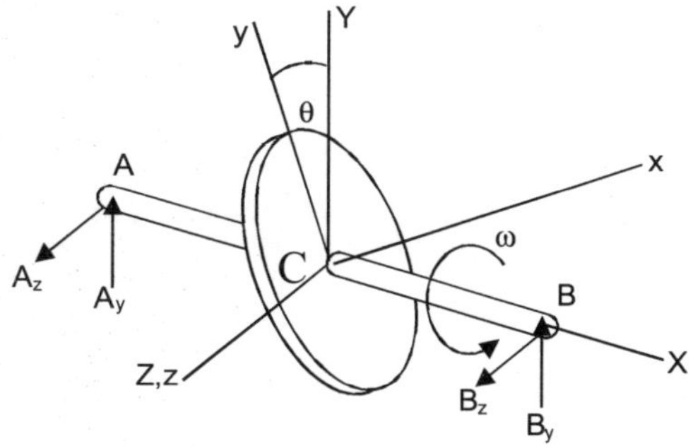

Fig.ED5.26 (a)

The angular velocity of disk is constant in magnitude and is always directed along the axis of the shaft AB. Since this vector is measured from X, Y, Z inertial frame of reference, for any position of x, y and z axes:

$$\omega_x = 30 \sin\theta,\ \omega_y = 30 \cos\theta \text{ and } \omega_z = 0.$$

For a disk, moment of inertia is $I_x = mr^2/4$ and $I_y = I_z = mr^2/2$. Then angular momentum with respect to the centroidal frame $Cxyz$ is:

$$H_C = I_x\omega_x \hat{i} + I_y\omega_y \hat{j} + I_z\omega_z \hat{k} \quad \text{...(1)}$$

where $\hat{i}, \hat{j}$ and $\hat{k}$ are unit vectors along x, y and z axis respectively. They are related to $\boldsymbol{I}, \boldsymbol{J}, \boldsymbol{K}$ unit vectors along X, Y and Z axis respectively according to the following equations:

$$\hat{i} = \sin\theta\, \boldsymbol{I} - \cos\theta\, \boldsymbol{J},\quad \hat{j} = \cos\theta\, \boldsymbol{I} + \sin\theta\, \boldsymbol{J}.$$

Substituting all in (1), we get

$$H_C = \frac{mr^2\omega}{4}\left[(1-\cos^2\theta)I + \sin\theta\cos\theta\, J\right] \quad \text{...(2)}$$

(*b*) Kinetic energy $T = \frac{1}{2}\omega \cdot H_C = \frac{1}{2}\left(I_x\omega_x^2 + I_y\omega_y^2 + I_z\omega_z^2\right) = \frac{mr^2\omega^2}{8}\left(1+\cos^2\theta\right)$ ...(3)

(*c*) The frame $CXYZ$ rotates with the disk with an angular velocity given by $\Omega = \omega = \omega \boldsymbol{I}$

Since ω = constant rate of change of angular momentum about C along XYZ frame is zero.

Hence $$\dot{H}_C = \left(\dot{H}_C\right)_{CXYZ} + \Omega \times H_C = 0 + \omega \boldsymbol{I} \times \frac{mr^2\omega}{4}\left[(1-\cos^2\theta)\boldsymbol{I} + \sin\theta\cos\theta\,\boldsymbol{J}\right]$$

$$= \frac{mr^2\omega^2}{4}[\sin\theta\cos\theta]\boldsymbol{k} \quad ...(4)$$

Complete MATLAB program for formulation based on equations (1) to (4) is given below:

```
%%%%%%%%%%%%%%%%%%%%
m=10; om=30;r=0.3;
theta=[0:10:90];
t=theta*pi/180;
Hc=0.25*m*r^2*om*sqrt((1-cos(t).^2)+(sin(t).*cos(t)).^2);
T=0.25*m*r^2*om^2*(1+cos(t).^2)/8;
Hcd=0.25*m*r^2*om^2* sin(t).*cos(t);
z=[theta;Hc;T;Hcd];
fprintf('Theta(deg)\tHc\t\tT\t\tHcd\n');
fprintf('%f\t%f\t%f\t%f\n',z);
subplot(3,1,1);plot(theta,Hc,'-+');ylabel('Hc(kg-m^2/s) of disk');
grid on;
subplot(3,1,2); plot(theta,T,'-p');ylabel('T(J) of disk'); grid on;
subplot(3,1,3); plot(theta,Hcd,'-*');ylabel('Hc-dot');
grid on;
%%%%%%%%%%%%%%%%%%%%
```

Output is shown below:

» Theta(deg)	Hc	T	Hcd
0.000000	0.000000	50.625000	0.000000
10.000000	1.645092	49.861735	34.629540
20.000000	3.167985	47.664000	65.082245
30.000000	4.464705	44.296875	87.685072
40.000000	5.465573	40.166485	99.711785
50.000000	6.146896	35.771015	99.711785
60.000000	6.535659	31.640625	87.685072
70.000000	6.703658	28.273500	65.082245
80.000000	6.746931	26.075765	34.629540
90.000000	6.750000	25.312500	0.000000

Figure ED5.26 (*b*) shows the plot of momentum and energy of the disk about *C*.

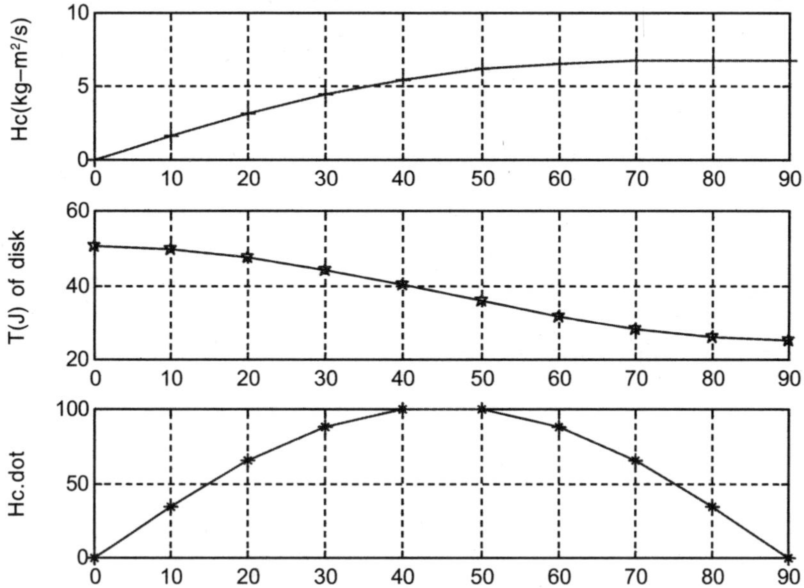

Fig. ED5.26(b) MATLAB output

ED5.27: The top of mass 0.5 kg shown in Fig.ED5.27 precessing about the vertical axis at a constant angle θ. If it spins with an angular velocity ω_s = 100 rad/s, plot the precessional velocity ω_p as a function of angle θ. Take the axial and transverse moments of inertia of the top as 0.45 mg-m^2 and 1.2 mg-m^2 respectively measured with respect to the fixed point *O*. Take *OG* = 5 cm.

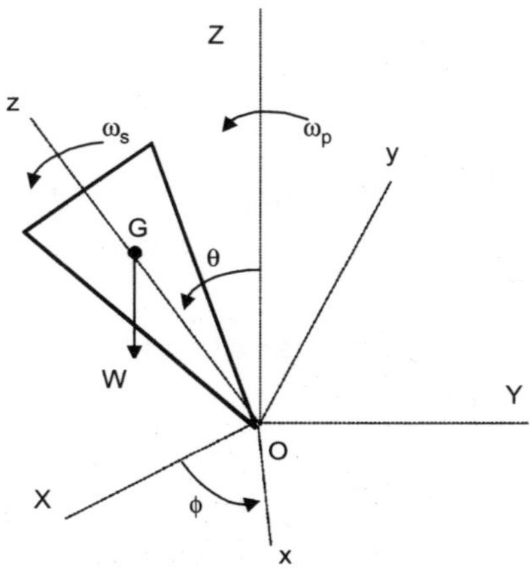

Fig. ED5.27

Solution: This example illustrates the use of MATLAB in computing motions of a spinning top. These are of particular interest in a device called gyroscope. Here body motion is analyzed using Euler's angles φ, θ, ψ. When θ and rate of change of angles φ and ψ are constant, this special case is called steady precision. Thus the moment about *x*-axis is

$$\Sigma M_x = -I\dot{\phi}^2 \sin\theta\cos\theta + I_z\dot{\phi}\sin\theta(\dot{\phi}\cos\theta + \dot{\psi})$$

$$\Rightarrow \qquad W \times OG \sin\theta = -1.2e-3 \sin\theta \cos\theta + 0.45e-3 \sin\theta\,(\cos\theta\,\dot{\phi}^2 + 100\,\dot{\phi})$$

$$\Rightarrow \qquad 0.5g \times 0.05 \sin\theta = -1.2e-3 \sin\theta \cos\theta + 0.45e-3 \sin\theta\,(\cos\theta\,\dot{\phi}^2 + 100\,\dot{\phi})$$

Here for any given θ, find using MATLAB solve command.

```
%%%%%%%%%%%%%%%%
m=0.5;g=9.81;OG=0.05;
eqn='0.5*9.81*0.05*sin(t)+1.2e-3*sin(t)*cos(t)*f^2-0.45e-3*sin(t)
*cos(t)*f^2-0.045*sin(t)*f=0'
solve(eqn,'f') % finding f in terms of t
%%%%%%%%%%%%%%%%%%%
```

Output is as follows:

eqn =

0.5*9.81*0.05*sin(*t*)+1.2e–3*sin(*t*)*cos(*t*)**f*^2– 0.45e–3*sin(*t*)*cos(*t*)**f*^2–0.045*sin(*t*)**f*=0

ans =

[1/2/cos(*t*)*(60+2*(900–327*cos(*t*))^(1/2))]
[1/2/cos(*t*)*(60–2*(900–327*cos(*t*))^(1/2))]

It is obvious to have two values for a quadratic equation. They practically represent low precision and high precision conditions. Usually low precision is more commonly observed.

A plot of the precision velocity as a function of angle θ is obtained as follows:

```
%%%%%%%%%%%%% This script plots precision velocity as a function of angle theta%%%%
theta=[0:10:80];
t=theta*pi/180;
   flow= 1./2./cos(t).*(60-2.*(900-327.*cos(t)).^(1./2));
   fhigh= 1./2./cos(t).*(60+2.*(900-327.*cos(t)).^(1./2));
subplot(2,1,1)
plot(theta,flow,'-*');ylabel('\omega_s(rad/s)'); grid on;
title('\bflow-precisional speed')
subplot(2,1,2)
plot(theta,fhigh,'-p');ylabel('\omega_s(rad/s)'); xlabel('Theta deg)');grid on;
title('\bfhigh-precisional speed')
%%%%%%%%%%%%%%%%%%%%%%%%%%%%%
```

Output is shown in Fig.ED5.27 (*a*)

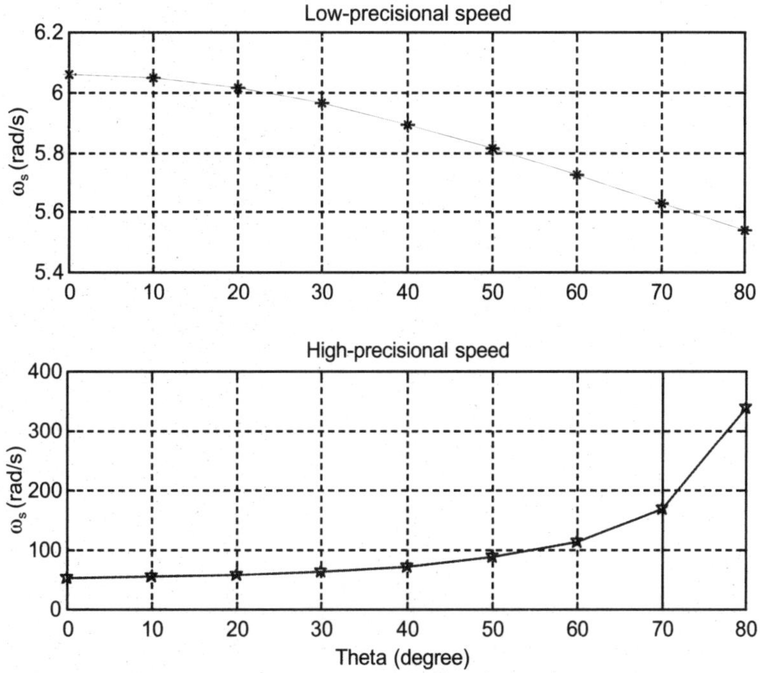

Fig. ED5.27(a) MATLAB output

5.3 SUMMARY

In this Chapter, we have presented a wide variety of illustrative worked examples of applications of MATLAB to engineering mechanics.

APPENDIX

APPENDIX—I

Centroids and Moments of Inertia

Following geometric properties of area elements are often used in Mechanics practice.

Geometry	Centroid	Area Moment of inertia
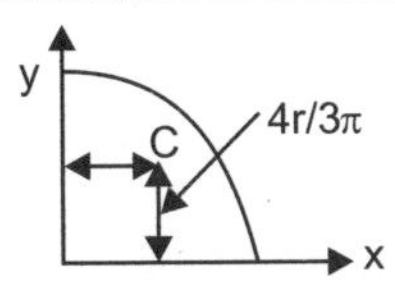 Quarter circle area	$\frac{4r}{3\pi}$ from either axis, where r is radius	$I_x = \frac{\pi r}{16}$ and $I_y = \frac{\pi r^4}{16}$
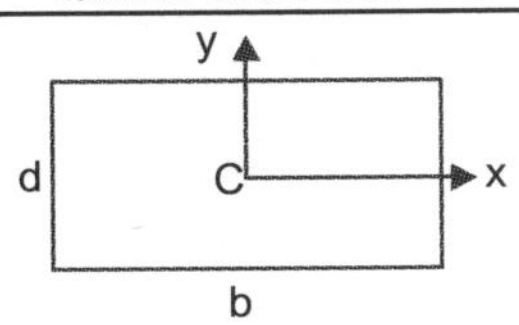 Rectangle	At the origin of coordinate frame	$I_x = \frac{bd^3}{12}$ and $I_y = \frac{db^3}{12}$
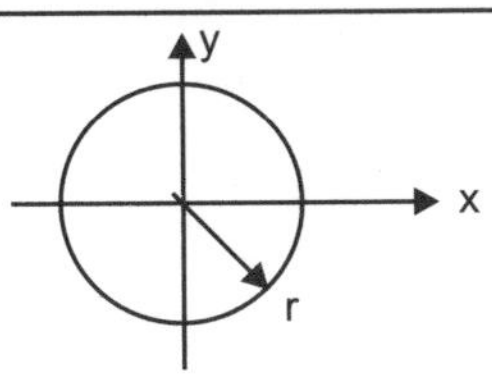 Circle	At the origin of coordinate frame	$I_x = \frac{\pi r^4}{4}$ and $I_y = \frac{\pi r^4}{4}$
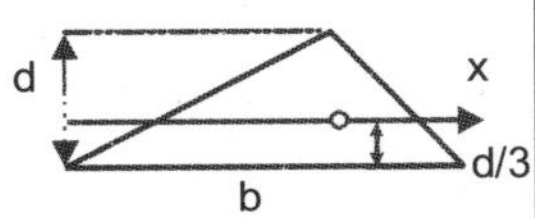 Triangle	At a height $d/3$ from base	$I_x = \frac{bd^3}{36}$

Name of Solid	I_{xx}	I_{yy}	I_{zz}
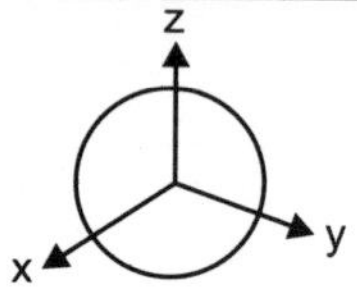 Solid sphere	$\frac{2}{5}mr^2$	$\frac{2}{5}mr^2$	$\frac{2}{5}mr^2$
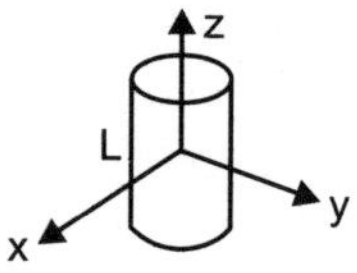 Cylinder	$\frac{1}{12}m(3r^2+L^2)$	$\frac{1}{12}m(3r^2+L^2)$	$\frac{1}{2}mr^2$
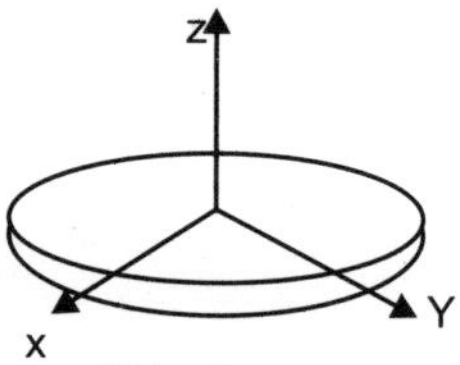 Thin circular disk	$\frac{1}{4}mr^2$	$\frac{1}{4}mr^2$	$\frac{1}{2}mr^2$
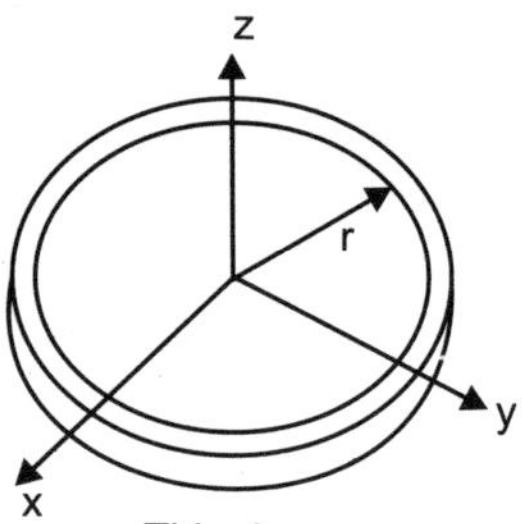 Thin ring	$\frac{1}{2}mr^2$	$\frac{1}{2}mr^2$	mr^2
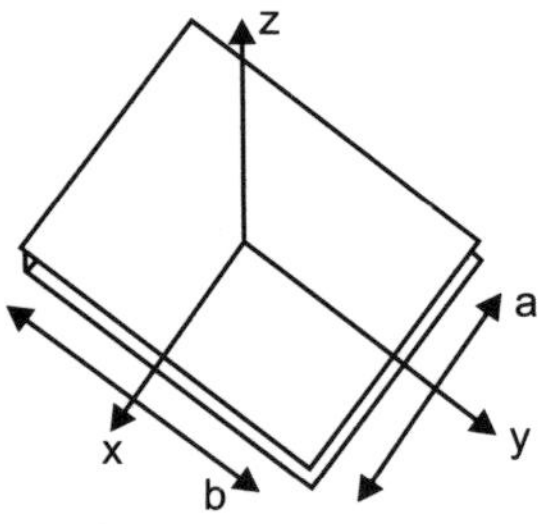 Thin rectangular plate	$\frac{1}{12}mb^2$	$\frac{1}{12}ma^2$	$\frac{1}{12}m(a^2+b^2)$
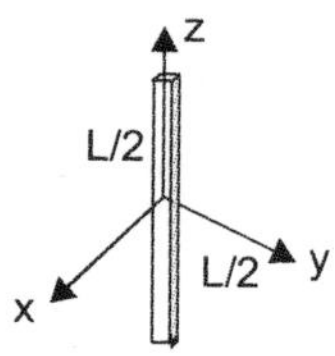 Slender rod	$\frac{1}{12}mL^2$	$\frac{1}{12}mL^2$	0

APPENDIX—II

Vibration Engineering–Application Of Dynamics

Important application of dynamics is vibration engineering. Vibrations are periodic motion of a body or system of connected bodies displaced from a position of equilibrium. The simplest type of vibrating motion is undamped free vibration. Free-vibrations are characterized by natural frequency of the system and corresponding mode of oscillation. Natural frequency is the number of oscillations made by the system in one second. It is an important dynamic parameter of a system, because it decides the condition of a violent vibrating state called 'resonance'. The natural frequency of a rigid body having one degree of freedom can be determined in three steps: (1) Draw free-body diagram.(2) Apply equations of motion. (3) Use kinematics to express body's accelerated motion in terms of second time derivative of position. In conservative systems, the energy method can be used as alternative approach for deriving equations of motions of vibrating system. The principle is based on concept that 'total energy will remain constant', i.e., $T+V=K$, or $\frac{d}{dt}(T+V)=0$. The expressions for T and V are written in terms of the displacement (also called degree of freedom) and its derivatives. In vibration engineering, often spring forces are considered as internal forces. In many cases damping is attributed to resistance created by the substances like water, oil or air in which system vibrates. The resistance to motion is directly proportional to body's speed and is called viscous damping force.

Fig. AII.1 shows a general one degree of freedom (also called lumped-parameter) model.

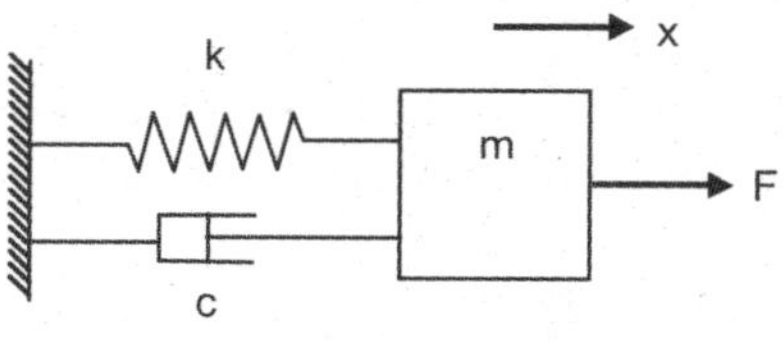

Fig. AII.1 One degree of freedom vibration model

The complete dynamic equation of motion for the forced vibrating system is given by

$$m\ddot{x}+c\dot{x}+kx=F=F_0\sin\Omega t$$

$$\text{or} \quad \ddot{x}+\frac{c}{m}\dot{x}+\frac{k}{m}x=\frac{F_0}{m}\sin\Omega t$$

Here $\sqrt{\frac{k}{m}}=\omega_n$ is called natural frequency and $\frac{c}{m}=2\xi\omega_n$, where ξ called damping ratio is a non-dimensional parameter taken for convenience. Since the equation is non-homogeneous, the general solution is sum of a complementary function x_c and particular integral x_p. The complementary function is determined by setting right hand side equal to zero and solving the homogeneous differential equation. The solution depends on the value of damping ratio ξ. If ≥ 1, the motion is aperiodic and when $\xi<1$, the motion is called underdamped vibration. The solution for such a state of motion is given as follows:

$$x_c = A\cdot e^{-(c/2m)t}\sin(\omega_d t+\theta)=A\,e^{-\xi\omega_n t}\sin(\omega_d t+x) \quad \text{for } \xi<1 \qquad \text{...(A1)}$$

$$=(A_1+A_2 t)\,e^{-\omega_n t} \quad \text{for } \xi=1 \qquad \text{...(A2)}$$

$$= A_1 e^{\left(-\xi+\sqrt{\xi^2-1}\right)\omega_n t} + A_2 e^{\left(-\xi+\sqrt{\xi^2-1}\right)\omega_n t} \text{ for } \xi<1 \quad ...(A3)$$

where $\omega_d = \omega_n \sqrt{1-\xi^2}$ is called damped natural frequency defined for under-damped system. A and θ or A_1 and A_2 are obtained from initial conditions of displacement and velocity. On the other hand particular integral describes the steady-state vibration of the system. It can be written as:

$$x_p = (c_1 \cos \Omega t + c_2 \sin \Omega t) = X_o \sin(\Omega t - \phi) \quad ...(A4)$$

where $$X_o = \frac{F_0/k}{\sqrt{\left(1-(\Omega/\omega_n)^2\right)^2 + (2\xi\Omega/\omega_n)^2}} \quad ...(A5)$$

is amplitude of steady state response and $\phi = \tan^{-1} \dfrac{2\xi(\Omega/\omega_n)}{1-(\Omega/\omega_n)^2}$ is the phase difference between the applied force and the resulting steady-state vibration of the damped system. Often the steady state response is written in non-dimensional form called 'magnification factor', which is the ratio of X_0 and static displacement F_0/k. Also the term Ω/ω_n is called frequency ratio. Force transmitted to the base can be written as:

$$F(t) = c(\dot{x} - \dot{y}) + K(x-y) = F_T \cos(\Omega t + \varphi) \quad ...(A6)$$

where $$\frac{F_T}{kY} = (\Omega/\omega_n)^2 \frac{\sqrt{1+(2\xi\Omega/\omega_n)^2}}{\sqrt{\left(1-(\Omega/\omega_n)^2\right)^2 + (2\xi\Omega/\omega_n)^2}}, \quad \varphi = \tan^{-1}(2\xi\Omega/\omega_n) \quad ...(A7)$$

Systems with multiple degrees of freedom can also be attempted on the same lines. Some applications of steady-state vibration analysis include unbalance response of rotors, base excited single-degree of freedom system and modelling of measuring devices such as seismometer and accelerometer. Brief introduction is given below:

Base excitation: Many times, force is applied at the base instead of at the mass for simulations. In such case, equation of motion is $m\ddot{x} + c(\dot{x} - \dot{y}) + k(x-y) = 0$ ...(A8)

where excitation function is $y(t) = Y \sin \Omega t$,

And the response: $x(t) = X\sin(\Omega t + \phi)$ ...(A9)

Response in non-dimensional form is called displacement transmissibility and can be written as:

$$\frac{X}{Y} = \frac{\sqrt{1+(2\xi\Omega/\omega_n)^2}}{\sqrt{\left(1-(\Omega/\omega_n)^2\right)^2 + (2\xi\Omega/\omega_n)^2}} \quad ...(A10)$$

Rotating unbalance: In rotor-dynamic systems, commonly disk are attached to rotating shafts. In such cases the eccentricity e of mass center of disk with the geometric center creates centrifugal forces and equations of motion are given as follows:

$$m\ddot{x} + c\ddot{x} + k\ddot{x} = m_0 e\Omega^2 \sin \Omega t \quad \text{...(A11)}$$

Here m_0 corresponds to mass of disk. Response is given as: $x(t) = X \sin(\omega t - \phi)$, where

$$\frac{mX}{m_0 e} = \frac{(\Omega/\omega_n)^2}{\sqrt{\left(1-(\Omega/\omega_n)^2\right)^2 + (2\xi\Omega/\omega_n)^2}} \text{ and } \phi = \tan^{-1}\frac{2\xi(\Omega/\omega_n)}{1-(\Omega/\omega_n)^2} \quad \text{...(A12)}$$

Measuring devices: They are also modelled as base excited systems. Thus equations of motion is:

$$m\ddot{x} = -c(\dot{x} - \dot{y}) - k(x-y) \quad \text{...(A13)}$$

Let the motion of mass relative to the base be $z(t) = x(t) - y(t)$

Then $$m\ddot{z} + c\dot{z} + kz = -my \quad \text{...(A14)}$$

If $y(t) = Y\cos \omega t$, equation becomes:

$$m\ddot{z} + c\dot{z} + kz = m\omega^2 Y \cos \omega t \quad \text{...(A15)}$$

The solution takes the form: $z(t) = Z\cos(\omega t - \phi)$, where

$$\frac{Z}{Y} = \frac{r^2}{\sqrt{(1-r^2)^2 + (2\xi r)^2}} \text{ and } \varphi = \tan^{-1}\frac{2\xi r}{1-r^2} \text{ for } r = \omega/\omega_n \quad \text{...(A16)}$$

The device becomes an accelerometer for low frequencies ($r < 0.2$) and a seismometer ($r > 3$) for high frequencies. Following illustrative examples demonstrate the application of MATLAB in simple vibration problems.

Example A.1

A motor of weight W is supported by four springs with a spring constant of 400 N/m each. The unbalance of the motor is found to be equivalent to a weight of 0.05 N located 5 cm from the axis of rotation. If the motor is constrained to move vertically, write a MATLAB program to calculate and plot the amplitude of the vibration and the maximum acceleration of the motor for motor speeds from 0 to 1200 rpm. The weight of the motor $W = 35$ N.

Solution: This is a problem from applications of dynamics (vibrations).

Amplitude of response due to harmonic load P_m is given by $x_m = \dfrac{P_m/k}{1-(\omega_f/\omega_n)^2}$ ...(1)

where $P_m = mr\omega_f^2 = (0.05/g) \times (5/100) \times \omega_f^2 = 2.54 \times 10^{-5}\omega_f^2$

$k = 4(400) = 1600$ N/m

Natural frequency of the system ω_n is found as follows:

$$\omega_n^2 = \left(\frac{k}{m}\right) = (1600 \times 9.81/35) = 453.02 \text{ or } \omega_n = 21.28 \text{ rad/s}$$

Substituting values into (1)

$$x_m = \frac{2.54 \times 10^{-5}\omega_f^2}{1600\left(1 - \dfrac{\omega_f^2}{453.02}\right)} \text{ m}$$

Acceleration $a_m = \omega_f^2\, x_m$ m/s^2

where $\omega_f = (\text{rpm})\left(\dfrac{2\pi}{60}\right)$

To plot the response, this forced frequency ω_f is varied over a range [0 – 400 rpm]

MATLAB program:

```
W=35;          % Weight of motor
g=9.81;        % Acceleration due to gravity
mW=W/g;        % Mass of motor in kg
ks=400;        % one spring stiffness
m=0.05/g;      % mass of unbalance in kg
r=5/100;       % radius of rotation in m
k=4*ks;        % total stiffness of all springs
omega_n2=k/mW; % square of natural frequency
omega_f=[0:10:400]; % variation of motor speeds in RPM
omeg=omega_f*2*pi/60;
Pm=m*r*omeg.^2;% Unbalance force due to rotation
xm=(Pm/k)./(1-omeg.^2/omega_n2);
am=omeg.^2.*xm; % acceleration in m/s^2
amp=xm*100;    % amplitude in cm
z=[omega_f;amp;am];
fprintf('omega_f(rpm)     Amplitude(cm.)     Acceleration(m/s^2)\n');
fprintf('\n');
fprintf('%5.0f                        %f                  %f\n',z);
figure(1)
plot(omega_f,amp, '-p')
xlabel('Motor Speed (rpm)')
ylabel('Amplitude of Vibration(cm)')
grid on;
figure(2)
plot(omega_f,am, '-+')
xlabel('Motor Speed (rpm)')
```

```
ylabel('Maximum Motor Acceleration(m/s^2)')
grid on;
```

Output is as follows:

omega_f(rpm)	Amplitude(cm.)	Acceleration(m/s^2)
0	0.000000	0.000000
10	0.000018	0.000000
20	0.000071	0.000003
30	0.000161	0.000016
40	0.000291	0.000051
50	0.000465	0.000128
60	0.000689	0.000272
70	0.000972	0.000523
80	0.001325	0.000930
90	0.001764	0.001567
100	0.002312	0.002535
110	0.003002	0.003983
120	0.003882	0.006131
130	0.005031	0.009324
140	0.006575	0.014131
150	0.008737	0.021558
160	0.011956	0.033564
170	0.017210	0.054544
180	0.027245	0.096803
190	0.053783	0.212917
200	0.319452	1.401275
210	–0.098265	–0.475221
220	–0.046061	–0.244475
230	–0.031473	–0.182581
240	–0.024628	–0.155565
250	–0.020663	–0.141619
260	–0.018081	–0.134036
270	–0.016269	–0.130064
280	–0.014931	–0.128370
290	–0.013904	–0.128227
300	–0.013091	–0.129206
310	–0.012434	–0.131037
320	–0.011892	–0.133541

330	–0.011438	–0.136597
340	–0.011053	–0.140117
350	–0.010722	–0.144039
360	–0.010436	–0.148316
370	–0.010185	–0.152911
380	–0.009965	–0.157797
390	–0.009770	–0.162952
400	–0.009595	–0.168359

Amplitude plots are shown in Fig. EA.1(a) and (b).

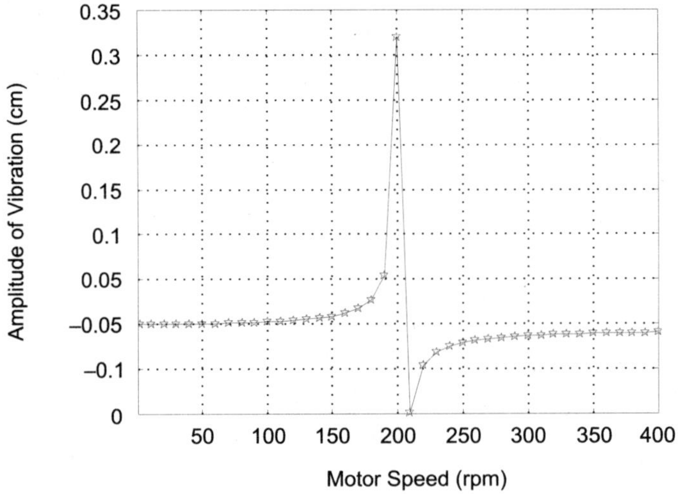

Fig. EA.1(a)

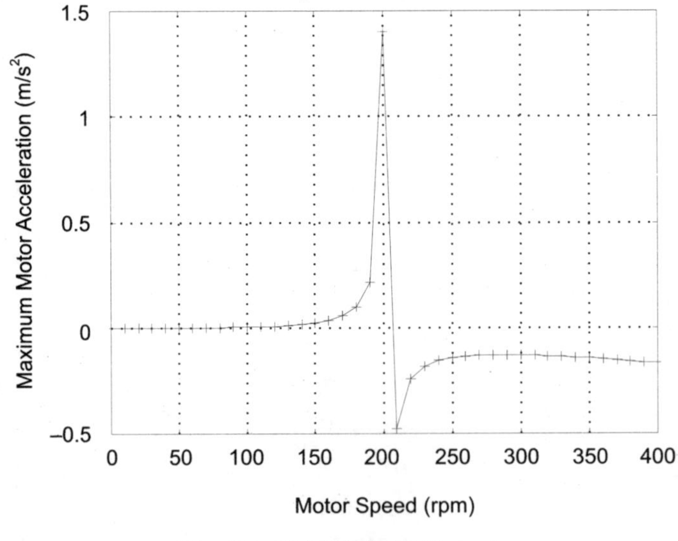

Fig. EA.1(b) MATLAB Outputs

It can be observed that maximum amplitudes occur at the speed equal to natural frequency ω_n in rpm.

Example A.2

Figure EA.2 shows an approximate model of a vehicle by the spring and dashpot system. The system moves with a speed of v over a road having sinusoidal cross-section of amplitude δ_m and wavelength ℓ. Write a MATLAB program to determine and plot the amplitude of the displacement of the mass relative to the road as a function of the speed v for damping factors $c/c_c = 0.1, 0.3$ and 0.7.

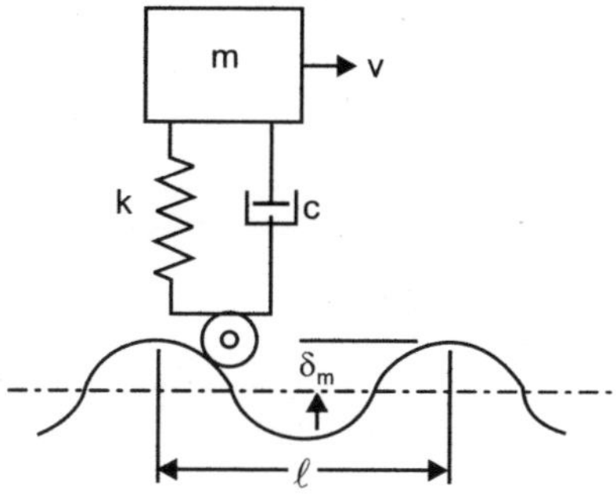

Fig. EA.2

Solution: Free body diagram is shown Fig. EA.2(a).

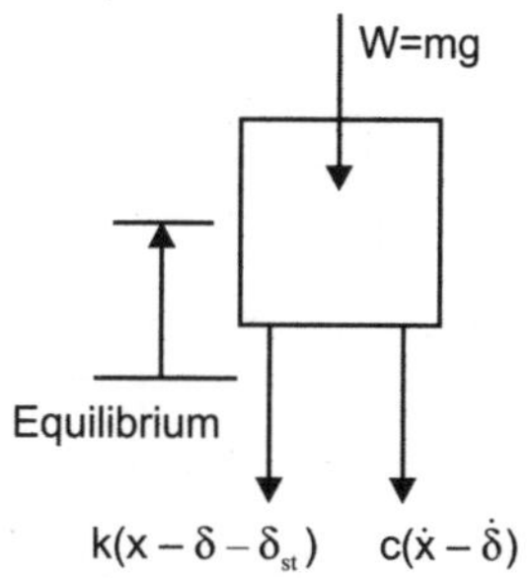

Fig. EA.2(a) Free-body diagram

Summing forces gives

$$+\uparrow \Sigma F = ma:$$

That is $-W - k(x - \delta - \delta_{st}) - c(\dot{x} - \dot{\delta}) = m\ddot{x}$

where $W = mg = k\delta_{st}$.(In static condition).

Let $u = x - \delta$, the relative displacement. Then

$$m\ddot{u} + c\dot{u} + ku = m\ddot{\delta}$$

where $\delta = \delta_m \sin \omega t$ and $\omega = 2\pi \frac{v}{l}$.

Thus $m\ddot{u} + c\dot{u} + ku = m\ddot{\delta} = p_m.\sin \omega t$

where $\quad p_m = m\omega^2\partial_m$.

For this equation, the amplitude ratio is given as:

$$\frac{u_m}{\delta_m} = \frac{(\omega/p)^2}{\sqrt{\left[1-(\omega/p^2)\right]^2 + (c/c_c)(\omega/p^2)}}$$

where $p = \sqrt{k/m}$ and $\omega = 2\pi \dfrac{v}{l}$.

Now a plot of non-dimensional frequency ratio and amplitude ratio can be drawn according to following MATLAB program for three values of c/c_c.

MATLAB program:

```
i=0;
for cR=0.1:0.3:0.9;
i=i+1;
Op=[0:0.5:5];                                  % Frequency ratio
den=sqrt((1-Op.^2).^2+(2*cR.*Op).^2);
umdm=Op.^2./den;                               % Amplitude ratio
plot(Op,umdm)
hold on;                           % In same plot to draw another curve
xlabel('Speed ratio (Omega/p)')
ylabel('Relative Displacement')
grid on;
end
legend('\zeta_1=0.1','\zeta_2=0.3','\zeta_3=0.9')
```

The output is shown in Fig. EA.2(b).

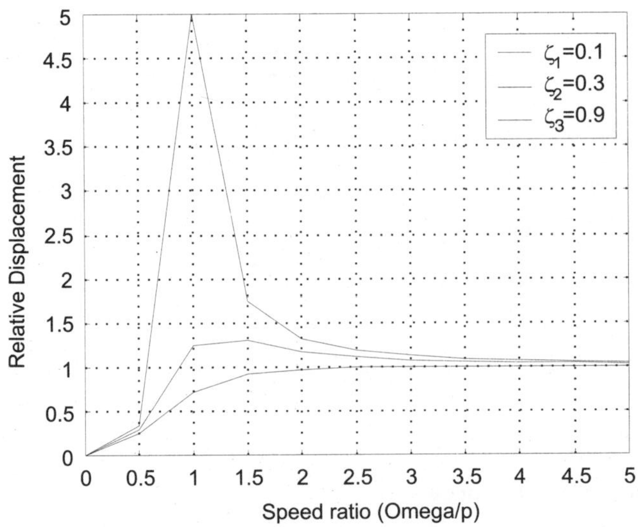

Fig. EA.2(b) MATLAB output

Example A.3

The 15 kg oscillator contains an unbalanced motor whose speed is N as shown in Fig. EA.3. The stiffness of the spring $k = 1100$ N/m. The oscillator is also restrained by a viscous damper whose piston is resisted by a force of 50 N when moving at a speed of 0.6 m/s. Determine (a) the viscous damping factor, (b) plot the magnification factor for motor speeds from 0 to 350, (c) the maximum value of the magnification factor and the corresponding motor speed.

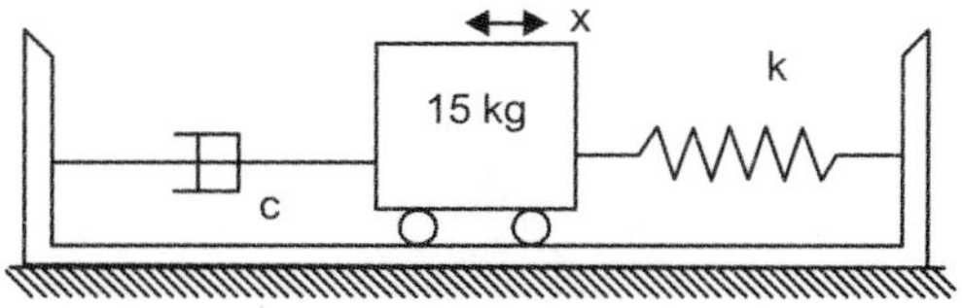

Fig. EA.3

Solution:

This is a single degree of freedom system. Here $m = 15$kg, $k = 1100$ N/m and $c = 50/0.6 = 83.34$ Ns/m

$$\text{Thus damping factor } \zeta = \frac{C}{2\sqrt{km}} = \frac{83.34}{2\sqrt{1100 \times 15}} = 0.3244 \text{ and } \omega_n = \sqrt{\frac{k}{m}} = \sqrt{\frac{1100}{15}} = 8.563 \text{ rad/s}$$

$$\text{Magnification factor is given by } M(\omega) = \frac{1}{\sqrt{(1-r^2)^2 + (2r\xi)^2}}, \text{ where } r = \omega/\omega_n = N/N_n$$

MATLAB code for drawing the plot is given below:

```
%%%%%%%%%%
k=1100;m=15;c=83.34;
wn=sqrt(k/m)*30/pi; %Natural frequency in rpm
zi=c/(2*sqrt(k*m));
wd=wn*sqrt(1-zi*zi);
w=0:5:350;
r=w./wn;
M=1./sqrt((1-r.^2).^2+(2*zi.*r).^2);
plot(w,M);
xlabel('speed in rpm');
ylabel('magnification factor');
title('Magnification factor Vs speed')
grid on;
%%%%%%%%%%%%%%%%%%%
```

The output is shown in Fig. EA.3(a).

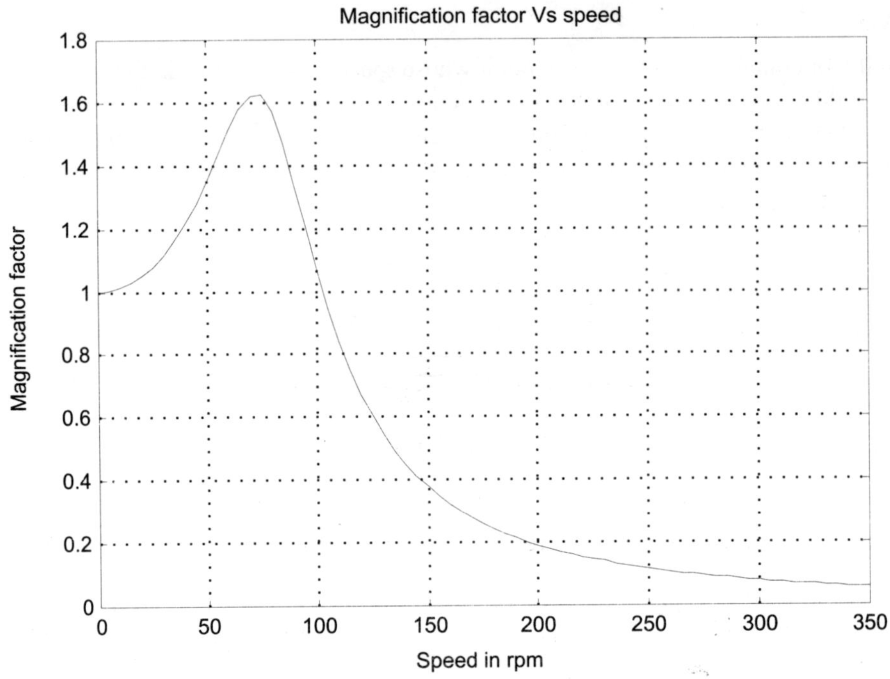

Fig. EA.3(a)

Maximum value occurs at a motor speed given by $N = (\omega_n \sqrt{1-2\zeta^2})\ 30/\pi$ rpm

$$= (8.563 \sqrt{1-2(0.3244)^2})\,30/\pi$$

$$= 72.67 \text{ rpm.}$$

Corresponding maximum magnification factor is $M_{max} = \dfrac{1}{2\zeta\sqrt{1-\zeta^2}} = \dfrac{1}{2\times 0.3244\sqrt{1-0.3244^2}}$

$$= 1.6297$$

Example A.4

The 10 kg body is moved 0.25 m to the right of the equilibrium position and released from rest at $t = 0$ as shown in Fig. EA.4. Plot the displacement as a function of time for four cases: $c = 10, 40, 50$ Ns/m. The stiffness of the spring is 40 N/m.

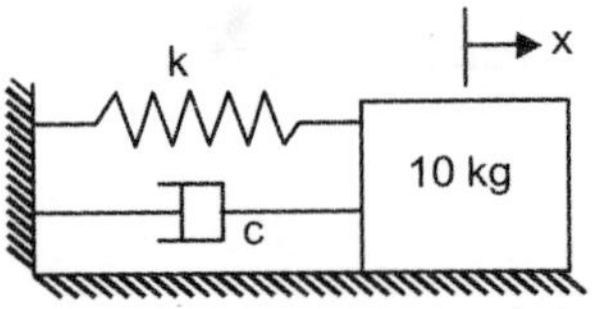

Fig. EA.4

Solution:

The equation of motion of the system is

$$-c\dot{x} - k\ddot{x} = m\ddot{x}$$

or $$\ddot{x} + \frac{c}{m}\dot{x} + \frac{k}{m}x = 0$$

The characteristic equation has two roots

$$s_{1,2} = -\frac{c}{2m} \pm \sqrt{\left(\frac{c}{2m}\right)^2 - \frac{k}{m}} = (-\zeta \pm \sqrt{\zeta^2 - 1})\omega_n$$

with $\frac{c}{2m} = \frac{k}{m} = \omega_n$ and $\zeta = \frac{c}{c_c} = \frac{c}{2m\omega_n}$

Case 1: when $\zeta < 1$, the solution for the motion is

$$x(t) = e^{-\zeta\omega_n t}(\mathrm{A}\cos(\sqrt{1-\zeta^2}\,\omega_n t) + \mathrm{B}(\sin\sqrt{1-\zeta^2}\,\omega_n t)) = Xe^{-\zeta\omega_n t}\sin(\omega_d t + \phi)$$

where $\omega_d = \sqrt{1-\zeta^2}\,\omega_n$

Case 2 : $\zeta = 1$, then solution is $x(t) = (\mathrm{A} + \mathrm{B}t)\,e^{-\omega_n t}$

Case 3 : $\zeta > 1$, then solution is $x(t) = C_1 e^{(-\zeta+\sqrt{\zeta^2-1})\omega_n t} + C_2 e^{(-\zeta-\sqrt{\zeta^2-1})\omega_n t}$

Thus first $\omega_n = \sqrt{\frac{k}{m}} = \sqrt{\frac{40}{10}} = 2$ rad/s and $\zeta = \frac{c}{C_c} = \frac{c}{2m\omega_n} = \frac{c}{2\times10\times2} = \frac{c}{40}$

When $c = 10$, $\zeta_1 = \frac{10}{40} = 0.4$, it belongs to case-1, with $\omega_d = 2\sqrt{1-0.4^2} = 1.4112$ rad/s

The solution is $x(t) = Xe^{-0.8t}\sin(1.4112t + \phi)$. Substituting initial conditions: at $t = 0$, $x(0) = 0.25$ m and $\dot{x}(0) = 0$, we find $0.25 = X\sin\phi$ and $\phi = 60.45$ degrees and $X = 0.2886$ m

So, it can be written as $x(t) = 0.2886\,e^{-0.8t}\sin(1.4112t + 60.45)$

When $c = 40$, $\zeta_2 = \frac{40}{40} = 1$, therefore solution becomes: $x(t) = (\mathrm{A} + \mathrm{B}t)\,e^{-2t}$

With given initial conditions: A = 0.25 and B = 0.5. Hence $x(t) = (0.25 + 0.5t)\,e^{-2t}$.

When $c = 50$, $\zeta_1 = 1.25$, it belongs to case-3, whose solution is

$$x(t) = C_1 e^{(-1.25+\sqrt{1.25^2-1})2t} + C_2 e^{(-1.25-\sqrt{1.25^2-1})2t} = C_1 e^{-1.8671t} + C_2 e^{-3.1328t}$$

The constants are given from initial conditions as $C_1 = 0.6189$ and $C_2 = 0.25 - 0.6189 = -0.3689$

MATLAB script for this problem can now be written as follows:

```
%%%%%%%%%%%%%%%%%%%%%%%%%%%%%%
t=0:0.01:5;
X=0.2886; phase=60.45*pi/180;
xa=X*sin(1.4112*t +phase).*exp(-0.8*t);
A=0.25;B=0.5;
xb=(A+B*t).*exp(-2*t);
C1=0.6189;C2=-0.3689;
xc=C1*exp(-1.867*t)+C2*exp(-3.1328*t);
x0=0.*t;
plot(t,xa,t,xb,t,xc,t,x0);
legend('\zeta_1=0.4','\zeta_2=1','zeta_3=1.25')
xlabel('time (sec)');
ylabel('x(m)');
%%%%%%%%%%%%%%%%%%%%%%%%%%%%%%%
```

The output is shown in Fig. EA.4(a).

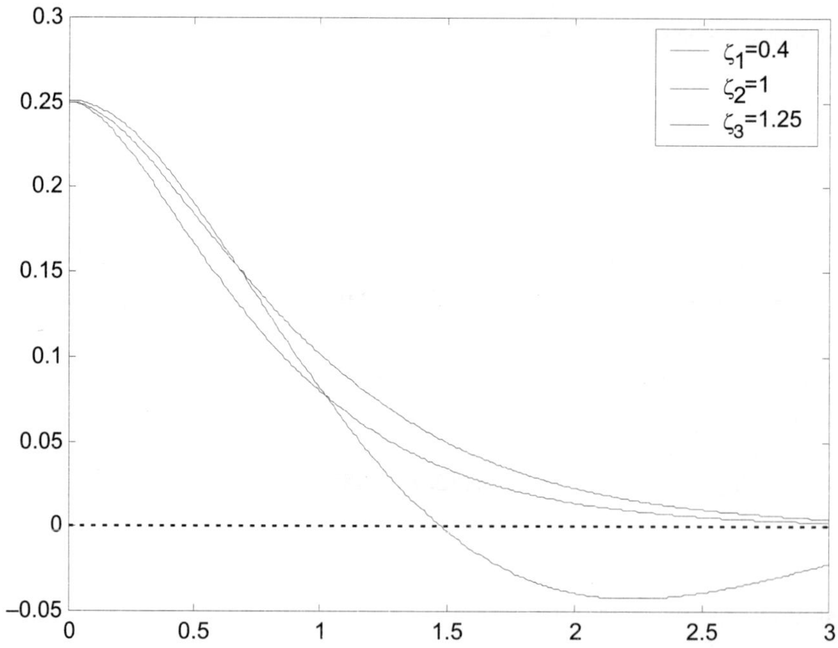

Fig. EA.4(a)

BIBLIOGRAPHY

There are several outstanding text and reference books on Engineering Mechanics: Statics and Dynamics, Mechanics of Materials, and MATLAB that merit consultation for those readers who wish to pursue these topics further. The following list is but a representative sample of the many excellent books.

Engineering Mechanics

Bedford, A., and Fowler, W., *Engineering Mechanics: Statics & Dynamics*, 4th ed, Prentice-Hall, New Jersey, 2005.

Beer, F.B., Russell Johnston, E., and Eisenberg, E.R., *Vector Mechanics for Engineers-Statics and Dynamics,* 8th ed, McGraw-Hill, New York, 2007.

Boresi, A., and Schmidt, R.J., Engineering *Mechanics: Statics,* 2nd ed, Brooks/Cole Publishing Company, New York, 2001.

Hibbler, R.C., *Engineering Mechanics: Statics & Dynamics,* Prentice-Hall, Englewood Cliffs, NJ, 1995.

Meriam, J.L., and Kraige, L.G., *Engineering Mechanics: Dynamics*, 5th ed, Wiley, New York, 2002.

Nelson, E.W., Best, C.L., and McLean, W.G., *Schaum's Outlines: Engineering Mechanics- Statics and Dynamics,* 5th ed, McGraw-Hill, New York, 1998.

Pytel, A., and Kiusalaas, J., *Engineering Mechanics: Statics*, 2nd ed, Brooks/Cole Publishing Company, New York, 1999.

Shelley, J.F., *Vector Mechanics for Engineers Volume I: Statics*, Schaum's Solved Problems Series, McGraw-Hill, New York, 1990.

Mechanics of Materials

Beer, F.B., Russell Johnston, E., and DeWolf, J.T., *Mechanics of Materials,* 3rd ed, McGraw-Hill, New York, 2001.

Cernica, J.N., *Strength of Materials*, Holt, Rinehart and Winston, Inc., New York, 1966.

Nash, W., *Schaum's Outlines: Statics and Mechanics of Materials,* McGraw-Hill, New York, 1992.

Nash, W., *Schaum's Outlines: Strength of Materials,* 4th ed, McGraw-Hill, New York, 1998.

Riley., W., Sturges, L., and Morris, D., *Mechanics of Materials*, Wiley, New York, 2007.

MATLAB

Chapman, S.J., *MATLAB Programming for Engineers,* 2nd ed., Brooks/Cole, Thomson Learning, Pacific Grove, CA, 2002.

Dukkipati, R.V., *MATLAB for Engineers,* New Age International (P) Ltd., New Delhi, India, 2007.

Dukkipati, R.V., *Solving Vibration Problems using MATLAB,* New Age International (P) Ltd., New Delhi, India, 2007.

Dukkipati, R.V., *Solving Engineering System Dynamics Problems with MATLAB,* New Age International (P) Ltd., New Delhi, India, 2007.

Etter, D.M., *Engineering Problem Solving with MATLAB,* Prentice-Hall, Englewood Cliffs, NJ, 1993.

Gardner, J.F., *Simulation of Machines using MATLAB and SIMULINK,* Brooks/Cole, Thomson Learning, Pacific Grove, CA, 2001.

Harper, B. D., *Solving Dynamics Problems in MATLAB,* 5th ed, Wiley, New York, 2002.

Harper, B. D., *Solving Statics Problems in MATLAB,* 5th ed, Wiley, New York, 2002.

Herniter, M.E., *Programming in MATLAB,* Brooks/Cole, Pacific Grove, CA, 2001.

Lyshevski, S.E., *Engineering and Scientific Computations Using MATLAB,* Wiley, New York, 2003.

Sigman, K., and Davis, T.A., *MATLAB Primer,* 6th ed, Chapman& Hall/CRCPress, Boca Raton, FL, 2002.

The MathWorks, Inc., *MATLAB: Creating Graphical User Interfaces, Version 1,* The MathWorks, Inc., Natick, 2000.

The MathWorks, Inc., *MATLAB: Function Reference,* The MathWorks, Inc., Natick, 2000.

The MathWorks, Inc., *MATLAB: Symbolic Math Toolbox User's Guide, Version 2,* The MathWorks, Inc., Natick, 1993-1997.

The MathWorks, Inc., MATLAB: *Using MATLAB Graphics, Version 6,* The MathWorks, Inc., Natick, 2000.

The MathWorks, Inc.,, *MATLAB: Using MATLAB, Version 6,* The MathWorks, Inc., Natick, 2000.

Vibrations

Dukkipati, R.V. and Srinivas, J., *Vibrations Problem Solving Companion,* Narosa Publishers, New Delhi 2007.

Dukkipati, R.V. and Srinivas, J., *Textbook of Mechanical Vibrations,* Prentice-Hall of India, New Delhi 2004.